PROTEIN DEPOSITION IN ANIMALS

Protein Deposition in Animals

P.J. BUTTERY, BSc, PhD
Department of Applied Biochemistry and Nutrition
University of Nottingham School of Agriculture

D.B. LINDSAY, MA, DPhil
ARC Institute of Animal Physiology
Babraham, Cambridge

BUTTERWORTHS
LONDON - BOSTON
Sydney - Wellington - Durban - Toronto

| United Kingdom | Butterworth & Co (Publishers) Ltd |
| London | 88 Kingsway, WC2B 6AB |

Australia	Butterworths Pty Ltd
Sydney	586 Pacific Highway, Chatswood, NSW 2067
	Also at Melbourne, Brisbane, Adelaide and Perth

| Canada | Butterworth & Co (Canada) Ltd |
| Toronto | 2265 Midland Avenue, Scarborough, Ontario, M1P 4S1 |

| New Zealand | Butterworths of New Zealand Ltd |
| Wellington | T & W Young Building, 77-85 Customhouse Quay, 1, CPO Box 472 |

| South Africa | Butterworth & Co (South Africa) (Pty) Ltd |
| Durban | 152-154 Gale Street |

| USA | Butterworth (Publishers) Inc |
| Boston | 10 Tower Office Park, Woburn, Massachusetts 01801 |

First published 1980

© The several contributors named in the list of contents, 1980

ISBN 0 408 10676 X

British Library Cataloguing in Publication Data

Easter School in Agricultural Science, *29th*
University of Nottingham, 1980
Protein deposition in animals.
1. Proteins in animal nutrition – Congresses
2. Protein metabolism – Congresses
I. Title II. Buttery, P J III. Lindsay, D B
636.089′2′398   SF98.P7   80-49869

ISBN 0–408–10676–X

Typeset by Scribe Design, Gillingham, Kent
Printed by Billing & Sons Ltd, London & Guildford

PREFACE

The 29th Easter School of the University of Nottingham, of which this book is the proceedings, discussed the factors controlling protein deposition in farm animals. The aim of the meeting was to mount a forum in which biochemists and physiologists could discuss with colleagues associated with the more practical aspects of animal production, the factors which influence protein production by animals. The book starts by discussing some fundamental aspects of protein synthesis and is followed by a consideration of the molecular control of protein breakdown. Two chapters then consider the measurement of whole-body protein metabolism and the integration of the metabolism of individual organs with the rest of the animal.

Two 'tissues', the muscle and the fetus, are singled out for detailed discussion in subsequent chapters, while another chapter attempts to describe the synthesis of egg proteins but shows clearly that much more work is required in this area. The factors which influence overall nitrogen retention by the animal are studied, as are the energy costs of protein deposition. Hormonal influences on protein deposition are considered from three different angles: first, a detailed discussion of hormone action; secondly, the way of manipulating growth with anabolic agents; and thirdly, the implications from a health point of view of current practice in the use of these anabolic agents. Two chapters, one on poultry and the other on ruminants, are concerned with predicting rates of protein deposition. The book ends by considering protein metabolism in a cold-blooded animal, the fish.

The meeting clearly indicated that there were numerous areas where additional work is required. The speakers were particularly asked to speculate on future developments in the area and much of this speculation appears in their papers.

The organizers would like to thank the speakers, the chairmen of sessions, and indeed all the staff of the University of Nottingham for their efforts in making the 29th Easter School a success. Particular mention should be made of Mrs Shirley Bruce and her staff who did so much to keep the organization of the conference running smoothly. In addition, the financial contributions by the concerns mentioned elsewhere were most gratefully received since they enabled the costs of the meeting to the delegates to be kept at a reasonable level.

P.J. BUTTERY
D.B. LINDSAY

ACKNOWLEDGEMENTS

The organizers wish to thank the staff at the University of Nottingham, the speakers and the chairmen (P.J. Buttery, D.J.A. Cole, Professor G.E. Lamming and Professor G.A. Lodge), who all contributed to the success of the meeting. The assistance of the following organizations is also gratefully acknowledged since without their help the meeting would not have taken place:

BOCM Silcock Ltd
BP Nutrition (UK) Ltd
Colborn Group Ltd
Imperial Chemical Industries Ltd
Pauls and Whites Foods Ltd
Pedigree Petfoods
Roussel Uclaf
Rumenco Ltd
Sun Valley Feed
Unilever Research Laboratory

CONTENTS

1

MECHANISM AND REGULATION OF PROTEIN BIOSYNTHESIS IN EUKARYOTIC CELLS

VIRGINIA M. PAIN
Department of Human Nutrition, London School of Hygiene and Tropical Medicine

and

MICHAEL J. CLEMENS
Department of Biochemistry, St George's Hospital Medical School, London

Summary

Protein biosynthesis comprises a series of complex processes involving three kinds of RNA and a large number of proteins. Messenger-RNA (mRNA) carries in its nucleotide sequence a code determining the order of insertion of amino acids into the polypeptide chain. Ribosomal-RNA, in combination with about 70 proteins, is formed into an organelle, the ribosome, which provides the correct structural alignment for the other protein synthetic components. Transfer-RNA (tRNA) exists as a number of different species, each specific for a particular amino acid, and is involved in activating and binding successive amino acids to the ribosome in the order directed by the structure of messenger-RNA (also bound to ribosomes). Protein factors involved in protein synthesis include many which function enzymically and others which have more structural significance.

The overall process, referred to as *translation*, is divided into three stages: (1) *Initiation*. A ribosome and a molecule of a specific initiator tRNA (Met-tRNA$_f$) bind to a particular site on the mRNA at the beginning of the coding sequence. Another aminoacyl-tRNA is then able to bind, and synthesis of the first peptide bond takes place. (2) *Elongation*. The ribosome moves relative to the messenger-RNA and a polypeptide chain is elaborated from amino acids in a specific sequence directed by the order of nucleotides in the messenger-RNA. (3) *Termination*. The ribosome reaches the end of the coding sequence on the messenger-RNA and is released together with the completed protein chain. In most tissues *in vivo* each molecule of messenger-RNA is translated simultaneously by several ribosomes, the entire structure being termed a *polyribosome* or *polysome*.

The overall rate of protein synthesis in eukaryotic cells and tissues can be regulated at two levels: (a) by the number of ribosomes available in the tissue, which determines the maximum rate of protein synthesis possible, and (b) by the activity of the ribosomes, i.e. the rate of protein synthesis per ribosome in the tissue. The activity of ribosomes appears, in most cases studied to date, to be regulated at the level of initiation of protein synthesis. Experiments which have led to this conclusion and possible mechanisms by which initiation may be regulated will be discussed. Technical difficulties have limited progress in this area of investigation with many animal tissues. Data are therefore presented which have been obtained with a model system of nutritional control (amino acid regulation of protein synthesis in Ehrlich ascites tumour cells in tissue culture), which can be subjected to more detailed analysis at the molecular level. Analogies will be drawn, where possible, with results obtained in various laboratories using normal tissues such as muscle and liver.

Introduction

During the 1970s we witnessed a very rapid development of knowledge of the mechanism of protein biosynthesis in animal cells. Even now our understanding of many details of the various chemical interactions is far from complete. It is apparent that, while the overall mechanism of protein synthesis in eukaryotes is analogous to that operating in bacteria, there are several stages which are considerably more complicated. Earlier reviews were concerned mainly with the bacterial process, but several recent articles have been devoted partly (Mazumder and Szer, 1977) or wholly (Pain, 1978; Revel and Groner, 1978; Pain and Clemens, 1980) to describing protein synthesis in eukaryotes.

In parallel with the characterization of the mechanism of protein biosynthesis, there have been developments in our understanding of the regulation of this process. Improved methods of measurement of rates of protein synthesis *in vivo* (see Chapter 3) permit the identification of situations in which physiological regulation occurs, and pinpoint fruitful areas for investigation at the subcellular level. At present, however, most of our knowledge of regulatory mechanisms is derived from model systems, e.g. reticulocytes and tumour cells in culture, in which very pronounced variations in the protein synthetic rate can be induced by the investigator. It is hoped that studies in these systems will provide information on potential sites of control of protein synthesis which will be applicable to tissues of normal animals.

In this chapter, the present knowledge of the mechanism of protein biosynthesis in animal cells is summarized, followed by a discussion of recent studies aimed at increasing our understanding of how the rate of this process may be controlled at the molecular level.

Mechanism of protein synthesis

TYPES OF RNA

Protein biosynthesis comprises a series of complex processes involving three kinds of RNA and a large number of proteins. The order of insertion of amino acids into a polypeptide chain is directed by the sequence of nucleotide bases in messenger-RNA (mRNA), which is transcribed from DNA in the nucleoplasmic region of the nucleus. Each amino acid is specified by a codon of three bases in the mRNA. Recent studies have revealed that most eukaryotic mRNAs contain certain characteristic features in addition to the coding sequence (*Figure 1.1*). At the 5' end they carry a 'cap' structure, i.e. a base-methylated guanosine residue joined 5'–5' to the next nucleotide by a trisophate linkage (see reviews by Rottman, 1976; Shatkin, 1976; Filipowicz, 1978; Revel and Groner, 1978; Pain and Clemens, 1980). The function of the cap is not yet certain but this structure may well play a role in regulating the initiation of protein synthesis (Filipowicz, 1978). Next to the 'cap' is an untranslated sequence, the length of which differs between species of mRNA, followed by the coding region which commences with the codon-specifying methionine (AUG). At the 3' end of the coding sequence is another untranslated region, and finally a segment of poly-adenylic acid residues, again untranslated. The presence of this poly(A) tract has been utilized extensively by biochemists in devising procedures for extracting mRNA from cells and tissues, but its physiological function is still obscure (Brawerman, 1976; Revel and Groner, 1978; Pain and Clemens, 1980). Some studies in cell-free systems suggest a direct role of poly(A) in regulating the rate of protein synthesis, but there are also indications that it may be involved in controlling the turnover of mRNA (e.g. Huez *et al.*, 1975).

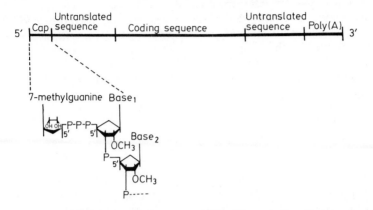

Figure 1.1   Generalized structure of messenger-RNA. The majority of eukaryotic mRNAs contain the five regions shown (see text). The relative lengths of the different parts of the molecule are not drawn to scale. The expanded portion shows the structure of the 5′−terminal cap

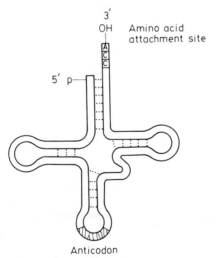

Figure 1.2   Characteristic structure of transfer-RNA, showing the position of the anticodon and the site for attachment of the specific amino acid

The genetic information represented by the sequence of bases in mRNA is decoded by transfer-RNA (tRNA) and converted into sequences of amino acids in proteins. The structure of tRNA (*Figure 1.2*) includes three features which enable it to carry out this function: (1) an anticodon triplet of bases for complementary binding to the codon in mRNA specifying a particular amino acid, (2) a site for attachment of that same amino acid at the 3′ terminus, and (3) a recognition site for the specific aminoacyl-tRNA synthetase enzyme which catalyses the binding of the amino acid. The formation of aminoacyl-tRNA involves the hydrolysis of ATP to AMP; hence, the preparation of each molecule of amino acid for protein synthesis requires the expenditure of two high-energy bonds.

The third type of RNA involved in protein synthesis is ribosomal-RNA (rRNA), which is synthesized in the nucleolus and becomes incorporated, together with about 70 proteins, into the ribosomes. Ribosomes consist of two subunits, termed by their sedimentation behaviour on sucrose-density gradients as 40S and 60S. The nature and possible roles of the protein components of ribosomes have been the subject of much recent investigation (Wool and Stöffler, 1974), and there is some evidence, as yet inconclusive, linking changes in protein synthetic activity of ribosomes with covalent modifications, e.g. phosphorylation, of certain ribosomal proteins.

THE RIBOSOME CYCLE

Figure 1.3 gives an overall scheme of the process of protein synthesis, often referred to as *translation*. This can be divided into three stages:

(1) *Initiation.* A ribosome and a molecule of a specific initiator tRNA (Met-tRNA$_f$) bind to a particular site on the mRNA at the beginning of the coding sequence.

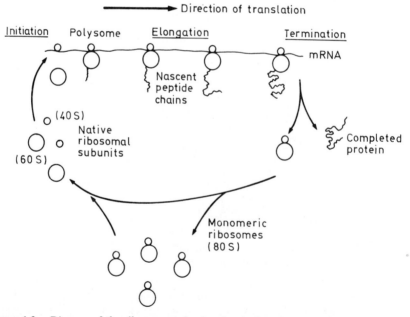

Figure 1.3　Diagram of the ribosome cycle, showing the incorporation of native ribosomal subunits into polysomes (peptide chain initiation), the transit of the ribosomes along the mRNA as the nascent polypeptide chains are extended (elongation), and the release of the ribosomes and completed protein (termination). (From Pain, 1978)

(2) *Elongation.* The ribosome moves relative to the mRNA and a polypeptide chain is elaborated from amino acids in a specific sequence directed by the genetic information encoded in the order of bases in the mRNA.

(3) *Termination.* The ribosome reaches the end of the coding sequence on the mRNA and is released together with the completed protein chain. The ribosome is then available for reattachment on the same or another molecule of mRNA. If it is not required immediately for another round of protein synthesis, however, it may remain in the cytoplasm, in an 'idling pool' of monomeric ribosomes unattached to mRNA.

In most tissues *in vivo*, as shown in *Figure 1.3*, each molecule of mRNA is translated simultaneously by several ribosomes. The entire structure is called a *polyribosome* or *polysome.*

The three stages of protein synthesis will now be described in more detail. The text concentrates mostly on the process of initiation, since this has been the subject of most dramatic expansion of knowledge in recent years and is thought to be an important site of regulation of overall rates of protein synthesis (see page 8).

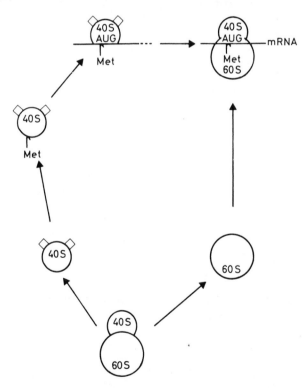

Figure 1.4   Expansion of the region of *Figure 1.3*, showing the initiation of protein synthesis. The diagram shows: (1) the dissociation of a monomeric (80S, i.e. 40S+60S subunits) ribosome into native 40S and 60S subunits. (2) the presence of extra proteins (initiation factors) on the native 40S subunits. (3) the formation of a complex between 40S subunits and the initiator tRNA (Met-tRNAf). (4) the binding of this complex to mRNA. The anti-codon of Met-tRNAf recognizes the initiation codon, AUG, on mRNA. (5) Addition of the 60S subunit and release of the initiation factors associated with the 40S subunit

Initiation of protein synthesis

Figure 1.4 represents an expansion of the part of *Figure 1.3* depicting the initiation of protein synthesis, showing extra details of the steps involved. Initiation in animal cells is somewhat more complicated than in bacteria, and is dependent on the catalytic action of several proteins known as *initiation factors*. At the present time the functions of six initiation factors appear to be fairly well defined and several other proteins have been suggested to be initiation factors on the basis of less complete evidence (Trachsel *et al.*, 1977; Safer and Anderson, 1978).

The first step in initiation is the dissociation of a monomeric ribosome into two subunits. This is associated with, and probably brought about by, the binding of one or more initiation factors to the 40S subunit. The amount of protein bound to these dissociated, or 'native', 40S subunits is consistent with the largest initiation factor, eIF-3, being involved at this stage. Native 60S subunits also

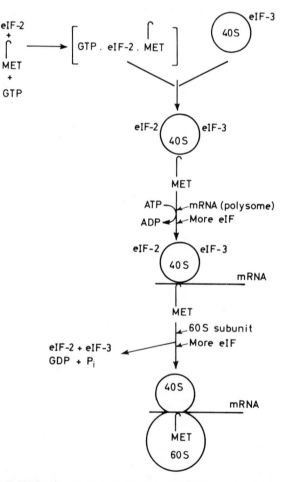

Figure 1.5   Detailed diagram of the mechanism of initiation of protein synthesis. For explanation, see text

carry an extra protein (van Venrooij and Janssen, 1976), but this has not been correlated with any of the initiation factors characterized so far.

The next step is the binding to the native 40S subunit of a molecule of initiator tRNA (Methionyl-tRNA$_f$). This is mediated by another initiation factor, eIF-2, which forms a ternary complex with Met-tRNA$_f$ and GTP (*Figure 1.5*). The entire 40S complex is then able to bind to the initiation site at the 5' end of the coding sequence of the mRNA (Kozak, 1978). The first codon of the sequence is always AUG, which is recognized by the anticodon of Met-tRNA$_f$. Binding of the 40S complex to mRNA requires the participation of at least three initiation factors and the hydrolysis of a molecule of ATP (see reviews by Safer and Anderson, 1978; Pain and Clemens, 1980).

The final step in the initiation pathway is the addition of the 60S ribosomal subunit to the complex. This occurs by a complicated reaction involving the release of the factors eIF-2 and eIF-3, previously bound to the 40S subunit, and hydrolysis of the GTP molecule which was originally associated with eIF-2 (Safer and Anderson, 1978). With the reassociation of the two ribosomal subunits on the mRNA, two binding sites for aminoacyl-tRNA are formed, one already occupied by Met-tRNA$_f$ and the other now available for binding of the aminoacyl-tRNA corresponding to the next codon on the mRNA.

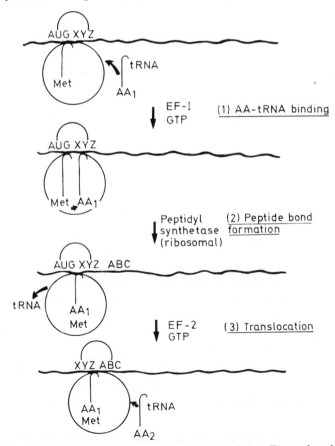

Figure 1.6   Mechanism of elongation of polypeptide chains. For explanation, see text

Elongation

Figure 1.6 shows details of the reactions by which the initial dipeptide is formed between methionine and the next amino acid, AA_1, corresponding to the codon 'XYZ' on the mRNA. First, AA_1-tRNA binds to the vacant tRNA site on the ribosome in a reaction dependent on elongation factor 1 (EF-1). This brings AA_1 into the correct position for formation of a peptide bond with the methionine attached to the initiator tRNA. A ribosomal protein, peptidyl transferase, catalyses the formation of the dipeptide, which remains attached to the second tRNA. Deacylated initiator tRNA is released. The cycle is completed by the shift of the ribosome relative to the mRNA by a distance equivalent to one codon. This process, known as *translocation*, requires another elongation factor, EF-2. A binding site is now available for the AA-tRNA corresponding to the next codon, ABC. This sequence of reactions is repeated until the ribosome has passed along the entire coding sequence of the mRNA and the complete protein has been synthesized. Since both the EF-1 catalysed binding of each AA-tRNA and the EF-2 catalysed translocation step involve the hydrolysis of a molecule of GTP, it follows that two further high-energy bonds are expended per amino acid incorporated.

Termination

At the $3'$ end of the coding sequence there is always a termination codon (UAA, UAG or UGA) which does not denote any amino acid. This constitutes a signal for the release of the ribosome and the completed protein chain from the mRNA. This release is mediated by a release factor (RF) which allows the ribosomal peptidyl transferase to hydrolyse the ester linkage between the nascent peptide chain and the last tRNA molecule.

Regulation of protein biosynthesis

Protein synthesis in eukaryotic cells is subject to both quantitative and qualitative regulation. Quantitative regulation, i.e. control of the overall *amount* of protein synthesized, is important in determining tissue size and its response to such factors as nutritional and hormonal state. Qualitative regulation, i.e. control of the *type* of proteins synthesized, is of great importance for the hormonal induction of enzymes and in the processes of development and differentiation. Most of the evidence available at present suggests that major changes in the pattern of proteins synthesized are exerted mainly at the level of relative abundance of different mRNAs (e.g. Ross, Ikawa and Leder, 1972; Feigelson *et al.*, 1975; Schimke *et al.*, 1975). The remainder of this chapter is devoted to discussion of mechanisms underlying the quantitative regulation of protein synthesis.

NUMBER AND ACTIVITY OF RIBOSOMES

Measurements of overall rates of protein synthesis *in vivo* have shown changes in various tissues in response to alterations in nutritional or hormonal conditions (see reviews by Waterlow, Garlick and Millward, 1978; Garlick, 1980; Garlick, this volume, Chapter 3; Henshaw, 1980; Pain, 1980).

Such changes can be analysed in terms of regulation at two levels: (a) by the number of ribosomes available in the tissue, which determines the maximum rate of protein synthesis possible at any one time, and (b) by the activity of the ribosomes, i.e. the rate of protein synthesis per ribosome in the tissue. Such analysis is a simple procedure, since a good approximation of the ribosome concentration in tissues can be obtained by measurement of total RNA (Hirsch, 1967; Young, 1970). In this way, several nutritional and hormonal deficiencies have been found to reduce both the number of ribosomes and the rate of protein synthesis per ribosome in muscle (e.g. Henshaw *et al.*, 1971; Millward *et al.*, 1973; Pain and Garlick, 1974; Flaim, Li and Jefferson, 1978). Where investigated, the ribosomal activity (protein synthesis per ribosome) tends to respond more acutely to nutritional or hormonal change than does the ribosome number (Millward, Garlick and James, 1973; Flaim, Li and Jefferson, 1978). Indeed, studies with isolated perfused muscle preparations have shown changes in protein synthesis rates which are far too rapid to be accounted for by alterations in ribosome number (Jefferson, Li and Rannels, 1977), and the speed of response of ribosomal activity to nutritional supply demonstrated in cells in culture is even more dramatic (see page 12).

Information on the mechanisms regulating the rate of synthesis and degradation of ribosomes is still very scarce, particularly in tissues of animals *in vivo*. It is possible that restriction of protein synthesis may have a secondary effect on ribosome production, by limitation either of the supply of newly synthesized ribosomal proteins or of a protein or proteins involved in the synthesis or processing of rRNA (see earlier review by Pain, 1978). Alternatively, production of rRNA may be regulated independently by nutritional or hormonal conditions. Studies with cells in culture suggest that the stability of ribosomes increases under conditions of rapid growth (Emerson, 1971; Weber, 1972; Abelson *et al.*, 1974), but reliable data on rates of degradation of ribosomes in intact animals await the development of suitable methodology.

REGULATION OF THE ACTIVITY OF RIBOSOMES

Standard procedures now exist for analysing changes in ribosomal activity in terms of effects on initiation, elongation and termination of protein synthesis. The technique which has been applied most widely is the investigation of polysome profiles on sucrose-density gradients. The tissue is homogenized and ribosomes are extracted under conditions designed to give optimum yield while minimizing the possibility of attack by endogenous and exogenous ribonucleases, to which the mRNA strands linking the ribosomes in polysomes are very vulnerable.

Figure 1.7 shows the patterns obtained when ribosomes from muscle of normal and diabetic rats are centrifuged on sucrose gradients. As originally demonstrated by Wool *et al.* (1968), the preparations from the diabetic animals contain less polysomes and more monomeric ribosomes than those from normal animals. Consideration of *Figure 1.3* helps us to interpret these results. It is clear that, provided the supply of mRNA does not become rate limiting, the distribution of ribosomes between polysomes and the idling pool of 80S monomers is determined by the relative rates of initiation and elongation. Thus, in the case of skeletal muscle of diabetic animals, there is an accumulation of 80S monoribosomes at the expense of polysomes, which suggests that the fall in the protein synthetic activity of ribosomes measured *in vivo* under these conditions (Pain and Garlick, 1974) is associated with a decrease in the rate of initiation *relative to* elongation. Conversely, the stimulation by insulin of the rate of protein synthesis per ribosome in the perfused hind-limb is accompanied by an increase in the proportion of ribosomes in polysomes (Jefferson *et al.*, 1974; *Figure 1.8*) which suggests that initiation is promoted relative to elongation.

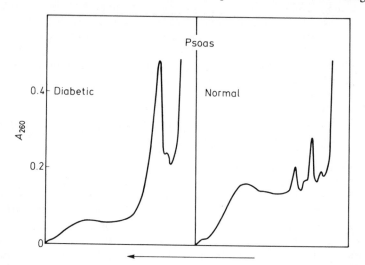

Figure 1.7   Polysome profiles from psoas muscles of alloxan-diabetic and normal rats. Post-mitochondrial supernatants were prepared in a buffer containing 0.25M KCl (see text). Samples were then layered on to exponential gradients of 15–68% sucrose dissolved in the same buffer, and centrifuged for 3.25 h at 284 000 × g_{max} in a Beckman SW 40 swing-out rotor. The gradients were then pumped through the flow cell of a recording spectrophoto-meter and the absorbance profiles at 260 nm were monitored. The direction of sedimenta-tion was from right to left. The profiles depict the distribution in the samples of ribosomal particles of different size classes, with the most rapidly sedimenting particles (large poly-somes) on the left and the lightest particles (ribosomal subunits) on the right. Under these conditions of high ionic strength, the monomeric ribosomes are dissociated into 40S and 60S subunits. (Results are taken from Jefferson *et al.*, 1974, in which a more complete description of the experimental procedure can be found)

Until relatively recently there has been a tendency by many workers to assume that a fall in the proportion of polysomes in a tissue *automatically* implies a decreased rate of protein synthesis and vice versa. This is not necessarily the case, however. For example, treatment of cells and tissues with low concentra-tions of the drug cycloheximide causes a partial inhibition of protein synthesis which is accompanied by an *increased* proportion of ribosomes in polysomes. This result indicates that the drug inhibits elongation to a greater extent than initiation. Thus, one can only use analysis of polysomes *in conjunction with* a measurement of protein synthesis/ribosome rather than as a substitute for it.

    There are now several examples of nutritional and hormonal deficiencies in which an impairment of initiation in muscle is indicated by the results of parallel measurements of rates of protein synthesis per ribosome in the intact tissue and proportion of ribosomes in polysomes, viz. starvation (D.E. Rannels *et al.*, 1978;

Li, Higgins and Jefferson, 1979), diabetes (Jefferson *et al.*, 1974) and hypo-physectomy (Flaim, Li and Jefferson, 1978). Treatment with high doses of glucocorticoids has a similar effect (S.R. Rannels *et al.*, 1978). These results indicate that it would now be appropriate to analyse in more detail the mechanisms underlying the regulation of initiation of protein synthesis in muscle.

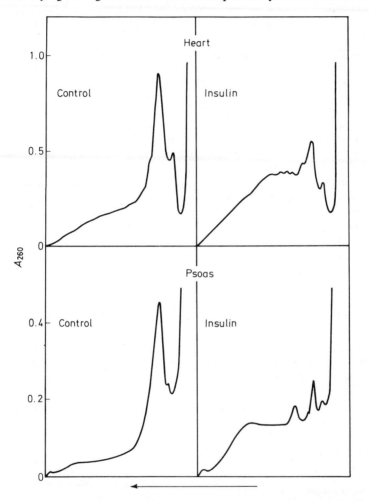

Figure 1.8   Effect of insulin on polysome profiles in the perfused heart and in the psoas muscle of the perfused hind-limb. Perfusion in the absence of insulin results in disaggregation of polysomes, indicating that initiation of protein synthesis is impaired relative to elongation. The polysome profile is preserved if insulin is present in the perfusate. (Results are taken from Jefferson *et al.*, 1974)

However, the extraction of ribosomes from this tissue presents serious technical difficulties which not only hinder the discovery of further details of regulation at the molecular level but also complicate the interpretation of polysome profiles. When muscle is subjected to fractionation into subcellular components by conventional techniques, myofibrils are precipitated and most of the ribosomes tend to co-precipitate with them. This is probably the result of two effects. First, it is likely that some ribosomes become trapped in the myofibrils as they are precipitated. Secondly, the ribosomes which are actually involved in the synthesis of the proteins of the contractile apparatus could well remain attached to the myofibrillar fraction by their nascent peptides. The second mode of attachment is potentially more serious as a factor in the interpretation of polysome profiles, as it would result in the selective loss of polysomes rather than monomeric ribosomes. Since contractile proteins represent over half the protein in muscle and are synthesized at about one-third to one-half the rate for muscle proteins as a whole (Waterlow, Garlick and Millward, 1978), the proportion of polysomal ribosomes which may be attached specifically to the contractile apparatus could be highly significant.

Some of these problems can be lessened by performing the extraction in a medium containing a high concentration of KCl (0.25–0.5M), which partially solubilizes the myofibrillar proteins. Complete solubilization can be achieved only by exhaustive extraction in such a medium, but this is not feasible for the preparation of polysomes in which rapid technique is necessary to avoid degradation by nucleases. Moreover, this treatment prevents further analysis of regulatory mechanisms by studies made in crude cell-free systems, as described below for other cell types, since a high concentration of KCl inhibits protein synthesis and removes initiation factors from ribosomes. At present, therefore, most of our information on mechanisms involved in regulation of initiation of protein synthesis is derived from experiments in model systems, e.g. reticulocytes and cells in tissue culture. The next section is a brief description of our work with one of these systems.

REGULATION BY AMINO ACID SUPPLY OF INITIATION OF PROTEIN SYNTHESIS IN EHRLICH ASCITES TUMOUR CELLS IN CULTURE

Protein synthesis in Ehrlich ascites tumour cells is exquisitely sensitive to the supply of essential amino acids in the culture medium (van Venrooij, Henshaw and Hirsch, 1970, 1972). Omission of a single essential amino acid results in a reduction of protein synthesis by 50–70% within minutes (*Figure 1.9*). This is accompanied by extensive disaggregation of polysomes (*Figure 1.10b*). Refeeding the missing amino acid results in reformation of polysomes and restoration of protein synthesis within 10 min (*Figure 1.10c*). Culture of the cells in the absence of exogenous glucose produces similar results, although less rapidly (*Figure 1.10d* and van Venrooij, Henshaw and Hirsch, 1970, 1972). These data suggest that initiation of protein synthesis is impaired relative to elongation in the deficient cells. Measurement of protein synthesis per polyribosomal ribosome, which gives an estimate of the rate of elongation, confirms that elongation is affected to a much lesser extent (van Venrooij, Henshaw and Hirsch, 1972).

Experiments designed to investigate the mechanism of these effects can be explained in terms of *Figure 1.5*. It can be seen that an early stage in initiation of protein synthesis is the formation of a complex between the native 40S ribosomal subunit and the initiator tRNA (Met-tRNA$_f$), which can be recovered from cell extracts by isolation of native 40S subunits from sucrose gradients. We estimated the amount of [Met-tRNA$_f$. 40S subunit] complexes in fed and lysine-starved Ehrlich cells by incubating the cells briefly in the presence of [^{35}S]-methionine and assaying the native 40S subunits for tRNA-bound ^{35}S radioactivity (Pain and Henshaw, 1975). The results are shown in *Table 1.1*. We found,

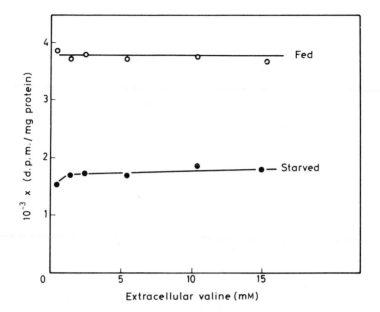

Figure 1.9   Incorporation of radioactivity from [³H]-valine into protein of fed and lysine-starved Ehrlich ascites tumour cells. Ehrlich cells were maintained in suspension culture in Eagles Minimal Essential Medium (MEM) containing 5% calf serum (Pain and Henshaw, 1975). Cells were separated from the medium by centrifugation and resuspended in serum-free medium, either complete MEM (fed) or MEM devoid of lysine (starved). After 5 min, samples were removed and [³H]-valine and unlabelled valine were added at various concentrations to give constant specific radioactivity. Incubation was continued for a further 20 min, after which protein was precipitated with 10% trichloroacetic acid and processed for estimation of radioactivity. This experiment was modelled on a technique used by Mortimore, Woodside and Henry (1972) to measure rates of protein synthesis in the perfused liver. At high concentrations of extracellular valine, the free pools of amino acids in the tissue become flooded and the intracellular specific radioactivity approaches that in the medium

Table 1.1.   ESTIMATED PROPORTION OF NATIVE 40S RIBOSOMAL SUBUNITS CONTAINING MET-tRNA$_f$

Incubation condition	*Percentage of 40S Subunits containing Met-tRNA*
Complete medium	10.7
Lysine-devoid medium	5.0
Lysine-devoid medium, then re-fed 10 min	8.3

Fed, starved and re-fed Ehrlich ascites wells were incubated for 15 min in the presence of [³⁵S]-methionine, as described in the legend to *Figure 1.10*. Native 40S ribosomal subunits were isolated from sucrose gradients subjected to prolonged centrifugation in order to give good resolution of these lighter particles. They were then purified by equilibrium density gradient centrifugation using caesium chloride. [³⁵S]-Met-tRNA bound to the subunits was measured as the radioactivity which was insoluble in cold, but soluble in hot (90°C) trichloroacetic acid. Further details are given in Pain and Henshaw (1975).

14

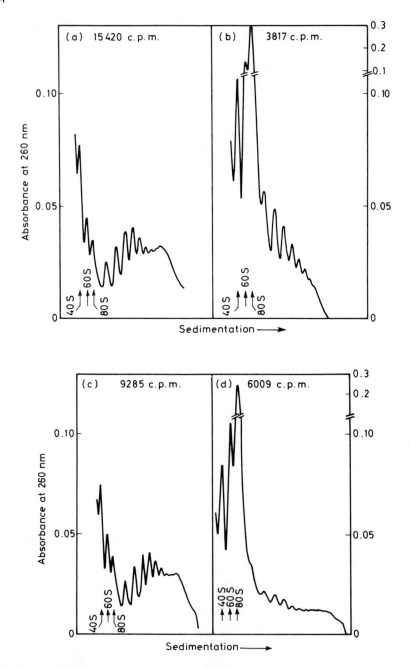

Figure 1.10   Polysome profiles of Ehrlich ascites tumour cells incubated under different nutritional conditons. The panels show profiles from postmitochondrial supernatants of (a) fed cells, (b) cells deprived of lysine for 75 min, (c) cells deprived of lysine for 65 min, then refed this amino acid for 10 min, and (d) cells deprived of glucose for several hours. Incorporation of radioactivity from [^{35}S]-methionine into 50 μl aliquots of cell suspension during the last 15 min of incubation is given in each panel. (Results are taken from Pain and Henshaw, 1975, in which further experimental details are given)

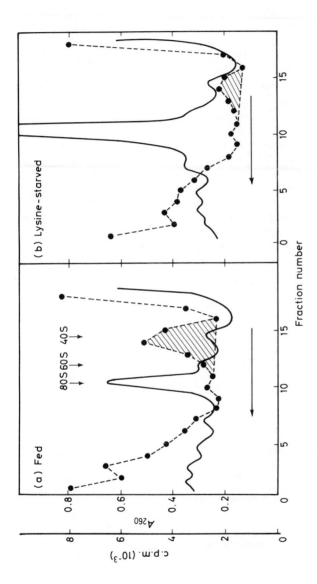

Figure 1.11 Formation of [Met-tRNA$_f$·40S subunit] complexes in cell-free protein synthesizing systems from fed and starved Ehrlich ascites tumour cells. Postmitochondrial supernatants were incubated for 2 min at 30°C with [^{35}S]-methionine in the presence of (final concentrations): 30mH Hepes, pH 7.5, 110 mM KCl, 1.5 mM Mg acetate, 6 mM 2-mercaptoethanol, 0.4 mM spermidine, 4% glycerol, 1 mM ATP, 0.25 mM GTP, 5 mM creatine phosphate, and 180 μg/ml creatine phosphokinase. The mixtures were subjected to sucrose density gradient centrifugation for sufficient time to give good resolution of ribosomal subunits, under which conditions most of the polysomes were pelleted. The gradients were fractionated, the RNA in the fractions was precipitated with cetyl trimethyl ammonium bromide, and radioactivity was determed. (a) Fed cell extract; (b) cell extract from cells deprived of lysine for 30 min. Optical density at 260nm (———); radioactivity (c.p.m./fraction) (● – – – ●)

in confirmation of earlier data (Smith and Henshaw, 1975), that only a small proportion of native 40S subunits contained [^{35}S]-methionine. This proportion was reduced substantially in the starved cells, even though the specific activity of the free pool of methionine in the cells was not altered (Pain and Henshaw, 1975). When the cells were re-fed with lysine there was some restoration of [Met-tRNA 40S] complexes within 10 min.

We are currently investigating this effect in further detail using postmitochondrial supernatants derived from fed and starved cells. Protein synthesis in cell-free extracts is much less active than that in the cells from which they are prepared (except in the case of reticulocytes), but the best retention of activity tends to be seen in rapidly derived crude extracts such as postmitochondrial supernatants, rather than in systems assembled from highly purified components. Since initiation of protein synthesis in cell-free systems is somewhat labile, we estimated formation of [Met-tRNA$_f$·40S] complexes by performing very brief (2 min) incubations of postmitochondrial supernatants in the presence of [^{35}S]-methionine and then isolating the native 40S ribosomal subunits on sucrose gradients. A typical result is depicted in *Figure 1.11*. It can be seen that labelling of the 40S subunits is reduced considerably in the starved cell extract. However, simple measurement of the radioactivity in the 40S peak tends to overestimate the extent of the effect, since extracts from amino acid starved cells tend to contain a higher concentration of methionine in the free amino acid pool. Hence the Met-tRNA$_f$ becomes labelled to a lower specific activity. We have therefore corrected the results obtained with several pairs of fed and starved cell extracts for the specific activity of the free methionine in the system, and a summary of

Table 1.2.   LEVELS OF [40S . Met-tRNA$_f$] INITIATION COMPLEXES IN EXTRACTS FROM FED AND LYSINE-STARVED EHRLICH ASCITES TUMOUR CELLS

Experiment	[40S . Met-tRNA$_f$] (pmol/A$_{260}$)		Inhibition (%)
	Fed	Starved	
1	0.044	0.022	50
2	0.034	0.011	68
3	0.130	0.067	48
4	0.094	0.022	77
5	0.065	0.005	92

Postmitochondrial supernatants were prepared from five sets of fed and lysine-starved Ehrlich ascites tumour cells, incubated with [^{35}S]-methionine and analysed as described in the legend to *Figure 1.11*. The free methionine concentration in each extract was measured on a Locarte amino acid analyser. The incorporation of [^{35}S]-methionine into the total 40S subunit peak could thus be converted to picomoles by correction for the specific radioactivity in the free amino acid pool.

the data is given in *Table 1.2*. While there are quantitative variations in the binding of Met-tRNA$_f$ to native 40S subunits between the different pairs of extracts, it is clear that in each experiment there is a substantial reduction in binding in the postmitochondrial supernatant derived from amino acid starved cells. Thus it is possible to obtain with cell-free systems results which reflect the difference observed in the intact cells.

We now have a situation where we can answer questions concerning the mechanism of the effect. For instance, it was thought possible that it may be mediated by the action of a cellular inhibitor, the activation of which could be induced by starvation. This followed by analogy from studies of the regulation by haem of initiation of protein synthesis in reticulocytes and lysates derived from them. Deficiency of haem results in a pronounced inhibition of initiation, associated with the activation of a protein kinase which phosphorylates the initiation factor (eIF-2) responsible for the binding of Met-tRNA$_f$ to 40S subunits (Farrell *et al.*, 1977; Clemens, 1980). We tested for the presence of a similar, dominant inhibitor in extracts from starved cells by performing a mixing experiment, i.e. estimating binding of Met-tRNA$_f$ to 40S subunits in a mixture of fed and starved extracts. The presence of a dominant inhibitor might be expected to reduce the extent of binding in the mixed extract to the level in the starved preparation. In contrast to this, the actual result showed binding to an extent close to that in the fed preparation, indicating that the starved extract may lack an active component.

Table 1.3. EFFECT OF ADDITION OF PURIFIED INITIATION FACTOR eIF-2 ON FORMATION OF [Met-tRNA$_f$.40S] COMPLEXES BY EXTRACTS FROM FED AND STARVED EHRLICH ASCITES TUMOUR CELLS

	[40S . Met-tRNA$_f$] (pmol/A$_{260}$)	
	Fed	Starved
No addition	0.077	0.022
+ eIF-2	0.077	0.072

The experiment was performed exactly as described in the legends to *Figure 1.11* and *Table 1.2*. Initiation factor eIF-2 (a highly purified preparation kindly supplied by Dr W.C. Merrick) was added at a final concentration of 44 µg/ml.

We have now found that addition of highly-purified eIF-2 to the cell-free system stimulates the ability of the starved extract to form [Met-tRNA$_f$.40S] complexes (*Table 1.3*). This does not occur with the fed extracts, and therefore the difference between them is eliminated.

INITIATION FACTOR ACTIVITY IN MUSCLE

In this final section is a brief description of some recent studies by Rannels and Jefferson and their colleagues on activity of the Met-tRNA$_f$ binding factor, eIF-2, in muscle. For the technical reasons outlined previously it is not feasible to prepare from muscle complete postmitochondrial supernatant systems active in protein synthesis. However, it is possible to isolate a postribosomal supernatant fraction, which, under the high salt conditions normally used for homogenizing muscle, contains initiation factors. Referring to *Figure 1.5*, it can be seen that before Met-tRNA$_f$ and eIF-2 bind to the 40S ribosomal subunit, they form a ternary complex together with GTP. Formation of this complex will occur *in vitro* in the absence of ribosomes, and can be measured easily since [^{35}S]-Met-tRNA$_f$ bound to the complex is retained by Millipore filters, whereas free initiator tRNA passes through.

Rannels and colleagues prepared postribosomal supernatants from muscle of fasted rats (D.E. Rannels *et al.*, 1978) and rats treated with a high dose of glucocorticoids (S.R. Rannels *et al.*, 1978), both conditions which lead to a substantial decline in the rate of protein synthesis in muscle. They found that both treatments reduced the ability of the supernatants to form a ternary complex with Met-tRNA$_f$ when this activity was expressed per gram of muscle (*Table 1.4*). However, there was no decline in activity expressed per gram of muscle RNA. Even the decline in activity per gram of muscle did not appear to be rapid enough to account for an acute effect on ribosomal activity. Moreover, reformation of polysomes in skeletal muscle from starved rats can be brought about by perfusion in the presence of insulin, in the absence of a change

Table 1.4.   Met-tRNA$_f$ BINDING ACTIVITY IN MUSCLE EXTRACTS PREPARED FROM STARVED OR CORTISONE-TREATED RATS

	[^{35}S]-Met-tRNA bound	
	10^{-3} · c.p.m./g muscle	10^{-3} · c.p.m./mg RNA
Experiment 1		
Fed	583 ± 42	490 ± 35
2 days starved	486 ± 8	528 ± 9
3 days starved	456 ± 6	512 ± 7
4 days starved	418 ± 9	464 ± 10
Experiment 2		
Normal	526 ± 50	446 ± 42
Cortisone	342 ± 32	398 ± 37

Results are mean ± S.E.M. for five or six observations.
Experiment 1 is taken from D.E. Rannels *et al.* (1978).
Experiment 2 is taken from S.R. Rannels *et al.* (1978). Rats were treated with 10 mg cortisone acetate/100 g body weight for 5 days.

in the Met-tRNA$_f$ binding activity measurable in the postribosomal supernatant (D.E. Rannels *et al.*, 1978). It therefore seems more likely that the slow change in binding activity in starved and glucocorticoid-treated rats reflects an adjustment in the cellular content of eIF-2 rather than the sort of effect on its state of activation which may be expected to play a role in the acute regulation of initiation.

Acknowledgements

This work was supported by the Medical Research Council. We are very grateful to Dr William C. Merrick for the provision of a highly purified preparation of initiation factor eIF-2.

References

ABELSON, H.T., JOHNSON, L.F., PENMAN, S. and GREEN, H. (1974). *Cell*, 1, 161–165

BRAWERMAN, G. (1976). *Progr. Nucleic Res. Mol. Biol.*, 17, 117–148

CLEMENS, M.J. (1980). In *The Molecular Biology of Gene Expression* (Clemens, M.J., Ed.), CRC Press, Boca Raton, Florida

EMERSON, C.P. (1971). *Nature, New Biol.*, 232, 101–106

FARRELL, P.J., BALKOW, K., HUNT, T., JACKSON, R.J. and TRACHSEL, H. (1977). *Cell*, 11, 187–200

FEIGELSON, P., BEATO, M., COLMAN, P., KALIMI, M., KILLEWICH, L.A. and SCHUTZ, G. (1975). *Rec. Progr. Hormone Res.*, **31**, 213–242

FILIPOWICZ, W. (1978). *FEBS Letts*, **96**, 1–11

FLAIM, K.E., LI, J.B. and JEFFERSON, L.S. (1978). *Am. J. Physiol.*, **234**, E38–E43

GARLICK, P.J. (1979). In *Comprehensive Biochemistry*, Vol. 19B, pp. 77–152 (Florkin, M., Stotz, E., Neuberger, A. and van Deenen, L.L.M., Eds), Elsevier–North-Holland, Amsterdam, in the press

HENSHAW, E.C. (1980). In *Molecular Biology of Gene Expression* (Clemens, M.J., Ed.), CRC Press, Boca Raton, Florida

HENSHAW, E.C., HIRSCH, C.A., MORTON, B.E. and HIATT, H.H. (1971). *J. Biol. Chem.*, **246**, 436–446

HIRSCH, C.A. (1967). *J. Biol. Chem.*, **242**, 2822–2827

HUEZ, G., MARBAIX, G., HUBERT, E., CLEUTER, Y., LECLERCQ, M., CHANTRENNE, H., DEVOS, R., SOREQ, H., NUDEL, U. and LITTAUER, U.Z. (1975). *Eur. J. Biochem.*, **59**, 589–592

JEFFERSON, L.S., LI, J.B. and RANNELS, S.R. (1977). *J. Biol. Chem.*, **252**, 1476–1483

JEFFERSON, L.S., RANNELS, D.E., MUNGER, B.L. and MORGAN, H.E. (1974). *Fedn Proc.*, **33**, 1098–1104

KOZAK, M. (1978). *Cell*, **15**, 1109–1123

LI, J.B., HIGGINS, J.E. and JEFFERSON, L.S. (1979). *Am. J. Physiol*, **236**, E222–E228

MAZUMDER, R. and SZER, W. (1977). In *Comprehensive Biochemistry* (Florkin, M., Neuberger, A. and van Deenen, L.L.M., Eds), Vol. 24, pp. 186–234, Elsevier–North-Holland, Amsterdam

MILLWARD, D.J., GARLICK, P.J., JAMES, W.P.T., NNANYELUGO, D.O. and RYATT, J.S. (1973). *Nature, Lond.*, **241**, 204–205

MORTIMORE, G.E., WOODSIDE, K.H. and HENRY, J.E. (1972). *J. Biol. Chem.*, **247**, 2776–2784

PAIN, V.M. (1978). In *Protein Turnover in Mammalian Tissues and in the Whole Body* (Waterlow, J.C., Garlick, P.J. and Millward, D.J.), Elsevier–North-Holland, Amsterdam

PAIN, V.M. (1980). In *The Molecular Biology of Gene Expression* (Clemens, M.J., Ed.) CRC Press, Boca Raton, Florida

PAIN, V.M. and CLEMENS, M.J. (1980). In *Comprehensive Biochemistry*, Vol. 19B, pp. 1–76 (Florkin M., Stotz, E., Neuberger, A. and van Deenen, L.L.M., Eds), Elsevier–North-Holland, Amsterdam

PAIN, V.M. and GARLICK, P.J. (1974). *J. Biol. Chem.*, **249**, 4510–4514

PAIN, V.M. and HENSHAW, E.C. (1975). *Eur. J. Biochem.*, **57**, 335–342

RANNELS, D.E., PEGG, A.E., RANNELS, S.R. and JEFFERSON, L.S. (1978). *Am. J. Physiol.*, **235**, E126–E133

RANNELS, S.R., RANNELS, D.E., PEGG, A.E. and JEFFERSON, L.S. (1978). *Am. J. Physiol.*, **235**, E134–E139

REVEL, M. and GRONER, Y. (1978). *A. Rev. Biochem.*, **47**, 1079–1126

ROSS, J., IKAWA, Y. and LEDER, P. (1972). *Proc. Natn. Acad. Sci. U.S.A.*, **69**, 3620–3623

ROTTMAN, F.M. (1976). *Trends. Biochem. Sci.*, **1**, 217–219

SAFER, B. and ANDERSON, W.F. (1978). *CRC Crit. Rev. Biochem.*, **5**, 261–290

SCHIMKE, R.T., McKNIGHT, G.S., SHAPIRO, D.J., SULLIVAN, D. and PALACIOS, R. (1975). *Rec. Progr. Hormone Res.*, **31**, 175–211

SHATKIN, A.J. (1976). *Cell,* **9**, 645–653

SMITH, K.E. and HENSHAW, E.C. (1975). *J. Biol. Chem.,* **250**, 6880–6884

TRACHSEL, H., ERNI, B., SCHREIER, M.H. and STAEHELIN, T. (1977). *J. Molec. Biol.,* **116**, 755–767

VAN VENROOIJ, W.J.W., HENSHAW, E.C. and HIRSCH, C.A. (1970). *J. Biol. Chem.,* **245**, 5947–5953

VAN VENROOIJ, W.J.W., HENSHAW, E.C. and HIRSCH, C.A. (1972). *Biochim. Biophys. Acta,* **259**, 127–137

VAN VENROOIJ, W.J.W. and JANSSEN, A.P.M. (1976). *Eur. J. Biochem.,* **69**, 55–60

WATERLOW, J.C., GARLICK, P.J. and MILLWARD, D.J. (1978). *Protein Turnover in Mammalian Tissues and in the Whole Body,* Elsevier–North-Holland, Amsterdam

WEBER, M.J. (1972). *Nature, New Biol.,* **235**, 58–61

WOOL, I.G., STIREWALT, W.S., KURIHARA, K., LOW, R.B., BAILEY, P. and OYER, D. (1968). *Rec. Progr. Hormone Res.,* **24**, 139–209

WOOL, I.G. and STÖFFLER, G. (1974). In *Ribosomes* (Normura, M., Tissiers, A. and Lengyel, P., Eds), Cold Spring Harbor Monograph, pp. 461–488

YOUNG, V.R. (1970). In *Mammalian Protein Metabolism,* Vol. IV (Munro, H.N., Ed.), pp. 585–674, Academic Press, New York

2

FACTORS CONTROLLING INTRACELLULAR
BREAKDOWN OF PROTEINS

R. JOHN MAYER
ROWLAND J. BURGESS
SUSAN M. RUSSELL
Department of Biochemistry, University of Nottingham Medical School

Summary

The presently known factors which control the intracellular breakdown of proteins are described. These factors are considered in terms of the correlates of protein structure and protein degradation; the roles of hormonal and nutritional factors; and the putative relationship between autophagic and non-autophagic processes. Studies on the integration of synthesis and degradation (turnover) of organelle proteins in steady-state conditions are described. Similarly, studies on the turnover of protein mixtures and specific proteins in non-steady-state conditions, i.e. in differentiating and developing cells, are discussed. A hypothesis which considers the mechanism and regulation of protein turnover in steady-state and non-steady-state conditions is presented.

Regulation of intracellular protein degradation

INTRODUCTION

The processes of protein synthesis and protein degradation play equally important roles in the control of protein deposition in animals. Whereas much is known about the mechanism and regulation of protein synthesis, comparatively little is understood concerning these features of protein degradation.

It is clear, however, that not only do different proteins have widely differing degradative rates, e.g. 0.2–150 h for various rat liver proteins (Dice and Goldberg, 1975) but also that these rates can be altered in response to changes in the physiological requirements of the cell.

In this chapter a discussion of the factors implicated in the regulation of protein degradation and a survey of the current knowledge of protein degradation in steady-state and non-steady-state conditions is presented. Putative models which may explain some of the observed phenomena are also given.

PROCESSES OF PROTEIN DEGRADATION

Three separate processes involving intracellular peptide-bond cleavage have been distinguished. First, the processing of proteins by a co-translational proteolytic mechanism, located in the rough endoplasmic reticulum, which removes 'signal' peptides from secretory proteins (Blobel and Dobberstein, 1975; Scheele,

Dobberstein and Blobel, 1978; Shields and Blobel, 1978). Signal sequences may also be implicated in the import of proteins into peroxisomes (Goldman and Blobel, 1978). Processing in this case would be post-translational. Secondly, the rapid degradation of a large proportion of newly synthesized protein. A large proportion of collagen is destroyed within fibroblasts (Bienkowski, Baum and Crystal, 1978) by means of an immediate co-translational or post-translational process. This process may serve to control the amount of the protein which is secreted. Thirdly, the post-translational degradation of proteins which is subjected to physiological regulation (physiological protein degradation).

FACTORS IMPLICATED IN PHYSIOLOGICAL PROTEIN DEGRADATION

Factors influencing the rate of degradation of proteins may be divided into two types: (a) those concerned with the properties of proteins as substrates for degradation; (b) those influencing the machinery of degradation.

Proteins as substrates for degradation

A number of protein properties have been correlated with degradative rates. *Table 2.1* gives a summary of these observations and suggests a possible molecular basis for each. These correlates have been discussed in detail by Goldberg

Table 2.1. CORRELATION OF MOLECULAR PROPERTIES OF PROTEINS WITH DEGRADATION RATES

Correlate	*Possible molecular basis*
Larger subunits are degraded more rapidly than small ones	Larger subunits contain more protease sensitive sites or may have lower stabilities
Acidic protein are degraded more rapidly than basic or neutral proteins	Proteases may bind to anionic sites or act on acidic residues. Charge may affect binding to membranes. Acidic proteins may have lower stabilities
Rapidly turning-over proteins show hydrophobic characteristics	Hydrophobic surface areas on proteins may allow binding to membranes and access to degradative systems
Thermostable proteins are destroyed slowly	Degradation requires partial or complete denaturation of the protein
In vitro susceptibility to endoproteases correlates with *in vivo* degradative rates	Degradation rates are determined by general conformational features of proteins

and Dice (1974), Dean (1975), Dice and Goldberg (1975), Goldberg and St. John (1976), Ballard (1977) and McLendon and Radany (1978). The potential for a protein to act as a substrate for degradation depends both on its susceptibility to proteolytic digestion and on its ability to interact with the degradative systems. In the former case proteins could contain protease-sensitive sites such as regions of loose folding or specific amino acid sequences which are recognized by proteases. Certain proteins could only reveal proteolytically sensitive sites when partially or completely denatured. The rate of degradation of these proteins will therefore be governed by their stability (McLendon and Radany, 1978). In the second case binding to, and transport across, membranes may be required for access to the degradative systems. Surface charge and regions of hydrophobic nature on the protein surface may influence this interaction with intracellular membranes.

The ligand binding of substrates, coenzymes and other factors may affect the degradation rate of a protein by altering its conformation, changing its stability or by masking protease-sensitive sites.

In all known cases degradation rates have been decreased by ligand binding. In rats, administration of tryptophan leads to decreased degradation of tryptophan oxygenase (Schimke, Sweeney and Berlin, 1965a, 1965b) and of pyruvate carboxylase (Ballard and Hopgood, 1973). Ligand stabilization of protein conformation may be reflected by reduced *in vitro* susceptibilities to proteases (Goldberg and Dice, 1974). Apoenzymes of pyridoxal- and NAD-requiring enzymes have a greater sensitivity to group specific proteases than their holoenzymes (Katunuma, Kito and Kominami, 1971a, 1971b). Litwack and Rosenfeld (1973) have shown a correlation between the relative rate of dissociation of cofactors from five liver enzymes and the half-lives of these enzymes.

Although ligand binding has not been shown to increase degradation, the sensitivity of haemoglobin to carboxypeptidase (Zito, Antonini and Wynan, 1964) and glycogen phosphorylase to trypsin (Graves *et al.*, 1968) are increased by the binding of oxygen and glucose, respectively.

Covalent modification of proteins is a possible regulatory mechanism of protein degradation and could act by changing protein conformation and stability, or by generating protease-sensitive sites or changing the affinity of proteins for endomembranes. Ekman *et al.* (1978) have demonstrated the proteolytic removal of sites on proteins which can be phosphorylated. Unphosphorylated sites which can be phosphorylated are twice as sensitive to proteolysis as phosphorylated sites. Phosphorylation may therefore stabilize against proteolysis. Dice *et al.* (1978) have shown that intracellular glycoproteins are degraded more rapidly than non-glycoproteins. This phenomenon does not occur in liver and muscle from starved and diabetic animals but is unchanged for glycoproteins in brain in these conditions.

Ashwell and Morell (1974) have shown that the presence of terminal sialic acid on oligosaccharides on circulatory glycoproteins is essential for continued circulation, since desialated proteins are rapidly removed. Exposure of terminal galactose residues is necessary for this removal from the circulation.

Mammalian cells possess the ability to selectively degrade abnormal proteins, often at rapid rates. Much of the evidence for this is supplied by experiments with amino acid analogues. Liver proteins containing canavanine in place of arginine are degraded at several times the rate of the normal proteins (Knowles *et al.*, 1975).

A number of abnormal haemoglobins are known to be unstable and to denature rapidly. These variants are degraded rapidly (Goldberg and St. John, 1976). Premature termination of protein chains brought about by puromycin produces peptides with half-lives as short as 10 min (McIlhinney and Hogan, 1974).

Degradation of abnormal proteins is probably brought about by immediate co-translational proteolysis, although some post-translational proteolysis and physiological protein degradation may occur.

Protein degradation systems

That more than one proteolytic system may exist in cells has been considered by many workers, but Ballard (1977) and Amenta, Sargus and Baccino (1977) have recently both proposed two independent pathways of protein degradation.

During autophagy, cell constituents are enclosed by membranes to form vacuoles which fuse with primary lysosomes to form autophagolysosomes in which proteins and other molecules are degraded. Knowles and Ballard (1976) estimate that autophagy accounts for approximately 30% of intracellular protein

Table 2.2.   FACTORS AFFECTING AUTOPHAGIC ACTIVITY

Stimulus	Response	Reference
Hormonal		
Insulin	Decreased vacuole formation in liver	Pfeifer *et al.* (1978)
Glucagon	Increased vacuole formation in liver	Arstila *et al.* (1972)
Chemical		
1. Respiratory inhibitors	Autophagic proteolysis 60% inhibited	Amenta *et al.* (1977)
2. Inhibition of protein synthesis by cycloheximide	Autophagic degradation inhibited	Amenta *et al.* (1977) Knowles and Ballard (1976)
3. Microtubular dissaggregators (vinblastine and colchicine)	Partial inhibition of autophagy	Amenta *et al.* (1977)
4. Proteolytic inhibitors	Inhibition of lysosomal proteases	Dean (1975) Dean and Barrett (1976)
5. Lysosomal weak bases	Degradation inhibited	Hopgood *et al.* (1977) Goldberg and St. John (1976)
Diurnal rhythms	Vacuole number increased by day, decreased by night in liver, pancreas and kidney	Pfeifer and Scheller (1971) Pfeifer (1976)
Nutritional		
Nutritional step-down of cultured cells	Transitory activiation of auto-phagic processes	Amenta *et al.* (1977) Knowles and Ballard (1976)

degradation. Ballard (1977) proposes that control of the rate of autophagic degradation is exerted by variations in the capacity of the system, i.e. in the number of vacuoles formed.

The autophagic degradative system cannot account for degradation of short-lived proteins because of both the short time periods involved and the insensitivity of the breakdown of short-lived proteins to inhibitors of autophagy. Ballard (1977) therefore defines a second degradative system, probably involving direct proteolytic breakdown, which is limited by the ability of each protein to act as a substrate for proteolysis (substrate limited degradation). Although the subcellular location(s) of such a system is unknown, it may involve membrane-linked processes in which enzymes become inactivated. Inactivation of enzymes may be a first step in their degradation (Knowles and Ballard, 1976).

Amenta, Sargus and Baccino (1977) distinguish between a process accounting for a steady basal rate of protein degradation and a promptly activated indiscriminate autophagic process. Selective lysosomal degradation and an alternative process may contribute to the basal rate of degradation.

Autophagic activity may be changed by a variety of factors (*Table 2.2*).

A rapid decrease in the number of autophagic vacuoles in hepatocytes is observed by electron microscopy following insulin administration to adult rats (Pfeifer, Werder and Bergeest, 1978). The decrease is probably mediated by inhibition of the formation of new vacuoles. Vacuole formation is increased in liver by the action of glucagon (Ashford and Porter, 1962; Arstila, Shelburne and Trump, 1972).

The activity of the autophagic system may be modulated in a variety of ways by different chemical agents. Depression of ATP levels by inhibitors of energy metabolism, such as cyanide or dinitrophenol, is known to reduce protein degradation (Simpson, 1953). Low ATP levels may decrease autophagic degradation by restriction of an active proton pump which is required for maintenance of the low lysosomal pH (Ballard, 1977). Additionally, energy may be required for synthesis of proteins necessary for vacuole formation. Inhibition of protein synthesis by cycloheximide decreases protein degradation by an equal amount to inhibition by insulin (Knowles and Ballard, 1976) suggesting that they affect the same system, i.e. autophagy. Disruption of microtubular function by colchicine and vinblastine is known to inhibit autophagy (Amenta, Sargus and Baccino, 1977). Inhibition of degradation by inhibitors of proteolytic enzymes such as pepstatin (Dean, 1975), antipain and leupeptin (Dean and Barrett, 1976) is probably due to action on lysosomal proteases.

Finally, protein catabolism may be reduced by the accumulation of weak bases, e.g. ammonia or methylamine (Hopgood, Clark and Ballard, 1977) and chloroquine (Goldberg and St. John, 1976) within lysosomes. Pfeifer and Scheller (1971) have observed a diurnal variation of the number of autophagic vacuoles in rat liver, pancreas and kidney convoluted tubule cells. The number decreases to a minimum during the rats' active night phase, i.e. when they eat, and increases to a maximum during their resting daytime phase. Reduction of the supply of nutrients in starved animals leads to an increase of the lysosomal enzymes in the soluble fraction of liver (Pontremoli *et al.*, 1973). Amenta, Sargus and Baccino (1977) and Knowles and Ballard (1976) report activation of the autophagic degradative system in cultured mammalian cells upon nutritional step-down by removal of serum from the growth medium.

Protein degradation in steady-state conditions

The amount of a protein within a cell is governed by the balance between the rate of synthesis and the rate of degradation of the protein. In steady-state conditions

$$\frac{dP}{dt} = 0 = K_S - K_D P \quad \text{(Schimke, 1973)}$$

and

$$P = \frac{K_S}{K_D}$$

where P is the protein mass, K_S the rate of synthesis, and K_D the rate of degradation. Control mechanisms which balance the opposing processes of protein synthesis and degradation may exist for individual proteins (heterogeneous or individual control of protein turnover) or for subgroups of proteins (homogeneous or unit control of protein turnover) within an organelle or subcellular compartment.

TURNOVER OF PROTEINS IN DIFFERENT CELL TYPES

Studies have been performed on the turnover of the same protein in several tissues and cell types.

The rate of degradation of catalase was similar when measured in several tissues of the mouse and budgerigar by the rate of disappearance of enzyme inhibited by aminotriazole (Crane, Holmes and Masters, 1978). There was, however, a considerably higher rate of degradation of the enzyme in the mouse intestine. This and other observations have generated the notion that proteins are degraded faster in intestinal cells (Cameron, 1971; Jones and Mayer, 1973; Crane, Holmes and Masters, 1978). Fritz *et al.* (1969) showed with a single-isotope technique that the rate of degradation of lactate dehydrogenase isoenzyme 5 was considerably different in the tissues of the rat (liver $t_{1/2}$ = 16 days, heart $t_{1/2}$ = 1.6 days, skeletal muscle $t_{1/2}$ = 31 days). Segal *et al.* (1969) determined the half-life of alanine aminotransferase in rat muscle to be 20 days and that in the liver to be 3 days. Both studies may be criticized in that re-utilization of the amino acid precursor may differ between tissues giving an artefactual variation in the degradation rate. However, such limited studies do suggest that a protein may have a different rate of turnover in different cell types.

TURNOVER OF PROTEINS IN DIFFERENT SUBCELLULAR COMPARTMENTS

It is clear that measurements on protein mixtures (*Table 2.3*) give the mean half-lives for proteins in the mixture. However, the half-lives of individual proteins in

Table 2.3.   MEAN HALF-LIVES OF PROTEINS IN VARIOUS SUBCELLULAR ORGANELLES OR FRACTIONS OF RAT LIVER

Fraction	Mean half-life of proteins	Reference
Cytoplasm	5.1 days	Arias *et al.* (1969)
	4–5 days	Schimke (1973)
Lysosome	7.1 days	Arias *et al.* (1969)
	4 days	Wang and Touster (1975)
	30 h	Segal (1975)
Nucleus	5.1 days	Arias *et al.* (1969)
Ribosomes	2–3 days	Buchanan (1961)
	5 days	Hirsch and Hiatt (1966)
Endoplasmic reticulum	2 days	Arias *et al.* (1969)
Plasma membrane	1.8 days	Arias *et al.* (1969)
Mitochondria	5 days	Aschenbrenner *et al.* (1970)

the mixture may show wide variation; for example, $t_{1/2}$ for cytosolic tyrosine aminotransferase is 1.5 h (Kenney, 1967) and $t_{1/2}$ for cytosolic arginase is 4–5 days (Schimke, 1964). The contribution from each individual protein to the mean rate of turnover is dependent not only on the turnover rate of the individual protein but also on its proportion in the total protein mass. The mean protein turnover rate for any particular subcellular compartment is therefore of limited information since it does not indicate how much heterogeneity of protein turnover rate exists within that compartment.

Cytoplasm

Certain structural properties of cytoplasmic proteins correlate with their rate of degradation (see page 22). These include correlations of subunit size, isoelectric point, hydrophobicity, ligand binding and thermodynamic parameters related to folding (see Goldberg and Dice, 1974 and Ballard, 1977 for reviews). Subunit size and isoelectric point have been shown to be linked in a general way for cytosolic proteins in liver and muscle to give a correlation with protein turnover rates. Thus the acidic proteins tend to turn over faster and are larger than the neutral and basic proteins (Duncan and Bond, 1977).

Lysosomes

The turnover of rat liver lysosomal enzymes has been measured with two radio-isomers of leucine as protein precursors in a double isotope technique (Arias, Doyle and Schimke, 1969; Dean, 1975; Wang and Touster, 1975). Membrane and non-membrane proteins obtained from lysosomes prepared from rats with or without treatment with Triton WR-1339 show a high degree of heterogeneity of turnover rates (Dean, 1975; Wang and Touster, 1975). No general correlation of degradation rate with subunit size was found for either membrane or non-membrane lysosomal proteins (Dean, 1975; Wang and Touster, 1975).

The turnover values obtained for lysosomal proteins are complicated both by the methods used to prepare the lysosomes and the problem of reutilization of the radiolabelled precursors. These complications may account for the different mean half-life for total lysosomal proteins obtained by Arias, Doyle and Schimke (1969) and by Segal (1975) who determined mean half-lives of lysosome proteins to be 7.1 days and 1.25 days, respectively.

The measurement of protein turnover in lysosomes may also be complicated by the role of lysosomes in the degradation of intracellular proteins (Dean, 1975; Dean and Barrett, 1976). Fragments derived from intralysosomal proteolysis may contribute to the heterogeneity observed for degradative rates of lysosomal proteins.

Nucleus

The mean half-life for rat liver nuclear proteins (*Table 2.3*) was determined with [^{14}C-guanidino]-arginine (Arias, Doyle and Schimke, 1969) which is re-utilized to some extent in liver. Rat liver chromosomal proteins and isolated nuclear acidic proteins show a subunit size correlation (Dice and Schimke, 1973). The rate of turnover of the histones which are small, basic proteins was very slow or undetectable. This is in agreement with observations on cytoplasmic proteins where small basic proteins have low rates of turnover. The mean half-life of the nuclear membrane proteins was estimated to be 2 days which is much shorter than that of the non-membrane nuclear proteins (Widnell and Siekevitz, 1967).

Ribosomal and endoplasmic reticulum proteins

Rat liver ribosomal proteins are destroyed heterogeneously and there is a systematic relationship between the size of a protein subunit and its relative rate of protein turnover; larger protein subunits have on the whole faster turnover rates (Dice and Schimke, 1972).

There appears to be a marked degree of heterogeneity in the rates of turnover of rat liver smooth microsomal and ribosomal proteins. Similar results were obtained for ribosome-depleted rough microsomes. A subunit size correlation was generally seen, except for cytochrome P-450 which is a major protein constituent of the endoplasmic reticulum (Arias, Doyle and Schimke, 1969; Dehlinger and Schimke, 1971).

Plasma membrane

The proteins of the plasma membrane have a mean turnover rate comparable to that observed for the endoplasmic reticulum and nuclear membrane (*Table 2.3*). Again, in rat liver plasma membrane a protein subunit size and degradation rate correlation has been described together with a degree of heterogeneity of turnover rates (Dehlinger and Schimke, 1971; Gurd and Evans, 1973). The real degree of heterogeneity of turnover of the plasma membrane proteins may however be much less than is observed, due to difficulties of preparing pure plasma membranes.

The plasma membrane is difficult to purify without significant contamination by membranes of lysosomes, endoplasmic reticulum and Golgi apparatus. Non-membrane cytoplasmic proteins may also be present as trapped vesiculated proteins (Tweto and Doyle, 1977). Gurd and Evans (1973) have found little heterogeneity in the turnover of plasma-membrane proteins, except for a fraction enriched in glycoproteins.

Plasma membrane proteins may be degraded coordinately at homogeneous rates. Proteins in different regions of the membrane may be degraded at different rates. Such homogeneity may be obscured in the experiments of Arias, Doyle and Schimke (1969), Dehlinger and Schimke (1971) and Gurd and Evans (1973). Proteins which contaminate cell membrane preparations (e.g. cytosolic proteins) may have considerable heterogeneity in their degradation rates. Such contamination may lead to the observed heterogeneity of degradation rates observed for plasma membrane proteins. Evidence in support of this proposal is outlined by Tweto and Doyle (1977). The arguments suggesting that the measured heterogeneity of turnover rates of plasma membrane proteins may obscure a true homogeneous or unit turnover rate of these proteins may also apply to reported measurements on the turnover of other membrane proteins.

Mitochondria

Studies on mitochondrial proteins support a model in which heterogeneous and homogeneous turnover occur within the same organelle. Studies on the haem labelling of rat liver cytochrome b, cytochrome c and cytochrome c oxidase suggested very similar rates of turnover for these inner membrane proteins (*Table 2.4*). This suggests the coordinated turnover of a large proportion of

Table 2.4. HALF-LIVES OF SPECIFIC RAT LIVER MITOCHONDRIAL PROTEINS

Mitochondrial subfraction	*Protein*	*Half-life*	*Reference*
Outer membrane	Cytochrome b_5	4.4 days	Druyan *et al.* (1969)
Intermembrane space	–	–	–
Inner membrane	Haem a	5.9 days	Aschenbrenner *et al.* (1970)
	Cytochrome c	5.8 days	Aschenbrenner *et al.* (1970)
	Cytochrome oxidase	5.7 days	Ip *et al.* (1974)
	Cytochrome b	5.5 days	Druyan *et al.* (1969)
	Cytochrome c	6.1 days	Druyan *et al.* (1969)
Matrix	Ornithine amino transferase	1.9 days	Ip *et al.* (1974)
	δ-Aminolevulinate synthetase	70 min	Marver *et al.* (1966)
	Ornithine amino transferase	0.7 days	Swick *et al.* (1968)
	Alanine amino transferase	1.0 days	Swick *et al.* (1968)

proteins in the inner mitochondrial membrane (Gross, Getz and Rabinowitz, 1969). Evidence from studies on outer membrane proteins (De Bernard, Getz and Rabinowitz, 1969; Druyan, De Bernard and Rabinowitz, 1969; Aschenbrenner *et al.*, 1970) indicates that their turnover rates are faster than those of the inner membrane. Gear (1970) has suggested that the turnover of some mitochrondrial proteins in different submitochondrial compartments may be coordinated. The turnover rates of the non-membrane mitochondrial proteins appear to be faster and more heterogeneous than the rates for membrane proteins (*Table 2.4*).

Studies on the mechanism of degradation of mitochondrial proteins may provide insight into mechanism(s) of degradative processes which occur in subcellular organelles.

MITOCHONDRIAL PROTEIN TURNOVER IN THE RAT LIVER

Measurement of protein turnover has many practical and theoretical difficulties. These have been outlined by several authors in great detail (Poole, 1971; Goldberg and Dice, 1974; Ip, Chee and Swick, 1974; Goldberg and St. John, 1976; Ballard, 1977; Tweto and Doyle, 1977; Burgess, Walker and Mayer, 1978). Some of these problems may be reduced considerably by the use of minimally re-utilizable precursors in the double-isotope procedure (Arias, Doyle and Schimke, 1969). By suitable choice of precursors (Burgess, Walker and Mayer, 1978) and subfractionation techniques the relative turnover of mitochrondrial proteins may be estimated accurately and the turnover characteristics of protein subgroups identified. The protein precursors used to study mitochondrial protein turnover by the double-isotope technique are $NaH[^{14}C]O_3$ and L-$[5-^3H]$-arginine.

Two subfractionation procedures have been carried out: an operationally defined subfractionation of mitochondria based on differential solubility of mitochondrial proteins (Walker, Burgess and Mayer, 1978) and a morphological fractionation by the method of Schnaitman and Greenawalt (1968).

Protein turnover in operationally defined submitochondrial fractions

Proteins were solubilized from mitochondrial preparations by water, salt, Tween 20 and Triton X-100. The Triton X-100 soluble proteins were further fractionated at DEAE-cellulose chromatography (Walker, Burgess and Mayer, 1978). There was heterogeneity in the turnover rates of proteins in each submitochondrial fraction, as shown by the variation in isotope ratios of the protein subunits separated by SDS-polyacrylamide gel electrophoresis. Heterogeneity of isotope ratios was greatest in fractions containing predominantly mitochondrial matrix and intermembrane space proteins (water-soluble fraction). There was less heterogeneity in isotope ratios for proteins in membrane fractions compared to the non-membrane proteins, particularly in the Triton X-100 soluble material not binding to DEAE-cellulose. These proteins also had very low isotope ratios indicating the low rates of degradation. The slow rate of protein turnover in the Triton X-100 fraction and low degree of heterogeneity may be a reflection of the long and similar half-lives of the integral inner membrane proteins (Druyan, De Bernard and Rabinowitz, 1969; Gross, Getz and Rabinowitz, 1969; Aschenbrenner *et al.*, 1970, Ip, Chee and Swick, 1974; Walker *et al.*, 1978). No subunit size correlation with rate of protein degradation was evident in any of the fractions (Walker, Burgess and Mayer, 1978).

Protein turnover in morphologically defined mitochondrial subcompartments

The method (*Figure 2.1*) gives fractions containing inner mitochondrial membrane proteins, outer membrane proteins, intermembrane space proteins and matrix proteins, as shown by marker enzyme measurements.

The rate of degradation of the proteins in mitochondrial subfractions is shown in *Figure 2.2*.

The degree of heterogeneity within each fraction (represented by mean ± S.D.) varies considerably. An estimate of the inherent experimental variation of isotope ratios is given by the error of the isotope ratios for proteins in mitochondria from control animals.

Clearly there is considerable heterogeneity in protein degradation rates in the outer membrane (4.4 X the control variation) and the intermembrane space fractions (2.7 X the control variation). The inner membrane plus some matrix proteins and matrix protein fractions turn over homogeneously: the variation

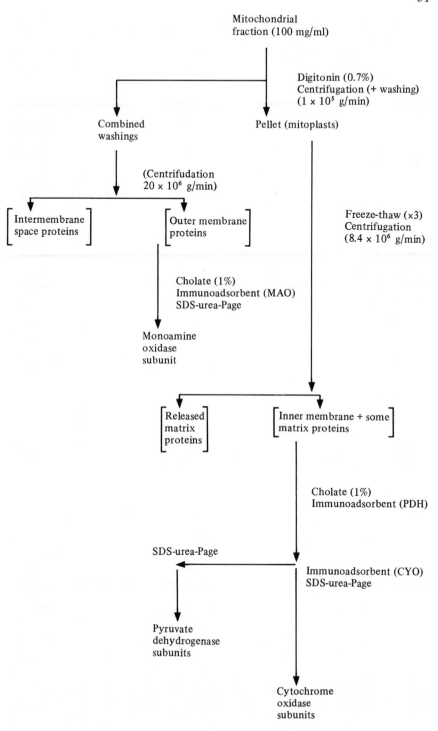

Figure 2.1   Mitochondrial fractionation procedure

32

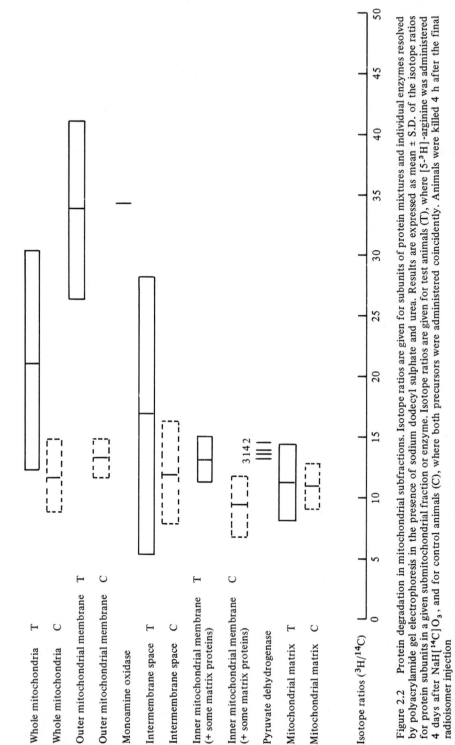

Figure 2.2 Protein degradation in mitochondrial subfractions. Isotope ratios are given for subunits of protein mixtures and individual enzymes resolved by polyacrylamide gel electrophoresis in the presence of sodium dodecyl sulphate and urea. Results are expressed as mean ± S.D. of the isotope ratios for protein subunits in a given submitochondrial fraction or enzyme. Isotope ratios are given for test animals (T), where [5-³H]-arginine was administered 4 days after NaH[¹⁴C]O₃, and for control animals (C), where both precursors were administered coincidently. Animals were killed 4 h after the final radioisomer injection

in isotope ratios for proteins from the test and control animals is very similar. The results in *Figure 2.2* also show the difficulty of drawing conclusions from the mean isotope ratios of the mitochondrial protein mixture. These mean isotope ratios give no information relevant to protein turnover in the mito-chondrial subcompartments. This criticism applies to interpretation of the measurements on the turnover of any protein mixture. No subunit size correla-tion with degradation rate was observed for any fraction.

Turnover of specific mitochondrial enzymes

As shown in *Figure 2.2*, three specific enzymes were immunoisolated from the submitochondrial fractions: monoamine oxidase, pyruvate dehydrogenase and cytochrome oxidase. The immunoisolated enzymes were further analysed by SDS–polyacrylamide gel electrophoresis and the isotope ratios determined for the subunits of each enzyme. Sufficient radioactivity was found in the protein subunits of monoamine oxidase and pyruvate dehydrogenase to determine accurately the isotope ratios for each subunit (*Figure 2.2*). Monoamine oxidase subunits are degraded at a relatively high rate compared to the proteins in other mitochondrial fractions. This result is similar to that obtained by Druyan, De Bernard and Rabinowitz (1969). The subunits of pyruvate dehydrogenase (which fractionated with the inner mitochondrial membrane proteins, *Figure 2.1*) have similar rates of degradation. The variation in isotope ratios between sub-units is clearly within the experimental error for the inner membrane fraction and is also similar to other rates of degradation of proteins in this fraction. The subunits of pyruvate dehydrogenase have very different molecular weights (36 000–74 000; Hamada *et al.*, 1975, 1976; Sakurai *et al.*, 1970) and yet all are apparently being turned over within the mitochondrion at similar or equal rates.

Studies on the turnover of other putative multienzyme complexes are subject to criticism. Experiments with fatty acid synthetase have shown that the 6–7 subunits observed on SDS–polyacrylamide gel electrophoresis turn over at different rates; the larger subunits turning over faster than the small subunits (Tweto, Dehlinger and Larrabee, 1972). However, there is strong evidence to suggest that fatty acid synthetase is composed of only two subunits of approx-imately equal molecular weight (Stoops *et al.*, 1978), the other smaller subunits being the result of *in vitro* proteolytic cleavage. A separate study by Hayes and Larrabee (1971) has shown that the component peptides of fatty acid synthetase turn over at approximately equal rates.

The heavy and light chains of myosin are degraded at similar or equal rates (LaGrange and Low, 1976) as are myosin and the actin (Rubinstein *et al.*, 1976).

A MODEL FOR THE TURNOVER OF MITOCHONDRIAL PROTEINS

A prerequisite of any model of mitochondrial protein turnover is the recognition of the existence of membrane and non-membrane proteins within the mito-chondrion. Membrane proteins are unlikely to exist free in the cytoplasm for any length of time (Tweto and Doyle, 1977), are specifically orientated in the membrane (Rothman and Lenard, 1977) and can move in the plane of the membrane (Singer and Nicholson, 1972).

Protein turnover in steady-state conditions implies a balance between synthesis and degradation. Proteins are synthesized and inserted into the mitochondrion and either degraded *in situ* or exported for subsequent degradation. The turnover of mitochondrial proteins is illustrated in *Figure 2.3*. This model can be included in a general protein turnover cycle (see *Figure 2.7*). Mitochondrial membrane proteins would be synthesized and inserted in their correct orientation in the endoplasmic reticulum membrane. Translocation would then occur either via a membrane continuum or via a vesicular route; thus the proteins

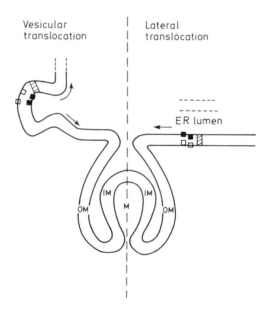

Figure 2.3   Mitochondrial assembly and disassembly. Proteins primarily on/in the luminal side of the ER (■), proteins primarily on/in the cytoplasmic side of the ER (□), transmembrane proteins (▨), vesicular (V), inner mitochondrial membrane (IM), matrix (M), outer mitochondrial membrane (OM), endoplasmic reticulum (ER)

retain their specific orientation (*Figure 2.3*). One feature of this model is that the two membranes of the mitochondrion are formed from an invaginated single membrane. The points of attachment of the mitochondrial outer membrane to the endoplasmic reticulum (Franke and Kartenbeck, 1971; Shore and Tata, 1977a, 1977b) or the inner membrane to the outer membrane (Hackenbrock and Miller, 1975) may not be permanent but the two membranes may continuously fuse and separate. Non-membrane proteins may be inserted into the mitochondrion by a similar mechanism involving temporary attachment of these proteins to the membrane during translocation (Kawajiri, Hirano and Omura, 1977). Degradation may be achieved by either reversal of the insertion mechanism and subsequent destruction in the Golgi endoplasmic reticulum lysosome complex, or by *in situ* proteolysis. Regulation of rates of protein insertion and removal may be achieved as shown in *Figure 2.7*.

Protein degradation in non-steady states: studies on differentiating and developing cells

INTRODUCTION

Cell differentiation implies the progressive restriction of developmental potential which in the final stages (terminal differentiation) is associated with the acquisition of the cell-specific proteins and enzymes which are involved in the production of the morphological and physiological characteristics of the phenotypically differentiated cell (Rutter, Pictet and Morris, 1973). Many studies on the turnover of enzymes and proteins have been carried out during terminal differentiation, since it is only during this period that sufficient material is available for the measurements of rates of protein synthesis and protein degradation. This problem is particularly acute when measurements on the turnover of specific proteins have been made and is of comparatively less importance for studies on the turnover of protein mixtures.

Protein accumulation in terminally differentiating cells may take place rapidly (e.g. in mammary gland) or slowly (e.g. in muscle). Protein accumulation (hypertrophy) usually occurs against a background of hyperplasia in many tissues. Hypertrophy occurs when protein synthesis exceeds protein degradation and the proteins synthesized in a particular cell type are a consequence of selective gene expression. However, the change in the amount of an enzyme or protein will depend not only on quantitative and temporal change in protein synthesis but also on any such changes in protein degradation during the phase of accumulation of the protein in the cell (Mayer and Paskin, 1978). Finally at least three protein subgroups may currently be recognized in terminally differentiating cells: these are cell-specific intracellular proteins, cell-specific secretory proteins and non-cell-specific proteins. It is to these protein subgroups that attention must be paid if the mechanism and regulation of protein turnover in differentiating cells is to be understood (Paskin and Mayer, 1977).

We will not understand the factors which control the breakdown of protein in differentiating cells until we recognize the protein turnover phenomena in specific cell types and collate the generality of these phenomena in many different species. Protein synthesis and degradation (protein turnover) must be seen against the developmental programme of specific cell types. Protein turnover must be considered with respect to the energy needs of a developing cell, the period over which the developmental change takes place, the degree of production of intracellular and extracellular cell-specific proteins in particular cell types, and any remodelling processes which may occur during differentiation. Furthermore, specific enzyme turnover characteristics may be related to the regulatory role of an enzyme: the accumulation profiles of key regulatory enzymes will determine flux through specific metabolic pathways.

The observed phenomena which are considered here concern the nature of protein degradation, the turnover characteristics of protein subgroups, and the turnover characteristics of protein mixtures and individual proteins. The generality of these observations is considered and a putative model for the mechanism and regulation of the phenomena suggested.

The nature of protein degradation

Recent advances in the study of the translation of proteins necessitates a clear definition of the presently recognizable types of proteolysis of proteins. Co-translational protein processing by limited proteolysis has been ascribed a key role in the secretion of secretory proteins (Blobel, 1978). It has also recently become apparent that a disproportionately large quantity (30–40%) of newly synthesized collagen is destroyed in fibroblasts to give fragments of around 500 daltons (Bienkowski, Baum and Crystal, 1978). This phenomenon may apply to non-secretory proteins produced in other cell types (e.g. in liver; Scornik and Bobtol, 1976), where 20% of newly synthesized intracellular protein is destroyed within three hours. Proteins are also destroyed post-translationally by processes of unknown mechanism which, along with protein synthesis, control the amounts of proteins in cells. In short, proteins may be co-translationally proteolytically processed; destroyed immediately co-translationally or post-translationally, or post-translationally destroyed at characteristic rates in defined physiological conditions.

Turnover characteristics of protein subgroups

The synthesis and degradation of proteins which are cell-specific and non-cell-specific may be expected not to be identical. Their turnover characteristics may be as shown in *Table 2.5*. Degradation applies here to the process of destruction

Table 2.5.   PROTEIN SUBGROUP TURNOVER DURING CYTODIFFERENTIATION

Subgroup	Rate of protein synthesis	Rate of protein degradation
Cell-specific intracellular proteins	Large (usually sustained) increase	Transient decrease during phase of most rapid protein accumulation
Cell-specific secretory proteins	Large (usually sustained) increase	Little degradation during accumulation, none during secretion
Non-cell-specific proteins	Some increase	Little or no change

of proteins post-translationally. Whether the interpretation proposed in *Table 2.5* is applicable to many developmental processes is the subject of the rest of this article. The production of cell-specific intracellular and secretory proteins is normally associated with the production and translation of increased amounts of specific messenger-RNA (e.g. Rosen and Barker, 1976; Farmer *et al.*, 1978). The extent, if any, of the modulation of rates of protein degradation during protein accumulation in developing cells is controversial. However, it must be noted that whatever modulation of protein degradation occurs in differentiating cells, protein degradation generally takes place in these cells often at high rates. Protein degradation probably has an important role in remodelling processes (Swick and Ip, 1974; Lodish, Small and Chang, 1975) as well as in the regulation of the amounts of enzymes and therefore metabolic flux (Paskin and Mayer, 1977). Finally, it might be expected that, by virtue of their packaging, secretory proteins would not be degraded post-translationally when they accumulate to some extent intracellularly during cytodifferentiation and presumably will not be degraded when secreted.

Turnover characteristics of protein mixtures

Studies of protein turnover have been carried out on protein mixtures and individual proteins. Both approaches are limited – the former since measurements are made on mixtures and therefore represent average turnover rates, the latter since measurement of the rate of synthesis and degradation of a single protein may not give sufficient information relevant to proteins in general in a cell in a specific physiological state. However, in spite of the limitations, studies on single proteins may be the only way to establish the molecular details of the mechanism of protein degradation. The comparative lack of progress in understanding protein degradation (relative to protein synthesis) is due to the difficulties of measuring the rate of degradation. Except in specially controlled circumstances, most authors prefer indirect estimation of protein degradation, i.e. the difference between protein accumulated and protein synthesized in a specified time rather than the direct estimation of protein degradation, i.e. by pulse-chase radioisotopic techniques. The isotopic decay techniques are fraught with difficulties due to isotopic re-utilization.

The turnover characteristics of protein mixtures in several developing tissues are shown in *Table 2.6.* The basic points concern the possible existence of general

Table 2.6.   TURNOVER CHARACTERISTICS OF PROTEIN MIXTURES DURING DEVELOPMENTAL PROTEIN ACCRETION

Tissue	*Rate of accretion*	*Rate of synthesis*	*Rate of degradation*	*Reference*
Postnatal rat skeletal muscle (0– 100 days)	Declines 6%–1% per day	Declines 36%–6% per day	Declines 24%–5% per day	Millward *et al.* (1975)
Newly hatched chicken muscle (1–2 weeks)	Greater at 2 weeks than 1 week	Approx. constant 20% per day	Possibly less at 1 week than 2 weeks	Maruyama *et al.* (1978)
Neonatal rat brain (0–30 days)	Greatest 2–10 days	Approx. constant 2.9% per hour	Approx. constant 1.8% per hour	Dunlop *et al.* (1978)
Neonatal rat brain (0–25 days)	Greatest 0–10 days	Approx. constant 2% per hour	Approx. constant 1% per hour	Berger *et al.* (1978)
Neonatal mouse liver (4 days)	Maximum at <4 days	Similar to adult approx. 36% per day	Half adult rate	Conde and Scornik (1977)
Regenerating adult mouse liver	Approx. 40% per day increase	Approx. 15% per day greater than adult	Half adult rate	Scornik and Bobtol (1976)

relationships between the rates of synthesis and degradation in different tissues. In muscle and brain the greatest rates of protein accretion are at developmental phases at which both synthesis and degradation rates are high and are either both declining (muscle) or initially approximately constant (brain). In contrast, protein accretion in neonatal (and regenerating) liver is achieved at the expense of a decreased degradative rate which is the most important regulatory factor determining the protein content of liver (Conde and Scornik, 1977). Although normal protein accretion in muscle may occur postnatally when degradation rates are high, protein degradation rates are further increased in nutritionally impaired chickens (Maruyama, Sunde and Swick, 1978). These authors propose an inverse relationship between protein degradation rate and growth rate in these conditions, i.e. lower degradation accompanying large protein accumulation which is similar to that described for 'catch-up' accretion in adult rats (Funabiki *et al.*, 1976).

Several comments need to be made on the data shown in *Table 2.6*. Protein degradation rates are computed by the difference between synthesis and accretion. Both measurements may be difficult. Accretion often depends on the difference between two large protein values and may therefore result in subsequent errors (Dunlop, Van Elden and Lajtha, 1978). If protein accumulation is accurate then the measurement of synthetic rate determines the extent of protein degradation. Measurements of synthetic rate by incorporation of radioactive precursors (all methods, *Table 2.6*) may be in error for several reasons including possible loss of specific proteins during protein sample preparation or, more importantly, overestimation of synthetic rate of bulk protein due to a rapidly synthesized protein subgroup(s) in the analysed mixture (Dunlop, Van Elden and Lajtha, 1978). These problems can be overcome by studies on specific proteins. Interestingly, however, studies (see *Figure 2.6*) on specific proteins in brain (S-100 protein) and liver (phosphoenol pyruvate carboxykinase) confirm the turnover characteristics of the protein mixtures shown in *Table 2.6*. This may indicate that the relationships between the synthesis and degradation of

Table 2.7.   PROTEIN DEGRADATION IN CELLS IN CULTURE

Cell type	Culture conditions	Protein degradation rate	References
HeLa Human conjunctiva Monkey kidney Mouse fibroblast (L)	Non-growing	Approx. 1% per hour	Eagle *et al.* (1958)
1. HeLa 2. KB	Growing	Approx. 1% per hour	Eagle *et al.* (1958)
Rat embryo fibroblasts	Growing	Decreases from 2.5% to 1.5% with increasing growth rate	Warburton and Poole, (1977)
Hepatocytes and 3T3 (fibroblasts) vs. Reuber H35 hepatoma and 3T3-SV40	Nutritional 'step-down'	Enhanced degradation decreased by 25–50% in transformed cells	Gunn *et al.* (1977)
Rat embryo fibroblasts	Nutritional 'step-down'	Enhanced degradation (increase 3–4 fold)	Amenta *et al.* (1977)

protein mixtures shown in *Table 2.6* are representative of the type of coordination of synthesis and degradation which may be expected for individual proteins. The data in *Table 2.6* do not indicate the extent of protein degradation in the earlier stages of cytodifferentiation, the extent to which proteins are degraded in different cell types or indeed the reason for the high rate of protein degradation in postnatal brain and muscle.

If growth rate of tissues is a factor in the regulation of protein degradation, then it might be expected that a relationship between protein degradation rate and growth occurs in cells in culture. Some studies on the degradation rate of protein mixtures in cultured cells are shown in *Table 2.7*. In non-growing normal and transformed cells the rate of protein degradation is approximately 1% per hour and in growing transformed cells the rate of degradation is approximately the same (Eagle, Fleischman and Oyama, 1958). Several authors have used nutritional 'step-down' conditions as an experimental probe of protein degradation. Enhanced proteolysis (measured over 2–4 h) of long-lived proteins occurs in nutritional step-down conditions (Gunn *et al.*, 1977; Amenta, Sargus and Baccino, 1977; Warburton and Poole, 1977). Transformed cells have a lower rate of enhanced proteolysis (Gunn *et al.*, 1977). Enhanced proteolysis has been equated mechanistically with lysosomal autophagy (Knowles and Ballard, 1976; Amenta, Sargus and Baccino, 1977). It is interesting that in cells in culture (Warburton and Poole, 1977) and in muscle in newly hatched chicks (Maruyama, Sunde and Swick, 1978) nutritional impairment of maximum growth results in increased protein degradation. A good correlation of the degree of impairment of growth and increased rate of protein degradation is found.

Turnover characteristics of specific proteins

As indicated in the previous section, studies of the rate of degradation of protein mixtures have increasingly relied on indirect estimation of degradation, i.e. the difference between protein synthesized and protein accumulated. Indirect estimation of the rate of degradation is preferable, since the alternative single radiochemical pulse-chase technique is beset with difficulties. Errors occur due to considerable isotopic re-utilization during the prolonged periods over which the loss of label from protein must be measured. Furthermore, pulse-chase methods are predominantly suited to steady-state conditions otherwise the measured rate of protein degradation only represents an average value over the prolonged period of measurement. Therefore, changes in the degradation rate are obscured. Studies on specific proteins are complicated by a further major difficulty which is that individual proteins constitute a very small proportion of total cell protein (e.g. often 0.001–1%) and therefore the incorporation of precursor label during a 'pulse' is very small – measuring the loss of label can therefore be hazardous. In short, indirect estimation of degradation rate is preferable for specific proteins.

Immunochemical methods are commonly used to show if changes in enzyme activity are due to changes in amount: relative or absolute rates of synthesis can be measured by incorporation of precursor label (e.g. amino acid) into immuno-isolated enzyme and rate of degradation computed from the relationship

$$\frac{dE}{dt} = K_S - K_D E \qquad (2.1)$$

where E is the enzyme amount, t the time, K_S the rate of enzyme synthesis, and K_D the rate of enzyme degradation.

The integrated form of this equation has been used to compute these parameters, particularly K_D (Schimke, 1970) but this assumes the time-independent nature of K_S and K_D. Recently this equation has been solved by a computer program based on a step-wise treatment of the calculus approach, i.e. equation (2.1) where k_S and k_D are time-dependent variables, i.e.

$$\frac{\Delta E}{\Delta t} = k_S - k_D E \qquad (2.2)$$

By this approach variations in the rate of enzyme degradation (k_D) can be accurately 'pinpointed' and profiles for transient changes in k_D estimated. Therefore, for the first time changes in degradation of individual proteins can be accurately estimated and the time axis of abrupt changes in degradation rate determined. Averages of degradation rates during developmental transitions and the concomitant difficulties of interpretation are therefore completely avoided (Paskin and Mayer, 1977; Mayer, 1978; Mayer and Paskin, 1978; Paskin and Mayer, 1978; Mayer, 1979).

Preliminary investigations on the synthesis and degradation of fatty acid synthetase and cytosolic protein in hormonally stimulated mammary explants in organ culture showed an apparent transient cessation in the rate of degradation of the enzyme and some decrease in the rate of degradation of the cytosolic protein (Speake, Dils and Mayer, 1975; Mayer, 1979). The rate of degradation of the enzyme apparently increased after hormone removal from the system, whereas an increase in the degradation rate of the cytosol protein mixture was somewhat equivocal (Speake, Dils and Mayer, 1975). Further studies showed that only in the presence of the hormone combination which gave optimal sustained tissue development (insulin, prolactin and cortisol) was the transient cessation in the degradation rate of fatty acid synthetase observed. These studies formed the basis for the development of the working hypothesis that a regulated programme of enzyme turnover exists in hormonally stimulated cell differentiation; the rate of synthesis and degradation of an accumulating enzyme are coordinated so that transiently decreased degradation accompanies increased synthetic rate (Speake, Dils and Mayer, 1976). The regulatory and energetic advantages of this model have been discussed (Paskin and Mayer, 1977).

Computation of the degradation profile for fatty acid synthetase by the method indicated in equation (2.2), in the presence of hormones and after hormone removal, is shown in *Figure 2.4*. The transient decrease in the rate of degradation of the enzyme is confirmed and the rapid increase in the rate of degradation of the enzyme on hormone withdrawal is clearly shown. Again as in the preliminary studies a hormonally regulated programme of enzyme accretion and diminution is seen: in the phase of most rapid enzyme accumulation an inverse relationship exists between enzyme synthetic and degradative rate. Therefore in cultured explants the turnover characteristics of a cell-specific intracellular protein are as indicated in *Table 2.5*. The degradation profile shown in *Figure 2.4* is a temporally compacted version of that seen for autophagy during pregnancy and lactation *in vivo* (Hollmann, 1974). However, whether data in *Figure 2.4* can be accounted for by autophagy alone remains to be determined.

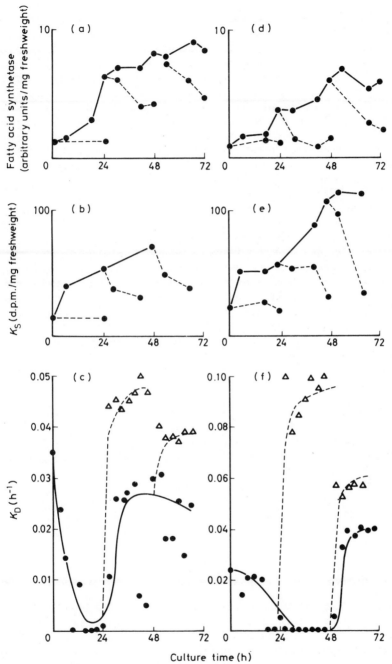

Figure 2.4 Turnover of fatty acid synthetase in hormonally stimulated explants of mammary gland from mid-pregnant rabbits. The results of two separate experiments are shown, (a)–(c) and (d)–(f). Culture with hormones (insulin, prolactin and cortisol,——), culture after removal of hormones (———). Enzyme degradation profile (K_D) determined as described in Paskin and Mayer (1978), based on pre-stimulation steady-state $t_{1/2}$ for fatty acid synthetase of 20 h, (a)–(c), and 29 h, (d)–(f)

In mammary gland the turnover characteristics of a cell-specific intracellular protein can be compared with the cell-specific secretory protein, casein. Autoradiographic studies (K. Al-Sarraj, D. White and R.J. Mayer, unpublished observations) with [³H]-mannosamine or [³H]-leucine have shown that casein granules can be seen in the lumina of the lobulo alveoli of mammary explants within 7 h of the radioisotopic pulse; subsequently, casein rapidly accumulates in the lumina which consequently become greatly distended. From its mode of synthesis and packaging (i.e. translocation into the endoplasmic reticulum, movement to the Golgi apparatus, and vesicular translocation to the plasma membrane) casein is always membrane-limited in the cell. It is rapidly secreted but a small luminal or vesicular pool exists. Two points need considering with regard to its degradation; namely, can it be destroyed intracellularly and extracellularly? It is very unlikely that casein is degraded extracellularly, particularly in view of the very low proportion of degradation products found in milk (D.T. Davies and D.G. Dalgleish, unpublished work). The results in *Figure 2.5* show that in mammary explants casein is not degraded from 36 h to 72 h in culture in the presence of hormones or after hormone removal. Furthermore no loss of [¹⁴C]-glucosamine label can be seen in the presence or absence of hormones – *Figure 2.5(b)* and *2.5(d)*. However, during the same period a considerable destruction of protein occurs in the cytosolic fraction of the cell and a parallel loss of [¹⁴C]-glucosamine label occurs – *Figure 2.5(a)* and *2.5(c)*. This observation confirms the previous results (Mayer, 1979) of considerable degradation of the cytosolic protein mixture in mammary explants in the presence of hormones. Since isotopic re-utilization is minimal for the cytosolic protein mixture in explants (Mayer, 1979) the lack of loss of label from casein can be confidently viewed as indicating no degradation of the protein in the presence or absence of hormones.

The results suggest that in the presence of hormones casein is rapidly secreted and not degraded. After hormone withdrawal casein is not degraded (although its synthetic rate falls gradually) and it is still secreted (results not shown). This indicates that a very small pool of compartmentalized intracellular casein exists, so small perhaps that any increased degradation of casein in this pool would not be detected by the pulse-chase technique used. The results are consistent with the model proposed in *Table 2.5* for a secretory protein with a small intracellular pool. Similarly the turnover characteristics of fatty acid synthetase and non-cell-specific cytosolic proteins also fit with the working hypothesis proposed in *Table 2.5*.

Generality of transient changes in degradation rate: analysis of data in literature

The extent to which transient cessation or decrease in the degradation rate of a cell-specific intracellular protein may accompany an increased synthesis rate during its accumulation in terminal differentiation, must be considered. Analysis of enzyme synthesis and accumulation profiles can only be carried out by calculations similar to those described by Paskin and Mayer (1978). Often the data on enzyme turnover which is found in the literature is insufficient for accurate analyses, but several examples have been studied. The results of the analyses (*Figure 2.6*) indicate the generality of the coordination of enzyme synthesis and degradation in several differentiating systems. In many cases transient decreases in degradation during enzyme accumulation are seen.

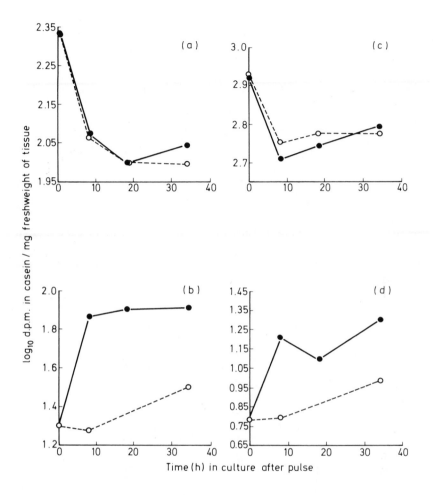

Figure 2.5   Degradation of casein in mammary explants in organ culture. Mammary explants from mid-pregnant rabbits were incubated for 18 h in the absence of hormones (to deplete the cells of mRNAs for cell-specific proteins) and then cultured for 38 h in the presence of insulin (5 μg/ml), prolactin (1 μg/ml) and cortisol (1 μg/ml). The explants were then incubated for 1 h at 37 °C under O_2/CO_2 (19:1) in M199 (glucose-free) containing 1.5 mM leucine, 5 mM sodium pyruvate, 20 μCi of L[4,5-^{3}H]-leucine and 14 μCi of D-[-^{14}C]-glucosamine. The explants were extensively washed, essentially as described by Speake, Dils and Mayer (1976), with Medium 199 containing 10 mM leucine and 10 mM glucosamine and subsequently cultured for a period of 34 h. At the times indicated, groups of 50 explants were taken, homogenized and sonicated for 5 min at 4 °C in 5 mM EGTA, pH 7.0 (1–1.5 ml) containing phenylmethane sulphonyl-fluoride (approx. 2 mM final conc.) and a particle-free supernatant prepared by centrifugation at 15 000 $g_{av.}$ for 60 min at 4 °C. Casein was immunoisolated from the supernatant by immunoaffinity chromatography (Al-Sarraj, White and Mayer, 1978). Incorporation into TCA-insoluble protein prepared from the supernatant was performed, essentially as described by Speake, Dils and Mayer (1975). The results are expressed as $\log_{10}$ radioactivity in TCA-precipitable protein, (a) and (c), and casein, (b) and (d), from protein precursor ([^{3}H]-leucine, (a) and (b)) and oligosaccharide precursor ([^{14}C]-glucosamine, (c) and (d)) vs. time in culture after the precursor pulse. Explants cultured in the presence (●) or the absence (○) of hormones

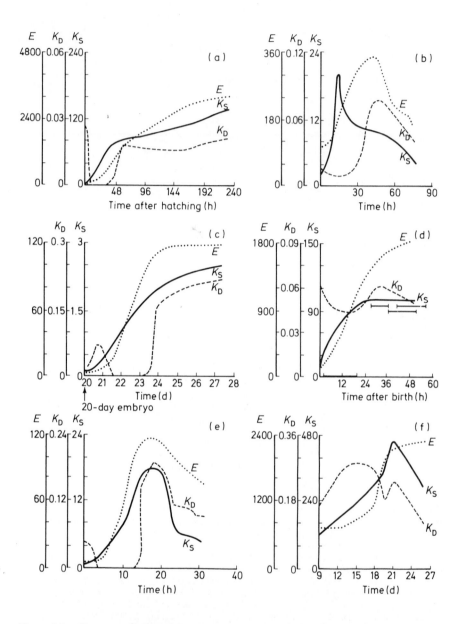

Figure 2.6 Turnover characteristics of enzymes in terminally differentiating cells. Protein degradation profiles (K_D) were calculated from protein accumulation profiles (E) and synthesis profiles (K_S) by the computer program described by Paskin and Mayer (1978). Data for malic enzyme (a) was taken from Silpanata and Goodridge (1971) and Goodridge and Adelman (1976); for glucose-6-phosphate dehydrogenase (b) from Smith and Barker (1977); for fatty acid synthetase (c) from Zehner, Joshi and Wakil (1977); for phosphoenolpyruvate carboxykinase (d) from Philippidis *et al.* (1972); for phenylalanine ammonia lyase (e) from Lamb, Merritt and Butt (1979); and for S-100 protein (f) from Stewart and Urban (1972). Bars represent the measured rate of degradation of phosphoenolpyruvate carboxykinase

It is interesting to speculate as to when transient decreases in protein degradation rates may occur in relation to developmental transitions undergone by various organisms. It is not inconceivable that when some extracellular demand is placed on the cell (hypertrophic or hyperplastic stimulus), cell-specific proteins are accumulated in the most efficient way by coordination of sustained increases in synthetic rates with transient decreases in degradation rates.

Mechanism of protein turnover in differentiating and developing tissues

Clearly, as indicated earlier, the mechanism(s) of protein degradation is unknown and its regulation incompletely understood. Protein synthesis and degradation are integrated in the turnover of organelle proteins – *Figure 2.7(a)*. The synthesis and degradation of some organelle subgroups of proteins are carefully coordinated (*Figure 2.7*). This may involve a cyclic series of events involving

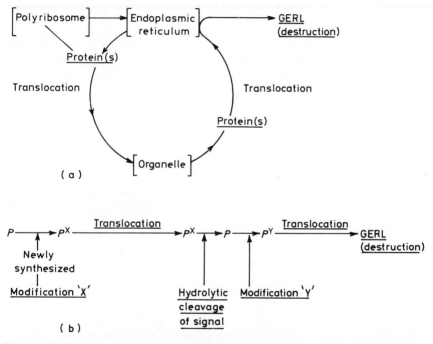

Figure 2.7   Putative mechanism and regulation of intracellular protein degradation: (a) protein turnover cycle; (b) regulation – X = peptide 'signal', covalent or non-covalent modification; Y = non-covalent or covalent modification regulated by transcriptional or translational signal molecule(s); P = protein(s); GERL = Golgi–endoplasmic–reticulum–lysosome

the synthesis and translocation of proteins to their functional sites and their subsequent return for ultimate incorporation and destruction in the Golgi–endoplasmic–reticular–lysosomal system. Such a putative process for protein degradation by protein cycling in steady-state conditions can be augmented by classical autophagy in times of physiological or pathological insult (e.g. Pfeifer, 1976). The alternative to such a cyclic process is that proteins are destroyed at or near their functional sites. However, the advantage of cycling is that protein synthesis and degradation can be carefully controlled at a morphological site, i.e. the polyribosome—endoplasmic reticulum. In this way careful coordination of the synthesis and degradation of protein(s) can be carried out.

If protein subgroups with different turnover characteristics can be identified in differentiating and developing cells (*Table 2.5*) then some selective mechanism must exist whereby some proteins may be destroyed faster than others. Such a mechanism may be as in *Figure 2.7(b)*, where the rate of protein modification (i.e. Y) controls the rate of protein degradation. The rate of modification would be enzymically catalysed and therefore could be regulated by signal molecule(s) generated at the transcriptional or translational level: the signal(s) would act in a negative feedback sense for proteins which are rapidly accumulating. Working hypotheses such as those in *Figure 2.7* stimulate the design of experiments by which we may eventually hope to understand what is the last unknown metabolic process, the mechanism and regulation of protein catabolism.

References

AL-SARRAJ, K., WHITE, D.A. and MAYER, R.J. (1978). *Biochem. J.*, **173**, 877—883

AMENTA, J.S., SARGUS, M.J. and BACCINO, F.M. (1977). *Biochem. J.*, **168**, 223—227

ARIAS, I.M., DOYLE, D. and SCHIMKE, R.T. (1969). *J. Biol. Chem.*, **244**, 3303—3315

ARSTILA, A.U., SHELBURNE, J.D. and TRUMP, B.F. (1972). *Lab. Invest.*, **27**, 317—323

ASCHENBRENNER, V., DRUYAN, R., ALBIN, R. and RABINOWITZ, M. (1970). *Biochem. J.*, **119**, 157—160

ASHFORD, T.P. and PORTER, K.R. (1962). *J. Cell Biol.*, **12**, 198—202

ASHWELL, G. and MORELL, A.G. (1974). *Adv. Enzymol.*, **41**, 99—128

BALLARD, F.J. (1977). In *Essays in Biochemistry*, Vol. 13, pp. 1—39 (Campbell, P.N. and Aldridge, W.N., Eds.), Academic Press, London and New York

BALLARD, F.J. and HOPGOOD, M.F. (1973). *Biochem. J.*, **137**, 259—264

BERGER, R., DIAS, Th. and HOMMES, F.A. (1978). *Proc. Eur. Soc. Neurochem.*, **1**, 590

BIENKOWSKI, R.S., BAUM, B.J. and CRYSTAL, R.G. (1978). *Nature, Lond.*, **276**, 413—416

BLOBEL, G. (1978). In *Gene Expression* (Clark, B.F.C., Klenow, H. and Zeuthen, J., Eds.), pp. 99—108, Pergamon Press, Oxford and New York

BLOBEL, G. and DOBBERSTEIN, B. (1975). *J. Cell Biol.*, **67**, 835—851

BOHLEY, P., KIRSCHKE, H., LANGER, J., WIEDERANDERS, B., ANSORGE, S. and HANSON, H. (1977). In *Intracellular Protein Catabolism* (Turk, V. and Marks, N., Eds.), Vol. II, pp. 108—110, Plenum, New York and London

BOSMAN, H.B., MYERS, M.W., DEHOND, D., BALL, R. and CASE, K.R. (1972). *J. Cell Biol.*, **55**, 147—160

BUCHANAN, D.L. (1961). *Archs Biochem. Biophys.*, **94**, 501—512

BURGESS, R.J., WALKER, J.H. and MAYER, R.J. (1978). *Biochem. J.*, **176**, 919—926

CAMERON, I.L. (1971). In *Cellular and Molecular Renewal in the Mammalian Body* (Cameron, I.L. and Thrasher, J.D., Eds.), pp. 45—85, Academic Press, New York

CONDE, R.D. and SCORNIK, O.A. (1977). *Biochem. J.*, **166**, 115—121

CRANE, D., HOLMES, R.S. and MASTERS, C.J. (1978). *Int. J. Biochem.*, **9**, 589—596

DEAN, R.T. (1975). *Nature, Lond.*, **257**, 414—416

DEAN, R.T. and BARRETT, A.J. (1976). In *Essays in Biochemistry*, Vol. 12, pp. 1–40 (Campbell, P.N. and Aldridge, W.N., Eds.), Academic Press,London and New York

DE BERNARD, B., GETZ, G.S. and RABINOWITZ, M. (1969). *Biochim. Biophys. Acta,* **193**, 58–63

DEHLINGER, P.J. and SCHIMKE, R.T. (1971). *J. Biol. Chem.,* **246**, 2574–2583

DICE, J.F. and GOLDBERG, A.L. (1975). *Archs Biochem. Biophys.,* **170**, 213–219

DICE, J.F. and SCHIMKE, R.T. (1972). *J. Biol. Chem.,* **247**, 98–111

DICE, J.F. and SCHIMKE, R.T. (1973). *Archs Biochem. Biophys.,* **158**, 97–105

DICE, J.F., WAKER, C.D., BRYNE, B. and CARDIEL, A. (1978). *Proc. Natn. Acad. Sci. U.S.A.,* **75**, 2093–2097

DRUYAN, R., DE BERNARD, B. and RABINOWITZ, M. (1969). *J. Biol. Chem.,* **244**, 5874–5878

DUNCAN, W.E. and BOND, J.S. (1977). *Fed. Proc.,* **36**, 918

DUNLOP, D.S., VAN ELDEN, W. and LAJTHA, A. (1978). *Biochem. J.,* **170**, 637–642

EAGLE, H., PIEZ, K.A., FLEISCHMAN, R. and OYAMA, V.I. (1958). *J. Biol. Chem.,* **234**, 592–597

EKMAN, P., HERMANSSON, V., BERGSTRÖM, G. and ENGSTRÖM, L. (1978). *Fedn Eur. Biochem. Soc. Letts,* **86**, 250–254

ERNSTER, L. and KUYLENSTIURNA, B. (1970). In *Membranes of Mitochondria and Chloroplasts* (Racker, E., Ed.), pp. 172–212, Van Nostrand Reinhold, New York

FARMER, S.R. HENSHAW, E.C., BERRIDGE, M.V. and TATA, J.R. (1978). *Nature, Lond.,* **273**, 401–403

FRANKE, W.W. and KARTENBECK, J. (1971). *Protoplasma,* **73**, 35–41

FRITZ, P.J., VESELL, E.S., WHITE, E.L. and PRUITT, K.M. (1969). *Proc. Natn. Acad. Sci. U.S.A.,* **62**, 558–565

FUNABIKI, R., WATANABE, Y., NISHIZAWA, N. and HAREYAMA, S. (1976). *Biochim. Biophys. Acta,* **451**, 143–150

GEAR, A.R.L. (1970). *Biochem. J.,* **120**, 577–587

GOLDBERG, A.L. and DICE, J.F. (1974). *Ann. Rev. Biochem.,* **43**, 835–869

GOLDBERG, A.L. and ST. JOHN, A.C. (1976). *Ann. Rev. Biochem.,* **45**, 747–803

GOLDMAN, B.M. and BLOBEL, G. (1978).*Proc. Natn. Acad. Sci. U.S.A.,* **75**., 5066–5070

GOODRIDGE, A.G. and ADELMAN, T.G. (1976). *J. Biol. Chem.,* **251**, 3027–3032

GRAVES, D.J., MANN, S., ANN, S., PHILIP, G. and OLIVEIRA, R.J. (1968). *J. Biol. Chem.,* **243**, 6090–6100

GROSS, N.J., GETZ, G.S. and RABINOWITZ, M. (1969). *J. Biol. Chem.,* **244**, 1552–1562

GUNN, J.M., CLARK, M.G., KNOWLES, S.E., HOPGOOD, M.F. and BALLARD, F.J. (1977). *Nature, Lond.,* **266**, 58–60

GURD, J.W. and EVANS, W.H. (1973). *Eur. J. Biochem.,* **36**, 273–279

HACKENBROCK, C.R. and MILLER, K.J. (1975). *J. Cell Biol.,* **65**, 615–630

HAMADA, M., HIRAOKA, T., KOIKE, K., OGASHARA, K., KANAKI, T. and KOIKE, M. (1976). *J. Biochem., Tokyo,* **79**, 1273–1285

HAMADA, M., OTSUKA, K.I., TANAKA, N., OGASHARA, K., KOIKE, K., HIRAOKA, T. and KOIKE, M. (1975). *J. Biochem., Tokyo,* **78**, 187–197

HARE, J.F. (1978). *Biochem. Biophys. Res. Commun.,* **83**, 1206–1215

HAYES, L.W. and LARRABEE, A.R. (1971). *Biochem. Biophys. Res. Commun.,* **45**, 955–963

HIRSCH, C.A. and HIATT, H.H.(1966). *J. Biol. Chem.*, **241**, 5936–5940

HOLLMANN, K.H. (1974). In *Lactation*, Vol. I, pp. 3–95 (Larson, B.L. and Smith, V.R., Eds.), Academic Press, New York

HOPGOOD, M.F., CLARK, M.G. and BALLARD, F.J. (1977). *Biochem. J.*, **164**, 399–407

IP, M.M., CHEE, P-Y. and SWICK, R.W.(1974). *Biochim. Biophys. Acta*, **354**, 29–38

JONES, G. and MAYER, R.J. (1973). *Biochem. J.*, **132**, 657–661

KATUNUMA, N., KITO, K. and KOMINAMI, E. (1971a). *Biochem. Biophys. Res. Commun.*, **45**, 76–81

KATUNUMA, N., KOMINAMI, E. and KOMINAMI, S. (1971b). *Biochem. Biophys. Res. Commun.*, **45**, 71–75

KAWAJIRI, J., HIRANO, T. and OMURA, T. (1977). *J. Biochem., Tokyo*, **82**, 1417–1423

KENNEY, F.T. (1967). *Science, N.Y.*, **156**, 525–528

KNOWLES, S.E. and BALLARD, F.J. (1976). *Biochem. J.*, **156**, 609–617

KNOWLES, S.E., GUNN, J.M., HANSON, R.W. and BALLARD, F.J. (1975). *Biochem. J.*, **146**, 595–600

LAGRANGE, B.M. and LOW, R.B. (1976). *Devl Biol.*, **54**, 214–229

LAMB, C.J., MERRITT, T.K. and BUTT, V.S. (1979). *Biochim. Biophys. Acta*, **582**, 196–212

LITWACK, G. and ROSENFELD, S. (1973). *Biochem. Biophys. Res. Commun.*, **52**, 181–188

LODISH, H.F., SMALL, B. and CHANG, H. (1975). *Devl Biol.*, **47**, 59–67

MARUYAMA, K., SUNDE, M.L. and SWICK, R.W. (1978). *Biochem. J.*, **176**, 573–582

MARVER, H.S., COLLINS, A., TSCHUDY, D.P. and RECHCIGL, M. (1966). *J. Biol. Chem.*, **241**, 4323–4329

MAYER, R.J.(1978). *Biochem. Soc. Trans.*, **6**, 505–509

MAYER, R.J. (1979). In *Vitamins and Hormones*, Vol. 36, pp. 101–163, Academic Press, New York

MAYER, R.J. and PASKIN, N. (1978).*In Regulation of Fatty Acid and Glycerolipid Metabolism*, pp. 53–62 (Dils, R. and Knudsen, J., Eds.), Pergamon Press, Oxford and New York

McILHINNEY, A. and HOGAN, B.L.M. (1974). *Fedn Eur. Biochem. Soc. Letts*, **40**, 297–301

McLENDON, G. and RADANY, E. (1978). *J. Biol. Chem.*, **253**, 6335–6337

MILLWARD, D.J., GARLICK, P.J., STEWART, R.J.C., NNANYELUGO, D.O. and WATER-LOW, J.C. (1975). *Biochem. J.*, **150**, 235–243

PASKIN, N. and MAYER, R.J. (1977). *Biochim. Biophys. Acta*, **474**, 1–10

PASKIN, N. and MAYER, R.J. (1978). *Biochem. J.*, **174**, 153–161

PFEIFER, U. (1976). *Verh. Dt, Ges. Path.*, **60**, 28–64

PFEIFER, U. and SCHELLER, H. (1971). *J. Cell Biol.*, **64**, 608–621

PFEIFER, U., WERDER, E. and BERGEEST, H. (1978). *J. Cell Biol.*, **78**, 152–167

PHILIPPIDIS, H., HANSON, R.W., RESHEF, L., HOPGOOD, M.F. and BALLARD, F.J. (1972). *Biochem. J.*, **126**, 1127–1134

PONTREMOLI, S., MELLONI, E., SALAMINO, F., FRANZI, A.J., DEFLORA, A. and HORECKER, B.L. (1973). *Proc. Natn. Acad. Sci. U.S.A.*, **70**, 3674–3678

POOLE, B. (1971). *J. Biol. Chem.*, **246**, 6587–6591

ROSEN, J.M. and BARKER, S.W. (1976). *Biochemistry*, **15**, 5272–5280

ROTHMAN, J.E. and LENARD, J. (1977). *Science, N.Y.*, **195**, 743–753

RUBINSTEIN, N., CHI, J. and HOLTZER, H. (1976). *Expl Cell Res.*, **97**, 387–393

RUTTER, W.J., PICTET, L.P. and MORRIS, P.W. (1973). *Ann. Rev. Biochem.*, **42**, 601–646

SAKURAI, Y., FUKUYOSHI, Y., HAMADA, M., HAYAKAWA, T. and KOIKE, M. (1970). *J. Biol. Chem.*, **244**, 1183–1187

SCHEELE, G., DOBBERSTEIN, B. and BLOBEL, G. (19′8). *Eur. J. Biochem.*, **82**, 593–599

SCHIMKE, R.T. (1964). *J. Biol. Chem.*, **239**, 3808–3817

SCHIMKE, R.T. (1970). In *Mammalian Protein Metabolism*, Vol. 4, pp. 178–228 (Munro, H.N., Ed.), Academic Press, New York

SCHIMKE, R.T. (1973). *Adv. Enzymol.*, **37**, 135–187

SCHIMKE, R.T., SWEENEY, E.W. and BERLIN, C.M. (1965a). *J. Biol. Chem.*, **240**, 322–331

SCHIMKE, R.T., SWEENEY, E.W. and BERLIN, C.M. (1965b). *J. Biol. Chem.*, **240**, 4609–4620

SCHNAITMAN, C. and GREENAWALT, J.W. (1968). *J. Cell Biol.*, **38**, 158–175

SCORNIK, O.A. and BOBTOL, V. (1976). *J. Biol. Chem.*, **251**, 2891–2897

SEGAL, H.L. (1975). In *Lysosomes in Biology and Pathology*, Vol. 4 (Dingle, J.T. and Dean, R.T., Eds.), North-Holland, Amsterdam

SEGAL, H.L., MATSUZAWA, T., HAIDER, M. and ABRAHAM, G.J. (1969). *Biochem. Biophys. Res. Commun.*, **36**, 764–770

SHIELDS, D. and BLOBEL, G. (1978). *J. Biol. Chem.*, **253**, 3753–3756

SHORE, G.C. and TATA, J.R. (1977a). *J. Cell Biol.*, **72**, 714–725

SHORE, G.C. and TATA, J.R. (1977b). *J. Cell Biol.*, **72**, 726–743

SILPANATA, P. and GOODRIDGE, A.G. (1971). *J. Biol. Chem.*, **246**, 5754–5761

SIMPSON, M.V. (1953). *J. Biol. Chem.*, **201**, 143–154

SINGER, S.J. and NICHOLSON, G.L. (1972). *Science, N.Y.*, **175**, 720–731

SMITH, E.R. and BARKER, K.L. (1977). *J. Biol. Chem.*, **252**, 3709–3714

SOTTOCASA, G., SANDRIA, G., PANFILI, E., DE BERNARD, B., GAZOTTI, P. and VASINGTON, F.D. (1972). *Biochem. Biophys. Res. Commun.*, **47**, 808–813

SPEAKE, B.K., DILS, R. and MAYER, R.J. (1975). *Biochem. J.*, **148**, 309–320

SPEAKE, B.K., DILS, R. and MAYER, R.J. (1976). *Biochem. J.*, **154**, 359–370

STEWART, J.A. and URBAN, M.I. (1972). *Devl Biol.*, **29**, 372–384

STOOPS, J.K., ARSELANIAN, M.J., AVNE, K.C. and WAKIL, S.J. (1978). *Archs Biochem. Biophys.*, **188**, 348–359

SWICK, R.W. and IP, M.M. (1974). *J. Biol. Chem.*, **249**, 6836–6841

SWICK, R.W., REXROTH, A.K. and STRANGE, J.L. (1968). *Fed. Proc.*, **27**, 462

TWETO, J. and DOYLE, D. (1977). In *The Synthesis, Assembly and Turnover of Cell Surface Components*, pp. 137–165 (Poste, G. and Nicholson, G.L., Eds.), Elsevier–North-Holland, Amsterdam

TWETO, J.M., DEHLINGER, P.J. and LARRABEE, A.R. (1972). *Biochem. Biophys. Res. Commun.*, **48**, 1371–1377

WALKER, J.H., BURGESS, R. and MAYER, R.J. (1978). *Biochem. J.*, **176**, 927–932

WANG, C-C. and TOUSTER, O. (1975). *J. Biol. Chem.*, **250**, 4896–4902

WARBURTON, M.J. and POOLE, B. (1977). *Proc. Natn. Acad. Sci. U.S.A.*, **74**, 2427–2431

WIDNELL, C.C. and SIEKEVITZ, P. (1967). *J. Cell Biol.*, **35**, 142a

ZEHNER, Z.E., JOSHI, V.C. and WAKIL, S.J. (1977). *J. Biol. Chem.*, **252**, 7015–7022

ZITO, R., ANTONINI, E. and WYNAN, J. (1964). *J. Biol. Chem.*, **239**, 1804–1812

3

ASSESSMENT OF PROTEIN METABOLISM IN THE INTACT ANIMAL

PETER J. GARLICK
Clinical Nutrition and Metabolism Unit, London School of Hygiene and Tropical Medicine

Summary

Methods of quantifying whole-body protein turnover were developed for measurements in human subjects, but can also be valuable tools for the study of nutrition or growth in experimental animals. Comparison between species indicates that the relationship between the rate of whole-body protein turnover and body weight is similar to that between fasting metabolic rate and body weight. Three techniques have been described using both ^{14}C- and ^{15}N-labelled amino acids. All give similar rates under normal dietary conditions and respond in the same way to changes in diet and growth rate. In animals the constant infusion of a ^{14}C-labelled amino acid will probably be found to be the most accurate and convenient method. Constant infusion also has the advantage of being the method of choice for studies of protein synthesis and breakdown in individual tissues of larger animals. The rate of synthesis is obtained by removing each tissue at the end of the infusion for measurements of the specific activity of the free and protein-bound amino acid. The rate of breakdown can also be derived if, in addition to synthesis, the growth rate of the tissue protein mass is also measured.

Introduction

Changes in the protein content of the body and tissues (e.g. during growth, malnutrition, disease and trauma) have traditionally been studied by nitrogen balance techniques both in man and in animals. Increasingly, investigators have become aware of the need to measure not only the net balance between dietary intake and nitrogen excretion, but also the rates of both synthesis and breakdown of body protein, for it is the imbalance between rates of synthesis and breakdown that results in the net gain or loss of body protein. However, the same net increase or loss of protein could be achieved by two distinct mechanisms; either by an alteration in protein synthesis, or by an opposite change in breakdown. Methods for measuring synthesis and breakdown must therefore be capable of distinguishing between these two mechanisms.

Except in certain special cases, rates of protein synthesis and breakdown in the live animal have been measured by administration of isotopically labelled amino acids. Three different approaches have been made, depending on the individual circumstances. In large animals, and particularly man, when it is not possible or convenient to take samples of tissue, the rates of whole-body protein synthesis and breakdown can be obtained by measuring the rate of disappearance of the administered label from the free amino acid pool. Alternatively, when tissue samples can be taken, the incorporation of label into the protein of that tissue can be determined in order to estimate its rate of synthesis. Finally, it is possible to wait until all the isotope has been incorporated into body protein, and then to observe its subsequent rate of loss from the tissues in order to determine not only the rate of protein breakdown, but also of synthesis. These approaches will now be considered in more detail.

Measurement of whole-body protein turnover

A historical account of the early methods that were used to measure whole-body protein synthesis and breakdown has been given by Waterlow, Garlick and Millward (1978). These methods frequently relied on complex models for whole-body protein metabolism (e.g. Oleson, Heilskov and Shonheyder, 1954) which with more recent techniques have largely been abandoned in favour of the very simple model shown in *Figure 3.1*. With this model the total free amino acid

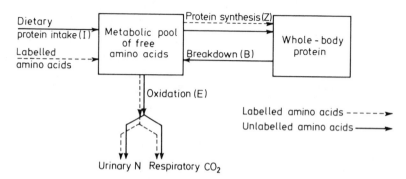

Figure 3.1   Model representing whole-body protein metabolism used to calculate rates of whole-body protein synthesis and breakdown from data obtained when labelled amino acids are given. ^{14}C- or ^{15}N-labelled amino acids are introduced into the metabolic pool by oral or intravenous administration. With a ^{14}C-labelled amino acid (e.g. [1^{14}C]-leucine), the scheme represents the metabolism of that single amino acid, its synthesis into protein and its oxidation with subsequent excretion of ^{14}CO$_2$ in the breath. The rates represented by the symbols, Z,B,I and E are therefore in the units of moles of that amino acid per unit time. The corresponding rates of protein synthesis and breakdown can then be calculated if the proportion of that amino acid in whole-body protein is known. With a ^{15}N-labelled amino acid such as [^{15}N]-glycine, the label enters the metabolic pool, where it is to some extent distributed among other amino acids, before it is either incorporated into protein or oxidized and excreted as urinary N. The ^{15}N is therefore used as a tracer for total N rather than a single amino acid, and the rates of Z,B,I and E are expressed in the units of gN per unit time

pool of the body is assumed to be homogeneous and is called the 'metabolic pool', and all the body proteins are combined to form a single protein pool. It is further assumed that any label that is introduced into the metabolic pool can only follow two pathways, which are incorporation into protein and oxidation to end products of metabolism. The sum of the rates of entry into these two pathways has been termed the flux, Q. Two different labels (^{14}C and ^{15}N) have commonly been used to measure the flux, but the technique is broadly similar whichever is chosen. Commonly the labelled amino acid is introduced into the metabolic pool by means of a continuous infusion. After a delay, the specific activity (or relative abundance of ^{15}N) of the free amino acid in the metabolic pool rises to a constant, or plateau value S. The flux Q is then calculated from the plateau value by the equation

$$i = Q.S \tag{3.1}$$

where i is the rate of infusion of isotope. This equation expresses the fact that, at plateau, the amount of isotope entering the metabolic pool by the infusion must equal the amount leaving for protein synthesis or for amino acid oxidation. It is assumed that any label which enters the protein pool remains there, and is not released back into the metabolic pool by protein breakdown during the course of the infusion. A further equation enables the rates of protein synthesis Z and breakdown B to be calculated from the flux Q if the rates of amino acid oxidation E and intake from food I are known, i.e.

$$Q = Z + E = I + B \tag{3.2}$$

The application of this basic technique with the two isotopes, ^{14}C and ^{15}N, will now be considered separately.

^{14}C-LABELLED AMINO ACIDS

Flux rates have been measured by infusion of a number of different ^{14}C-labelled amino acids. The best amino acid is probably $[1^{14}C]$-leucine, because its metabolism approximates most closely to the simple model shown in *Figure 3.1* (James *et al.*, 1974). The first step in leucine oxidation is its transamination, followed by decarboxylation and loss of the ^{14}C into the bicarbonate pool. The production of $^{14}CO_2$ in the breath is therefore a measure of the proportion of the flux which is oxidation (E) of leucine; the remaining part of flux must be synthesis into protein, since there are no other known pathways of leucine metabolism. Uniformly labelled amino acids (e.g. $[U^{14}C]$-tyrosine) are not satisfactory because it is possible for labelled metabolic products to be diverted into synthesis of glucose, fat and hormones as an alternative to conversion to $^{14}CO_2$ (James *et al.*, 1976). However, even with $[1^{14}C]$-leucine a correction factor has to be applied, because a certain amount ($\sim$20%) of the $^{14}CO_2$ produced is fixed by other pathways, possibly into bone (James *et al.*, 1976).

Equation (3.1), which is used to calculate the flux, requires a value for S, the specific radioactivity of leucine in the metabolic pool. It has been shown that in man the specific activity of free leucine in plasma reaches a plateau within 8 h, and this value is maintained for up to 30 h (Golden and Waterlow, 1977). A plateau of specific activity of respiratory CO_2 is also maintained during this time (Golden and Waterlow, 1977).

With infusion of $[1^{14}C]$-leucine, therefore, the model of *Figure 3.1* represents the metabolism of leucine. Rates of flux, synthesis, breakdown and oxidation, expressed in moles of leucine in unit time, can be calculated if the specific activity of plasma leucine, the rate of production of respiratory $^{14}CO_2$ and the rate of intake of leucine from the food are measured. Corresponding rates for protein can be calculated if it is assumed that leucine is a fixed proportion (e.g. 8%, O'Keefe, Sender and James, 1974) of body protein.

LABELLING WITH ^{15}N

^{15}N-labelled amino acids can be used in a similar way to ^{14}C, with measurements of the abundance of ^{15}N (cf. specific activity) in the amino acid in plasma (e.g. Halliday and McKeran, 1975). However, because of the differences in metabolism between the carbon and nitrogen parts of an amino acid, special methods have been devised using ^{15}N which require measurement to be made not on the plasma, but on the urine. The important distinctions between ^{15}N and ^{14}C are:

(1) ^{14}C is a useful tracer for leucine metabolism because it cannot be transferred to another compound without irreversible degradation of the leucine molecule. ^{15}N, by contrast, can be transferred reversibly to other amino acids (e.g. by transamination). This means that the label cannot, in general, be used to trace the behaviour of a single amino acid, but instead can be used as a tracer for total amino N.
(2) The metabolic pool is not only the precursor of protein synthesis, but also the precursor for oxidation of amino acids. The end products of oxidation of amino N are excreted in the urine and can be used to reflect the ^{15}N abundance in the metabolic pool. By contrast, ^{14}CO$_2$, which is the end product of oxidation of the carbon, cannot be used in this way because unlabelled CO$_2$ is also produced from non-amino acid sources.

These two characteristics of labelling with ^{15}N raise two important practical questions: which ^{15}N-labelled amino acid should be given, and which urinary end product should be used for measurements of ^{15}N abundance in the metabolic pool? The first of these questions, which amino acid, has mostly been answered by economic considerations, since [^{15}N]-glycine is the only amino acid which could, until recently, be obtained at a reasonable price. The methods described here have therefore used [^{15}N]-glycine. A discussion of studies in which other ^{15}N-labelled amino acids were given can be found in Waterlow, Garlick and Millward (1978). The most commonly used method is that of Picou and Taylor-Roberts (1969) in which [^{15}N]-glycine is given by constant infusion or by regular oral doses. Urinary urea is used as the end product to reflect the abundance of ^{15}N in the metabolic pool. A plateau of ^{15}N abundance in urinary urea is reached in about 25 h in children (Picou and Taylor-Roberts, 1969), but may take as much as 3 days in adults. The flux (in gN/day) is then calculated from equation (3.1) using this plateau value. Protein synthesis and protein breakdown can also be calculated from equation (3.2) if amino acid oxidation (i.e. nitrogen excretion) and dietary intake of N are also determined.

The disadvantage of this method is the long time taken to reach plateau, which makes it unsuitable for measuring acute changes. The reason for the delay before plateau is reached is that newly synthesized urea first enters the body urea pool, which has a half-life of the order of 10 h, before it is excreted. It was therefore suggested that urinary ammonia might be a good alternative to urea, since the plateau of ^{15}N abundance in ammonia is reached far more rapidly, generally in about 12 h (Waterlow, Golden and Garlick, 1978). The choice between these two end products, urea and ammonia, is difficult to make on theoretical grounds since neither can perfectly reflect the ^{15}N abundance in the metabolic pool. Urea is synthesized from amino acids in the liver, and urinary

ammonia from amino acids, mainly glutamine, in the kidney. In practice, however, it appears from comparisons between the different methods that, in general,
both urea and ammonia can be used to give results which are of practical value
(see below).

A further modification to the method, which simplifies its application, is to
give the [^{15}N]-glycine as a single dose rather than by constant infusion or
repeated dose (Waterlow, Golden and Garlick, 1978). This does not result in a
plateau of isotope abundance in urinary end products such as urea or ammonia,
but instead the cumulative excretion of isotope reaches a plateau. In theory,
this plateau will only be reached at infinite time. However, particularly when
ammonia is used as end product, it is possible to define a time t when there
is, effectively, complete elimination of isotope from the metabolic pool, but
before significant recycling of isotope from protein breakdown has occurred.
With urea as end product this point is more difficult to define because of the
delaying effect of the urea pool. The flux can then be calculated from the
following, which is completely analogous to equation (3.1) for calculation of
flux from constant infusion:

$$d = Q \cdot \int_{0}^{t} S \, dt \qquad (3.3)$$

where d is the dose of isotope given, and the integral is the area under the curve
for ^{15}N abundance against time in urinary ammonia. This integral is most easily
computed if a single collection of urine is made between times zero and t, since
it is then equal to the ^{15}N abundance of ammonia in that sample multiplied by
t. Further discussion of the single-dose method has been given in papers by
Garlick and Waterlow (1977), Waterlow, Garlick and Millward (1978), Waterlow,
Golden and Garlick (1978) and Garlick (1979).

COMPARISONS BETWEEN METHODS

Each of the methods of assessing whole-body protein synthesis and breakdown
has potential sources of error which have been discussed in detail by Waterlow,
Garlick and Millward (1978). One way of evaluating these is by comparison of
results obtained by different techniques. Three such comparisons are given in
Table 3.1. Overall, the agreement between all methods is quite good. There is
very close agreement between single dose and constant infusion of [^{15}N]-glycine,
but ammonia seems to give a lower rate of synthesis than urea. This difference
could, however, have resulted from the time periods chosen for collections in
comparisons of urea with ammonia (24–30 h). It is possible that this time period
was not long enough to have reached plateau with urea and far longer than is
necessary for plateau in ammonia. In adults the agreement between [1^{14}C]-
leucine and [^{15}N]-glycine with either end product is very good.

Different methods should not only agree in normal subjects receiving normal
diets; it is important that all methods should be capable of showing the same
change in synthesis when the conditions of measurement are altered. Thus,
Waterlow, Golden and Garlick (1978) showed that in spite of the difference in
results obtained with urea and ammonia after [^{15}N]-glycine administration in
recovered children (*Table 3.1*), the same differences between malnourished,
recovering and recovered children were observed when either end product was

Table 3.1.   COMPARISON OF DIFFERENT METHODS FOR MEASURING WHOLE-BODY PROTEIN SYNTHESIS

Subjects	Units	$[1^{14}C]$-leucine infusion	$[^{15}N]$-glycine			
			single dose		constant infusion	
			urea	NH_3	urea	NH_3
Children	g/kg body wt/ day	—	6.1±0.6	4.1±0.5	6.9±0.5	3.6± 0.3
Elderly adults	g/kg body wt/ day	2.7±0.3	—	—	3.3±0.1	2.3±0.2
Obese adults	g/day	287±23	—	299±31	—	287±41

The table includes results from three studies. The study on children (Waterlow, Golden and Garlick, 1978) included six subjects who were each given both repeated hourly doses of $[^{15}N]$-glycine with collections of urine from 24 h to 34 h, and also a single dose of $[^{15}N]$-glycine with collections of urine for the subsequent 24 h. With both methods, rates of protein synthesis were calculated from the excretion of isotope both in urea and ammonia. In the study on elderly adults (Golden and Waterlow, 1977) each patient was given a 30 h constant infusion of both $[1^{14}C]$-leucine and $[^{15}N]$-glycine. Rates of protein synthesis were calculated from the specific activity of free leucine in the plasma and from the ^{15}N abundance in both urea and ammonia. The study on obese adults (Garlick, Clugston and Waterlow, 1980) used a different set of patients for each of the three methods, infusion of $[1^{14}C]$-leucine for 12 h, repeated dose of $[^{15}N]$-glycine for 12–14 h and single dose of $[^{15}N]$-glycine with collection of urine for the following 12 h. Rates of protein synthesis have been quoted in the units of g/kg body wt per day, except in the obese in whom the units of g/day were more appropriate.

used. Synthesis in the malnourished children was lower, and in recovering children on a very high food intake synthesis was higher than in the recovered. Another such comparison is shown in *Table 3.2*. Protein synthesis was measured by one of three methods in obese subjects receiving a normal diet (8.0 MJ, 70 g protein), and then again by the same method after a further three weeks on a low-energy diet (2.1 MJ) containing either zero or 50 g protein. The table shows the rate of synthesis observed when the low-energy diet was given, expressed as

Table 3.2.   THE EFFECT OF LOW-ENERGY DIETS WITH AND WITHOUT PROTEIN ON WHOLE-BODY PROTEIN SYNTHESIS MEASURED BY THREE METHODS

Method	Protein synthesis on low-energy diet as percent of rate on normal diet	
	Diet	
	50 g protein	protein-free
$[1^{14}C]$-leucine, infusion	87 ± 3	64 ±7
$[^{15}N]$-glycine, repeated dose	111 ±11	58 ±6
$[^{15}N]$-glycine, single dose	—	50 ±7

Data of Garlick, Clugston and Waterlow (1980). Obese subjects were given a normal diet (8.0 MJ, 70 g protein) for three days and the rate of whole-body protein synthesis was measured by one of three methods. For the following three weeks all subjects received a low-energy diet (2.1 MJ/day), but for some this diet contained 50 g protein and for others it was protein-free. After three weeks the rate of whole-body protein synthesis was measured again by the same technique. The values in the table are the rates obtained after three weeks on the low-energy diet, expressed as percent of the rates obtained on the normal diet (± S.E.M.).

a percent of the rate seen with the normal diet. From both infusion of $[1^{14}C]$-leucine and repeated dose of $[^{15}N]$-glycine with measurements on urinary ammonia, we arrive at the same conclusion. The low-energy, protein-containing diet had little effect on protein synthesis, and the low-energy, protein-free diet caused a substantial decrease in synthesis. A similar decrease with the protein-free diet was also observed with the single-dose $[^{15}N]$-glycine method. We have found the single-dose method to be particularly convenient. It is not only very simple to use, but can be repeated many times at regular intervals. *Figure 3.2* is an illustration of how this method was used to measure the variations in protein synthesis during a succession of changes of diet in a single patient.

We therefore conclude that each of the methods described can provide valid and useful results under the conditions that have so far been tested. This is perhaps surprising, in view of the simple model that was used to represent whole-body protein metabolism, with all its complexities, and there is still a need for more work on validation of these methods under different sets of conditions.

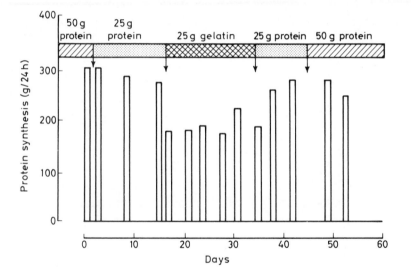

Figure 3.2 The rate of whole-body protein synthesis in a single patient given diets containing constant energy (2.1 MJ) but changing levels of protein. Protein synthesis was estimated from the excretion of ^{15}N in urinary ammonia after a single dose of $[^{15}N]$-glycine. (Data of Garlick *et al.* 1980)

Measurement of the rate of protein synthesis in individual tissues

Measurement of whole-body protein synthesis is useful when it is not possible or convenient to remove samples of tissue (e.g. in man), but it is more satisfactory if rates of synthesis in individual tissues can also be obtained. This is generally done by measuring the amount of incorporation of isotope into tissue protein after administration of a labelled amino acid. This technique also requires measurement of the time course of specific activity of the free amino acid at the site of protein synthesis. Most investigators have used the intracellular free amino acid for this purpose. Because the intracellular pool is compartmented, however, improved accuracy can be obtained by making measurements on aminoacyl tRNA (for discussion see Waterlow, Garlick and Millward, 1978). With the following two methods of administration of isotope, errors arising from this source are minimized.

CONSTANT INFUSION OF A TRACER DOSE OF LABELLED AMINO ACID

If a tracer dose of a labelled amino acid is given as a single injection, the specific activity of the free amino acid in the tissues rapidly rises and then falls. In order to determine the rate of protein synthesis it is necessary to measure the time course of these changes by killing numbers of animals at different times. The constant infusion method was introduced by Waterlow and Stephen (1968) to avoid this need. As described above, when a labelled amino acid is infused the specific activity of the free amino acid in the blood rapidly rises to a plateau value. In the intracellular pool a plateau is also reached, but its value is some-what lower, and it is attained a little later than in the plasma (*Figure 3.3*). The

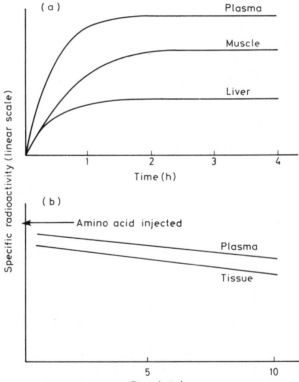

Figure 3.3   Schematic representation of the specific radioactivity of the free amino acid in plasma and tissues during a constant infusion of a labelled amino acid lasting several hours (a), or during the first 10 min following the single intravenous injection of a massive dose of labelled amino acid (b)

constant infusion method depends on the assumption that it is possible, knowing the time course for rise to plateau in the plasma, to predict the time course for the intracellular amino acid without the need to remove the tissue for measurement, except at the end of the infusion. It can be shown that this time course approximates quite closely to a single exponential function, whose rate constant λ_i can be predicted (Waterlow and Stephen, 1968; Garlick, Millward and James, 1973), i.e.

$$S_i = S_{i\max} \left[1 - \exp\left(-\lambda_i t\right)\right] \tag{3.4}$$

If the infusion is continued for some time after plateau is reached the value for λ_i need only be approximate since it has relatively little influence on the rate of synthesis obtained (Garlick, 1978). The value of λ_i depends on the tissue and the amino acid infused. With tyrosine infusion, λ_i in liver can be taken to equal to λ_p, which is the measured rate constant for the rise to plateau in the plasma. In muscle, the appropriate value for λ_i is Rk_s, where R is the ratio of protein bound to free amino acid in the tissue and k_s is the fractional rate of muscle protein synthesis (for further explanation of the fractional rate see pages 64 and 65).

Once the time course for free amino acid specific activity has been determined, an equation can be derived which enables the fractional rate of protein synthesis (k_s) to be calculated from the specific activities of the amino acid in the protein (S_B) and in the free amino acid pool (S_i) at the end of the infusion (Garlick, Millward and James, 1973). With infusion of [^{14}C]-tyrosine in the rat, the appropriate equation for liver protein is:

$$\frac{S_B}{S_i} = \frac{\lambda_p}{(\lambda_p - k_s)} \cdot \frac{[1 - \exp(-k_s t)]}{[1 - \exp(-\lambda_p t)]} - \frac{k_s}{(\lambda_p - k_s)} \tag{3.5}$$

and for muscle protein

$$\frac{S_B}{S_i} = \frac{R}{(R-1)} \frac{[1 - \exp(-k_s t)]}{[1 - \exp(-Rk_s t)]} - \frac{1}{(R-1)} \tag{3.6}$$

Probably the most convenient amino acid for infusion is tyrosine, because there is a very simple method for assay of its specific activity (Garlick and Marshall, 1972). In addition, its rise to plateau in plasma and tissues is very rapid, being complete in about 0.5 h in the rat (Garlick, Millward and James, 1973) and in about 3 h in animals as large as the pig (Garlick, Burk and Swick, 1976) and man (James *et al.*, 1976). However, in principle there is no reason why other amino acids should not be used if a suitable assay procedure is available. An amino acid analyser is accurate but tedious and costly, and is unable to distinguish the L isomer from the D, which may occasionally be present as an impurity in the infusate (Waterlow and Stephen, 1968; Nicholas, Lobley and Harris, 1977).

The major source of error with this technique is the use of the total free amino acid of the tissue as a measure of the specific activity at the site of protein synthesis. There are some who argue that it would be more appropriate to use the free amino acid of the plasma (for a discussion see Waterlow, Garlick and Millward, 1978). In most tissues, e.g. muscle, heart, brain, this source of error is in any case not very serious since the specific activity in the tissue is usually about 75% or more of that in the plasma. With some other tissues, however, this source of error may be a problem. This is particularly so in gut mucosa in which the intracellular specific activity is only about 25% of that in the plasma (*Table 3.3*). With tissues such as this it would be unwise to place any reliance on results obtained by constant infusion since other methods are more appropriate (see below).

Table 3.3.   FRACTIONAL RATES OF PROTEIN SYNTHESIS IN LIVER AND MUSCLE ESTIMATED BY TWO METHODS

| | *Constant infusion of [^{14}C]-tyrosine* precursor | | *Large dose of [1^{14}C]-leucine* precursor | |
	plasma	tissue	plasma	tissue
Liver	21 ± 1	59 ± 5	77 ± 6	87 ± 7
Jejunum	50 ± 17	218 ± 92	104 ± 13	136 ± 15

Data from McNurlan, Garlick and Tomkins (1979). After constant infusion of [^{14}C]-tyrosine for 6 h, rates of protein synthesis in liver and jejunal mucosa were calculated from the specific activity of tyrosine in protein and that of free tyrosine in either the plasma or the tissue. Because the specific activity of free tyrosine at these two sites differed appreciably, the rates calculated by the two methods differed by a factor of 3 or 4. When [1^{14}C]-leucine was given as a large dose (100 μmol/100 g body wt) and incorporation measured 10 min afterwards, rates of protein synthesis calculated from the specific activity of free leucine in either the plasma or the tissue were much more similar.

INJECTION OF LARGE DOSES OF LABELLED AMINO ACIDS

Scornik (1974) has described a method which appears to eliminate the need for measurement of free amino acid specific activity. Mice were given different doses of labelled leucine at the same specific activity, and the incorporation into liver proteins measured 5 min later. A graph was plotted of 1/dose against 1/incorporation, which was seen to be a straight line. The line was then extrapolated to zero on the 1/dose axis, a point corresponding to the incorporation at infinite dose. This intercept value was then used to calculate the rate of protein synthesis, on the assumption that at infinite dose the specific activity at the site of protein synthesis would be the same as that of the injected amino acid.

The difficulty with this method, as with other techniques in which unphysiological doses of amino acids are given, is the possibility that the rate of protein synthesis might be altered by the injected amino acid. With liver, measurement of the incorporation into protein of tracer doses of labelled lysine and glycine has failed to reveal any effect of leucine on protein synthesis (Scornik, 1974; McNurlan, Garlick and Tomkins, 1979). In muscle, however, leucine has been cited as a regulator of protein synthesis (Buse and Reid, 1975; Fulks, Li and Goldberg, 1975) and it would not be wise to use this amino acid for studies on this tissue. Unfortunately, our own attempts to validate this technique with other amino acids (valine and phenylalanine) in a number of tissues have failed because we were not, in general, able to obtain straight lines when 1/dose was plotted against 1/incorporation (P.J. Garlick, E. Mengheri and M.A. McNurlan, unpublished results). We have therefore preferred to use the conventional approach of measuring the specific activity of the free amino acid in the tissue.

When a very large dose of an amino acid is injected into an animal the specific activity of that amino acid in the tissues very rapidly rises to a value close to that in the plasma (Henshaw *et al.*, 1971; see illustration in *Figure 3.3*). With 100 μmol/100 g body wt of $[1^{14}C]$-leucine, the specific activity in liver and gut falls slowly but linearly between 2 and 10 min after injection (McNurlan, Garlick and Tomkins, 1979). The mean specific activity (S_i) over the first 10 min, therefore, can readily be computed from measurements at these two time points only. From the specific activity of the protein at 10 min (S_B) the rate of protein synthesis can then be calculated very simply from the formula

$$\frac{S_B}{\bar{S_i}} = k_s t \qquad\qquad (3.7)$$

The fall in free amino acid specific activity after a large dose is linear and is therefore more easy to measure, with fewer animals, than the rapid changes which take place after single injection of a tracer dose of label (e.g. Haider and Tarver, 1969; Peters and Peters, 1972). Nevertheless, this technique requires more animals than constant infusion. There are, however, several advantages. The specific activity of the free amino acid in the tissues rises close to that in the plasma in both liver and intestinal mucosa, so that the difficulty of defining the specific activity at the site of protein synthesis no longer introduces a significant error (*Table 3.3*). The single injection with measurement of incorporation over 10 min is not only more convenient than constant infusion in small animals; this short time period does not allow for any turnover of labelled protein during the time of incorporation. With constant infusion for 6 h, gut protein became so highly labelled that recycling of label prevented the attainment of a plateau in the free amino acid pool (McNurlan, Garlick and Tomkins, 1979). The 10 min incorporation period also means that labelled plasma proteins synthesized in the liver will still be retained in that tissue. Hence the rate of synthesis in liver protein obtained by the large-dose method is about 30% higher than that obtained by constant infusion, which does not include the plasma protein component (*Table 3.3*).

Measurement of tissue protein breakdown

DECAY OF LABELLED PROTEINS

The principle of measuring the loss of isotope from labelled protein is very simple. The decay of total label in the protein, when plotted on semi-log axes, should be a straight line whose gradient is the fractional rate of breakdown; similarly, decay of specific activity gives the fractional rate of synthesis (see Waterlow, Garlick and Millward, 1978). In general, however, this technique is not very useful for measurements of total tissue proteins for the following reasons:

(1) Recycling of label from protein breakdown must be eliminated if valid rates of turnover are to be obtained. Whereas this is possible with certain labels in some tissues, e.g. carbonate labelling of liver proteins (Millward, 1970; Swick and Ip, 1974), in general it is difficult to be sure that all recycling has been eliminated in a tissue such as muscle. Slowly turning-over tissues like muscle

are constantly receiving labelled amino acids from the more highly labelled tissues such as gut and liver, which makes the prevention of recycling more difficult.

(2) Unlike single purified proteins, mixed proteins from whole tissues do not decay linearly on semi-log axes. The curves obtained can be analysed in a way which allows for the heterogeneity of turnover rates in the tissue (Garlick, Waterlow and Swick, 1976), but the analysis requires that decay be measured over a very long time period (i.e. greater than 3 X the mean half-life). For rat liver and muscle this would mean experiments lasting >6 days and >20 days, respectively. For larger animals with slower turnover rates the time scale would probably be prohibitive.

BREAKDOWN ESTIMATED FROM SYNTHESIS AND GROWTH

The problems encountered with measurements of decay in mixed proteins do not cause difficulty when synthesis is measured from incorporation of label (see above and Waterlow, Garlick and Millward, 1978). In the absence of a suitable direct method of measurement, therefore, breakdown rates of whole-tissue proteins can be estimated from the difference between the rate of synthesis and the rate of growth, i.e.

$$k_s - k_d = k_g \tag{3.8}$$

where k_s, k_d and k_g are the fractional rates of synthesis, breakdown and growth. This method has been used successfully in rat muscle (Turner and Garlick, 1974; Garlick *et al.*, 1975; Millward *et al.*, 1975) and liver (Scornik and Botbol, 1976). The sources of error for this method have been discussed in detail previously (Waterlow, Garlick and Millward, 1978; Garlick, Fern and McNurlan, 1979; Garlick, 1979). The main difficulty arises from the time scale of measurements. Synthesis is measured over a period of a few minutes or hours, while growth, particularly in muscle, must be measured over a period of several days to obtain accurate values. Hence it must be assumed that the synthesis rate measured at a particular time of day is representative of the whole day. Whereas diurnal variations in synthesis in muscle have been shown to occur, their amplitude is small (Garlick, Millward and James, 1973) and there appears to be no reason why breakdown rates calculated by this method should not be accurate.

Applicability of methods to studies in different species

Methods of measuring whole-body protein turnover were mostly developed for use in humans because of the difficulty of obtaining samples of tissue. However, they are equally applicable for use in both small experimental animals, such as rats and mice, and in larger animals. The advantage of these techniques is that they can be repeated many times in the same individual, for example during development, since only urine samples or blood and respiratory CO_2 are required. With animals there is no particular hazard to the use of ^{14}C-labelled amino acids, which are easier to measure than stable isotopes such as [15]N. In addition, we know more about the metabolism of carbon compared with nitrogen and are therefore more able to assess the magnitude of possible errors with, for example, infusion of $[1^{14}C]$-leucine. Constant infusion of $[1^{14}C]$-leucine has been applied successfully to the measurement of whole-body protein turnover and leucine oxidation rate in both rats (Sketcher, 1976) and pigs (Reeds *et al.*, 1978).

A comparison of adults of different species, ranging in body weight from 20 g to over 600 kg is shown in *Table 3.4*. The values shown are flux rates calculated

Table 3.4.   WHOLE-BODY PROTEIN TURNOVER IN VARIOUS SPECIES

Species	*Protein turnover* (g/kg body wt per day)	(g/kg$^{0.75}$ body wt per day)	*Reference*
Adult:			
mouse (20 g)	38.4	14.5	Garlick and Marshall (1972)
mouse (42 g)	43.4	19.5	Garlick, Trayhurn and James, unpubl.
rat (510 g)	20.5	17.4	Millward, unpubl.
rabbit (3.6 kg)	18.0	24.8	Nicholas, Lobley and Harris (1977)
dog (10.2 kg)	12.1	21.5	Everett and Sparrow, unpubl.
sheep (67 kg)	5.3	15.9	Bryant and Smith, unpubl.
man (77 kg)	5.7	16.7	James *et al.* (1976)
cow (628 kg)	3.7	18.7	Lobley, Reeds and Pennie (1978)
Immature:			
rat (37 g)	78.9	34.5	Millward, unpubl.
rat (116 g)	45.0	25.4	Millward, unpubl.
pig (76 kg)	9.0	26.6	Garlick, Burk and Swick (1976)

The values shown are rates of protein flux obtained during constant infusion of labelled tyrosine on the assumption that tyrosine comprises 3% of whole-body protein.

from constant infusion of labelled tyrosine. Tyrosine is not the best amino acid for studies of this kind, nor was $^{14}CO_2$ production measured in most cases, so that protein synthesis rates could not be calculated from flux rates. However, the tyrosine flux enabled the widest range of comparison of results obtained by the same technique to be made. From the table it can be seen that the flux rate is much slower in the smaller species, when expressed per kilogramme of body weight. In this respect, protein turnover parallels other metabolic processes, e.g. oxygen consumption. Brody (1945) and Kleiber (1961) showed that fasting metabolic rate in adults of a very wide range of species was related to body weight raised to a power close to 0.75. In *Table 3.4* the rates of flux have also been expressed per kg$^{0.75}$. The range of values obtained, 14.5–24.8 g protein/kg$^{0.75}$ per day, is very small, considering the 30 000-fold range of body weights. Nor is there any systemic change in the value with increasing body weight. This shows that protein turnover bears a similar relationship to fasting metabolic rate in all the species listed. If it is assumed that four high-energy phosphate bonds are consumed per peptide bond synthesized, and that the fasting energy metabolism is 70 kcal/kg$^{0.75}$ (Kleiber, 1961), we may conclude that about 15–20% of fasting metabolic rate goes towards the synthesis of body protein. In the immature animals shown in *Table 3.4*, flux rate expressed per kg$^{0.75}$ is higher in all cases than any of the values for adults.

As described above, measurements of rates of protein synthesis and breakdown in individual tissues can be made by a variety of techniques. Most of these methods have been developed using rats, and only constant infusion appears to have been at all widely used in other species, e.g. mouse (Garlick and Marshall, 1972), rabbit (Nicholas, Lobley and Harris, 1977), fowl (Laurent *et al.*, 1978; Maruyama, Sunde and Swick, 1978), dog (Everett, Taylor and Sparrow, 1977), pig (Garlick, Burk and Swick, 1976; Edmunds, Buttery and Fisher, 1978; Simon *et al.*, 1978), sheep (Buttery *et al.*, 1975), cattle (Lobley, Reeds and Pennie, 1978) and man (Halliday and McKeran, 1975). The advantage of the constant infusion over single-dose methods is that specific activity time curves do not have to be measured by killing animals at several points in time. Instead it is only necessary to obtain tissue samples at the end of the infusion in order to calculate the rate of synthesis in each individual animal. Hence, the constant infusion is particularly suited to measurements in large animals. The single injection of a large-dose technique, which is certainly more accurate than constant infusion for rapidly turning-over tissues such as liver and gut, is more suitable for studies in the smaller species, such as rats. The requirement for larger numbers of individual animals would restrict the use of the latter technique with large animals.

Some examples of rates of protein synthesis and breakdown in liver and skeletal muscle obtained by constant infusion of labelled amino acids in a variety of species are shown in *Table 3.5*. As with flux rate, the fractional rate of synthesis (i.e. the fraction or percent of the tissue protein renewed per day) is lower

Table 3.5.   PROTEIN SYNTHESIS IN LIVER AND MUSCLE OF VARIOUS SPECIES

Species	Amino acid infused	Muscle		Liver		Reference
		k_s (%/day)	% of whole body	k_s (%/day)	% of whole body	
Adult:						
mouse (36 g)	[^{14}C]-tyrosine	9	18	87	16	Garlick, Trayhurn and James, unpubl.
rat (716 g)	[^{14}C]-tyrosine	5	28	54	20	Lo, Bates and Millward, unpubl.
rabbit (3.62 kg)	[^{3}H]-tyrosine	2	13	32	14	Nicholas *et al.* (1977)
sheep (40 kg)	[^{3}H]-lysine	2	50	10	17	Buttery *et al.* (1975)
man (67 kg)	[^{15}N]-lysine	2	53	–	–	Halliday and McKeran (1975)
cow (628 kg)	[^{3}H]-tyrosine	0.9	18	14	8	Lobley *et al.* (1978)
Immature:						
rat (108 g)	[^{14}C]-tyrosine	13	19	67	10	Garlick *et al.* (1975)
pig (24 kg)	[^{3}H]-lysine	6	62	37	14	Edmunds *et al.* (1978)
pig (76 kg)	[^{14}C]-tyrosine	4	42	23	10	Garlick *et al.* (1976)

All results were obtained by constant infusion of labelled amino acids. Total protein synthesis in liver and muscle was calculated by multiplying the fractional rate of synthesis by the protein content of that tissue; these values were then expressed as percent of whole-body synthesis, which was calculated from the amino acid flux as in *Table 3.4*. When the proportion of the infused amino acid in protein was not given in the original publication, tyrosine was assumed to be 3% and lysine 7% of whole-body protein.

in both liver and muscle of the larger species. In immature animals the values of k_s are somewhat higher than in adults of the same body weight. This would allow for growth. In muscle, the effect of age and growth is particularly pronounced; both synthesis and degradation are higher in younger, more rapidly growing individuals (Millward *et al.*, 1975; Arnal, Ferrara and Fauconneau, 1976).

It has often been suggested that in smaller animals protein synthesis in muscle would contribute a smaller proportion of whole-body synthesis, and in liver a higher proportion (e.g. Munro, 1969). This comparison requires that the fractional rates of synthesis shown in *Table 3.5* be converted to numbers of grammes of protein synthesized per organ. The values for synthesis in tissues as a percent of whole-body synthesis were therefore calculated by multiplying the fractional rates k_s by the amount of protein in that tissue, and dividing by the rate of whole-body synthesis. The results are disappointing in that there is no consistent change in the proportion of whole-body synthesis in either liver or muscle with increasing body weight, in spite of the large variability between species. Similarly, there is no clear difference between immature and adult animals. However, these measurements were made in a number of laboratories, using two different amino acids for infusion. In particular, it appears that infusion of labelled lysine gives higher values than tyrosine. This may have resulted from low values for flux with lysine. Simon *et al.* (1978) have shown that the flux of lysine is very much lower than that of leucine, even though their proportions in protein are very similar. In addition, the food intakes in the various studies may not have been strictly comparable; variations in protein synthesis with dietary intake of energy and protein have been demonstrated (e.g. Garlick *et al.*, 1975). Also the rate of protein synthesis in the whole body and in muscle has been shown to depend upon the time of day with respect to meals (Garlick *et al.*, 1978, 1973). More measurements therefore need to be done, preferably using the same technique throughout, and under controlled conditions of food intake, before the differences between species indicated by the data in *Table 3.5* can be accepted as valid.

Acknowledgements

I am indebted to Professor Waterlow and colleagues at the Clinical Nutrition and Metabolism Unit for constant discussion and advice, to those who contributed unpublished data, and to the Royal Society for the award of a J. Sainsbury Research Fellowship.

References

ARNAL, M., FERRARA, M. and FAUCONNEAU, G. (1976). In *Nuclear Techniques in Animal Production and Health,* pp. 393–401, International Atomic Energy Agency, Vienna

BRODY, S. (1945). *Bioenergetics and Growth,* Hafner, New York

BUSE, M.G. and REID, S.S. (1975). *J. Clin. Invest.,* **56**, 1250–1261

BUTTERY, P.J., BECKERTON, A., MITCHELL, R.M., DAVIES, D. and ANNISON, E.F. (1975). *Proc. Nutr. Soc.,* **34**, 91A

EDMUNDS, B.K., BUTTERY, P.J. and FISHER, C. (1978). *Proc. Nutr. Soc.,* **37**, 32A

EVERETT, A.W., TAYLOR, R.R. and SPARROW, M.P. (1977). *Biochem. J.,* **166**, 315–321

FULKS, R.M., LI, J.B. and GOLDBERG, A.L. (1975). *J. Biol. Chem.,* **250**, 290–298

GARLICK, P.J. (1978). *Biochem. J.,* **176**, 402–405

GARLICK, P.J. (1979). In *Comprehensive Biochemistry,* Vol. 19B (Florkin, M., Ed.), Elsevier, Amsterdam

GARLICK, P.J. and MARSHALL, I. (1972). *J. Neurochem.,* **19**, 577–583

GARLICK, P.J. and WATERLOW, J.C. (1977). In *Stable Isotopes in the Life Sciences,* pp. 323–333, International Atomic Energy Agency, Vienna

GARLICK, P.J., BURK, T.L. and SWICK, R.W. (1976). *Am. J. Physiol.,* **230**, 1108–1112

GARLICK, P.J., CLUGSTON, G.A. and WATERLOW, J.C. (1980). *Am. J. Physiol.,* **238**, E235–E244

GARLICK, P.J., FERN, E.B. and McNURLAN, M.A. (1979). In *Proceedings of the 12th F.E.B.S. Meeting, Dresden,* Vol. 53 (Rapoport, S. and Schewe, T., Eds), pp. 85–94, Pergamon Press, Oxford

GARLICK, P.J., MILLWARD, D.J. and JAMES, W.P.T. (1973). *Biochem. J.,* **136**, 935–945

GARLICK, P.J., WATERLOW, J.C. and SWICK, R.W. (1976). *Biochem. J.,* **156**, 657–663

GARLICK, P.J., MILLWARD, D.J., JAMES, W.P.T. and WATERLOW, J.C. (1975). *Biochim. Biophys. Acta,* **414**, 71–84

GARLICK, P.J., CLUGSTON, G.A., SWICK, R.W., MEINERTZHAGEN, I.H. and WATERLOW, J.C. (1978). *Proc. Nutr. Soc.,* **37**, 33A

GOLDEN, M.H.N. and WATERLOW, J.C. (1977). *Clin. Sci. Mol. Med.,* **53**, 277–288

HAIDER, M. and TARVER, H. (1969). *J. Nutr.,* **99**, 433–445

HALLIDAY, D. and McKERAN, R.O. (1975). *Clin. Sci. Mol. Med.,* **49**, 581–590

HENSHAW, E.C., HIRSCH, C.A., MORTON, B.E. and HIATT, H.H. (1971). *J. Biol. Chem.,* **246**, 436–446

JAMES, W.P.T., GARLICK, P.J., SENDER, P.M. and WATERLOW, J.C. (1976). *Clin. Sci. Mol. Med.,* **50**, 525–532

JAMES, W.P.T., SENDER, P.M., GARLICK, P.J. and WATERLOW, J.C. (1974). In *Dynamic Studies with Radioisotopes in Man,* pp. 461–472. International Atomic Energy Agency, Vienna

KLEIBER, M. (1961). *The Fire of Life,* Wiley, New York

LAURENT, G.J., SPARROW, M.P., BATES, P.C. and MILLWARD, D.J. (1978). *Biochem. J.,* **176**, 393–401

LOBLEY, G.E., REEDS, P.J. and PENNIE, K. (1978). *Proc.Nutr. Soc.,* **37**, 96A

MARUYAMA, K., SUNDE, M.L. and SWICK, R.W. (1978). *Biochem. J.,* **176**, 573–582

McNURLAN, M.A., GARLICK, P.J. and TOMKINS, A.M. (1979). *Biochem. J.,* **178**, 373–379

MILLWARD, D.J. (1970). *Clin. Sci.,* **39**, 591–603

MILLWARD, D.J., GARLICK, P.J., STEWART, R.J.C., NNANYELUGO, D.O. and WATERLOW, J.C. (1975). *Biochem. J.,* **150**, 235–243

MUNRO, H.N. (1969). In *Mammalian Protein Metabolism,* Vol. III, Ch. 25 (Munro, H.N., Ed.), Academic Press, New York

NICHOLAS, G.A., LOBLEY, G.E. and HARRIS, C.I. (1977). *Brit. J. Nutr.,* **38**, 1–17

O'KEEFE, S.J., SENDER, P.M. and JAMES, W.P.T. (1974). *Lancet,* **ii**, 1035–1037

OLESON, K., HEILSKOV, N.C.S. and SHONHEYDER, F. (1954). *Biochim. Biophys. Acta,* **15**, 95–107

PETERS, T. Jr. and PETERS, J.C. (1972). *J. Biol. Chem.*, **247**, 3858–3863

PICOU, D. and TAYLOR-ROBERTS, T. (1969). *Clin. Sci.*, **36**, 283–296

REEDS, P.J., FULLER, M.F., LOBLEY, G.E., CADENHEAD, A. and McDONALD, J.D. (1978). *Proc. Nutr. Soc.*, **37**, 106A

SCORNIK, O.A. (1974). *J. Biol. Chem.*, **249**, 3876–3883

SCORNIK, O.A. and BOTBOL, V. (1976). *J. Biol. Chem.*, **251**, 2891–2897

SIMON, O., MÜNCHMEYER, R., BERGNER, H., ŻEBROWSKA, T, and BURACZEWSKA, L. (1978). *Brit. J. Nutr.*, **40**, 243–252

SKETCHER, R.D. (1976). *Ph.D. thesis*, University of London

SWICK, R.W. and IP, M.M. (1974). *J. Biol. Chem.*, **249**, 6836–6841

TURNER, L.V. and GARLICK, P.J. (1974). *Biochim. Biophys. Acta,* **349**, 109–113

WATERLOW, J.C. and STEPHEN, J.M.L. (1968). *Clin. Sci.,* **35**, 287–305

WATERLOW, J.C., GARLICK, P.J. and MILLWARD, D.J. (1978). *Protein Turnover in Mammalian Tissues and in the Whole Body,* North-Holland, Amsterdam

WATERLOW, J.C., GOLDEN, M.H.N. and GARLICK, P.J. (1978). *Am. J. Physiol.,* **235**, E165–E174

4

INTEGRATION OF WHOLE-BODY AMINO ACID METABOLISM

E.N. BERGMAN
R.N. HEITMANN
Department of Physiology, Biochemistry and Pharmacology, New York State College of Veterinary Medicine, Cornell University

Summary

Techniques will be described for measuring free amino acid inter-organ transport and integration of nitrogen metabolism by several tissues of fed, fasted and acidotic sheep. Net addition to, or removal from, the blood was calculated from veno-arterial concentration differences and rates of whole blood and plasma flow across the portal-drained viscera, liver, kidneys and hindquarters. In some experiments, net portal blood addition (or appearance) of free amino acids was compared with the total amino acids disappearing from the lumen of the small intestine, i.e. between the pylorus and terminal ileum.

Comparisons of free amino acid transport in both whole blood and plasma showed that there was at least some blood cell transport of glutamine, glutamate, taurine, glycine and leucine. In fact, blood cells, per unit volume, transported even more glutamine, glutamate and taurine than did plasma. Plasma thus is believed to reflect amino acid transport but, in most cases, it probably underestimates total transport to the extent of the packed cell volume.

Of all the amino acids hydrolysed in the intestinal lumen, only alanine and serine were found to appear in the portal blood in amounts equal to that disappearing from the intestine. Glutamate and aspartate had the highest rates of intestinal disappearance but yet appeared in the blood in only negligible amounts. All other amino acids had net rates of portal appearance of only 40–80% of their intestinal disappearance. The highest percentages of portal appearance occurred when the sheep were fed a 20%, as compared with a 15%, protein diet. Considerable gut epithelial metabolism must thus occur or else many amino acids are absorbed as small polypeptides.

In the body, the liver was found to remove large quantities of amino acids. Also, alanine and glutamine cycles especially seemed important in nitrogen transport between tissues. Both are continuously released by muscle and removed by liver with large increases occurring during fasting. The gut also adds alanine to the blood during feeding but glutamine is removed. The kidneys of fed sheep add glutamine to the blood but, during fasting and acidosis, glutamine is removed for urinary ammonia production. The urea cycle amino acids also are involved in a nitrogen transport cycle since the kidneys, and to some extent the muscles, remove citrulline and ornithine and add arginine. The liver does the reverse by removing arginine and adding ornithine, citrulline and urea. Glycine is added to the blood by both kidneys and muscle, especially during fasting, and is removed by liver. The above amino acid movements between different tissues thus emphasize nitrogen carrier systems for maintenance of acid-base balance and for avoidance of NH_3 toxicity. Also, their carbon skeletons are used mainly by liver for synthesis of glucose and other compounds.

Introduction

The amounts and kinds of amino acids participating in the metabolism of different tissues of the intact animal are only incompletely understood. In

69

ruminants, the significance of fermentation in the reticulorumen has led to considerable research on rates of amino acid degradation and disappearance within the digestive tract, but this gives little or no information on actual availability or metabolism by different organs. Further, the animal must integrate and change its metabolism to adapt to different nutritional, physiological or stressful situations. Notable among the latter are starvation, and even acidosis which can occur under conditions of high grain feeding.

Some tissues that undoubtedly have high requirements for amino acids even in the adult animal are the gut epithelium, liver, muscle, kidneys, and, of course, the mammary gland and developing fetuses. Intestinal epithelial cells are known to have an extremely rapid rate of turnover and large quantities of amino acids must surely be required for their formation. Further, virtually unknown quantities of dietary amino acids are metabolized by the gut epithelium during the process of absorption (Bergman and Heitmann, 1978; Neame and Wiseman, 1958). The liver of the intact animal also requires amino acids, not only for plasma protein synthesis, but for ureogenesis and gluconeogenesis, especially if glucose precursors are in short supply (Bergman, 1973). Muscle represents a considerable storehouse for amino acids simply because of its large mass. Its daily protein turnover for normal 'wear and tear' is known to be considerable and it also can remove or release free amino acids in net amounts. Muscle thus must have metabolic and homeostatic activities, as well as having the important function of providing for locomotion (Daniel, 1977; Lindsay and Buttery, 1980). Further, the kidneys not only require amino acids for their protein turnover, as in the case of muscle, but they also require amino acids, particularly glutamine, for ammonia production and maintenance of acid-base balance (Baruch *et al.* 1975). Considerable bodily integration and inter-organ transport of free amino acids must thus occur in all animal species and this transport undoubtedly becomes altered by diet, and by conditions of fasting, acidosis, lactation or pregnancy.

The purpose of this chapter is to summarize recent and current research, primarily in the authors' laboratory, on amino acid metabolism simultaneously occurring in the digestive tract, liver, kidneys and hindquarters of adult fed, fasted and acidotic sheep. The methods depend upon measurement of rates of blood and plasma flow across each tissue, measurement of veno-arterial amino acid concentration differences and, in some instances, whole-body rates of blood amino acid turnover and rates of amino acid disappearance from the gut lumen. Emphasis will be placed on those amino acids primarily involved in nitrogen transport for the integration of metabolism by the above various organs. Additionally, amino acid transport in plasma will be compared with that occurring in whole blood. Comparisons will be made with non-ruminant species where possible.

Experimental techniques

Most of the studies to be reviewed were performed using multicatheterized and continuously fed adult sheep. Polyvinyl catheters were surgically implanted into the portal, hepatic and renal veins, lower vena cava and aorta to measure veno-arterial amino acid concentration differences across the portal-drained viscera (digestive tract plus pancreas and spleen), liver, kidneys and hindquarters (Katz and Bergman, 1969; Kaufman and Bergman, 1974; Heitmann and Bergman,

1978). Experiments could be performed only after about one week and when the animals were fully recovered from the effects of surgery. Results on three groups of sheep will be reviewed in this paper: fed, fasted, and fed plus acidotic (*Table 4.1*). Fed sheep were given 33 g of pelleted feed via an automatic feeder

Table 4.1. EXPERIMENTAL ANIMALS AND BLOOD ACID-BASE PARAMETERS

Sheep*	pH	HCO₃ (mmol/l)	Base excess (mmol/l)
Fed (5)	7.52	22	0.0
Fasted (4)	7.42†	20	−3.1
Acidotic (5)	7.35†	12†	−11.3†

* Sheep (50–60 kg) were fed 800 g of pelleted feed per day, fasted for 3 days, or fed plus made acidotic by administering 20–30 g NH₄Cl for 3 days.
† Significantly different from fed sheep ($P < 0.05$). Summarized from Heitmann and Bergman (1978).

each hour throughout each 24 h period (800 g/day) and this feeding regime was started at least four days prior to any experiment. Fasted sheep were given only water and salt for 72 h. For the production of acidosis, fed sheep were drenched daily for 3 days with sufficient $NH_4 Cl$ to maintain the blood base excess at −10 to −15 mEq/l.

Rates of whole-blood flow through the portal vein, liver and kidneys were measured by infusion of P-aminohippuric acid (PAH), as the indicator substance, into a mesenteric vein. Portal and hepatic flows were calculated by measuring the downstream dilution of the PAH (Katz and Bergman, 1969). Renal flow through the two kidneys' was calculated by simultaneously measuring PAH excretion in the urine and renal veno-arterial PAH concentration differences (Bergman *et al.*, 1974). Blood flow through the hindquarters was calculated at the end of each experiment by infusing PAH into the caudal aorta and measuring jugular–lower vena caval concentration differences. Jugular and general arterial PAH concentrations are identical since PAH is not excreted or metabolized by the head (Heitmann and Bergman, 1978). Plasma flows were obtained by subtracting that flow represented by the packed cell volume. Net tissue metabolism of amino acids thus is calculated by multiplying either the whole-blood or plasma flow by the respective whole-blood or plasma amino acid veno-arterial difference.

Amino acid fluxes in plasma vs. whole blood

The use of plasma, instead of whole blood, previously had been widely accepted and assumes that blood cells make no contribution to net movements of amino acids for tissue metabolism (Pitts, de Haas and Klein, 1963; Wolff, Bergman and Williams, 1972). Consequently, due to problems inherent in sample preparation of whole blood, most of the data on amino acid fluxes available to us today have been obtained by using plasma. More recent work in dogs and human beings, however, has shown that erythrocytes can indeed be involved, at least for a few specific amino acids (Elwyn *et al.*, 1972; Felig, Wahren and Raf, 1973; Aoki *et al.*, 1976; Drewes, Conway and Gilboe, 1977). Comparisons of plasma and whole-blood fluxes and concentrations of all major amino acids are therefore necessary.

Amino acid fluxes (removal or addition) across the above four tissues have been measured simultaneously in both plasma and whole blood of sheep. The tissue flux rates were calculated as mmol/h and data for plasma and whole blood were correlated *(Table 4.2)* using the formula $Y = mX + b$, where Y and X are

Table 4.2. CORRELATIONS ($Y = mX + b$) OF TISSUE AMINO ACID METABOLISM (mmol/h), AS MEASURED IN PLASMA (Y) AND WHOLE BLOOD (X) OF SHEEP

Amino acid	Slope (m)	Intercept (b)	S. D.
Alanine	0.86	0.02	1.16
Glutamine	0.45*	−0.37	1.35
Glutamate	0.44*	0.02	0.82
Glycine	0.52†	−0.24	1.57
Arginine	0.81	−0.11	0.37
Citrulline	0.39	0.10	0.60
Ornithine	0.71	0.04	0.56
Leucine	0.85†	0.01	0.29
Isoleucine	0.78	−0.02	0.26
Taurine	0.04*	−0.10	0.29

$N = 64-75$ (see *Figure 4.1*).
* Significantly lower than (one-packed cell volume) or 0.77 ($P < 0.05$). All intercepts were not significantly different from zero. Summarized from Heitmann and Bergman (1980).
† Significantly lower than 1.0 ($P < 0.05$).

plasma and whole-blood flux rates, m is the slope, and b the intercept. Detailed data for leucine, as an example, are expressed in graphical form in *Figure 4.1.* There were no significant differences in slopes or intercepts for any particular amino acid between tissue sites; e.g., data for the portal-drained viscera gave the same slope and intercept as that for the liver and other tissues. Further, animal treatment (fed, fasted or acidotic) also had no significant effect on slope or intercept. Tissue site and treatment can be ignored, therefore, on the assumption that blood cells are handled by similar mechanisms in all tissues and for all treatments.

It is noteworthy that all slopes were less than 1.0 and all intercepts did not differ significantly from zero *(Table 4.2; Figure 4.1)*. The slope is a function of amino acid transport rate and direction. If blood cells were not involved in amino acid transport, the slopes should be equal to 1.0. However, if the cells, per unit volume, were transporting amino acids in the same direction and at the same rate as plasma, the slope would be less than 1.0 and actually equal to (one-packed cell volume) or 0.77 for these experiments. Further, if the blood cell transport rate exceeded that of plasma, the slope would be significantly less than 0.77. Indeed, five of the ten observed amino acids had slopes statistically less than 1.0 and therefore were partially being transported by blood cells and always in the same direction with that of plasma. However, only glutamine, glutamate and taurine had slopes significantly less than 0.77 and these three amino acids must have been transported by the blood cells at a rate greater than that of plasma.

Measurements of amino acid transport in plasma must thus underestimate true amino acid fluxes. Previous workers (Felig, Wahren and Raf, 1973; Chaisson *et al.*, 1975) have reported 75% and 93% of alanine transport in human beings to be by way of plasma. Data on sheep (*Table 4.2*) agrees with this, in that 86% of alanine transport occurs via plasma. Similarly, Drewes, Conway and Gilboe (1977) found that plasma measurements especially underestimate glutamate transport in dogs (44%) and, again, data on sheep (*Table 4.2*) for both glutamate

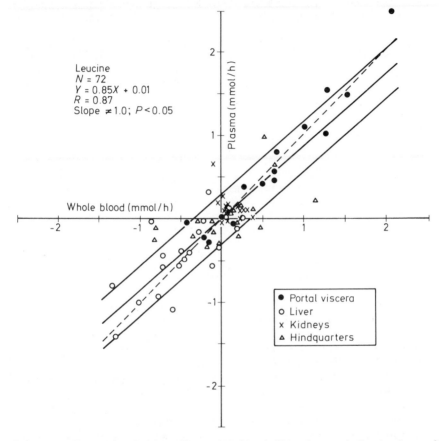

Figure 4.1   Regression of plasma, *Y*, vs. whole blood, *X*, leucine metabolism by tissues of sheep. Dashed line represents a slope of 1.0 and an intercept of 0.0. Solid lines represent the calculated regression line ± 1 S.D.

and glutamine are in good agreement (56%). In conclusion, plasma surely reflects amino acid transport but it does seem to underestimate total transport to about the extent of the packed cell volume and, in the cases of glutamine, glutamate and taurine, perhaps even more.

Arterial concentrations of amino acids

Table 4.3 summarizes values obtained for sheep of comparable size and breed. In comparison to dog, pig or human blood, sheep blood has higher concentrations of glycine and the urea cycle amino acids, arginine, citrulline and ornithine (Elwyn *et al.*, 1972; Felig, Wahren and Raf, 1973; Aoki *et al.*, 1976; Chavez and

Bayley, 1976). However, glutamine and alanine, the most glucogenic amino acids in sheep (Brockman and Bergman, 1976; Heitmann and Bergman, 1978), are lower than in dogs, pigs and humans, probably reflecting the continuous process of gluconeogenesis unique to ruminants.

Fasting clearly had a more profound effect on amino acid concentrations than did acidosis (*Table 4.3*). Glutamine levels were depressed but glycine and the branched chain amino acids, leucine, isoleucine and valine, were all elevated

Table 4.3.   ARTERIAL CONCENTRATIONS OF MAJOR AMINO ACIDS IN SHEEP

Amino acid	Plasma concentrations			Whole-blood–plasma (19) μM
	fed (9)	fasted (8) μM	acidotic (10)	
Alanine	89	94	88	−5
Glutamine	225	173*	176*	−30†
Glutamate	59	56	58	45†
Glycine	303	559*	514*	100†
Arginine	117	81*	102	−29†
Citrulline	149	65*	167	−4
Ornithine	76	65*	100	11†
Leucine	74	126*	82	10†
Isoleucine	59	89*	61	−4
Valine	131	184*	168	−1

*   Significantly different from fed sheep ($P < 0.05$).
†   Whole blood significantly different from plasma ($P < 0.05$). Summarized from Heitmann and Bergman (1980) and from Joo *et al.* (1976).

during fasting. This phenomenon is well documented in the literature (Baird, Heitzman and Hibbitt, 1972; Adibi, 1976). Branched-chain amino acid concentrations are dependent on the type of fasting. Complete fasting temporarily elevates these amino acids but primarily a protein deficiency decreases their concentrations. The three urea cycle amino acids, arginine, citrulline and ornithine, all decreased during fasting. As large amounts of ammonia are produced during ruminal fermentation of protein, and ammonia production decreases during fasting, the concentrations of the urea cycle amino acids are probably an index of the quantities of ammonia being detoxified in the liver.

It also is of interest to compare plasma amino acid concentrations with that of whole blood (*Table 4.3*). There were no significant differences due to treatment and thus all three treatments were combined to obtain whole-blood–plasma differences. Glutamate, glycine, ornithine and leucine were significantly higher in whole blood as compared to plasma, while glutamine and arginine were lower. It is of especial interest that the glutamine–glutamate couplet behaved oppositely. Whereas glutamine was mostly in plasma, glutamate was concentrated in the erythrocytes. Further, and as stated earlier, at least one-half of both glutamine and glutamate transport was handled by the erythrocytes and only about 45% by the plasma.

Net tissue metabolism of amino acids

NET GUT AND LIVER METABOLISM

Although considerable research has been done on protein hydrolysis and amino acid disappearance within the digestive tract, only little information is available on amino acids actually absorbed into the portal blood. *Table 4.4*, however, summarizes some recent data in sheep fed two different diets. The amounts of amino acids disappearing between the pylorus and ileocaecal junction (small intestine) were compared with net amounts appearing in the portal plasma.

Of all the amino acids hydrolysed in the intestinal lumen, only alanine and serine were found to appear in the portal blood in amounts nearly equal (72–103%) to that disappearing from the intestine (*Table 4.4*). Further, even

Table 4.4. COMPARISON OF AMINO ACIDS DISAPPEARING FROM THE SMALL INTESTINE WITH THAT APPEARING IN PORTAL BLOOD

Amino acid	Intestinal disappearance*		Net portal appearance†	
	20% prot.	15% prot.	20% prot.	15% prot.
	(g/day)		(% of disappearance)	
Alanine	4.5	2.7	86	83
Glutamine			<0	<0
Glutamate	7.3	5.2	6	−10
Aspartate	7.0	4.9	−2	9
Glycine	4.5	2.7	64	32
Serine	2.8	1.5	72	103
Arginine	3.4	2.3	58	32
Ornithine	6.0	1.9	46	18
8 EAA + cysteine	31.6	19.0	44	33

Sheep were fed 800 g/day high protein (20%) or 650 g/day medium protein (15%) diets at hourly intervals. Each value represents the mean of 4–5 experiments. Calculated from Tagari and Bergman (1978).

* Pyloric-ileal differences $\times$ ingesta flow.

† Portal-arterial differences $\times$ plasma flow.

though glutamate and aspartate were present in the digesta in greater amounts than any other amino acid, only negligible amounts were found to be absorbed into the blood. All other amino acids had net rates of portal plasma appearance of only about one-third to two-thirds of their rates of intestinal disappearance. Even if these plasma appearance rates are corrected to that of whole blood, still only 40–80% of their intestinal disappearance could be accounted for. The highest percentages of portal appearance also occurred when the sheep were fed a 20%, as compared with a 15%, protein diet. Considerable gut epithelial metabolism must thus occur for energy purposes, transaminations or even subsequent ammonia production. Additionally, many amino acids could be conjugated by the gut mucosae into protein membranes of chylomicrons or into enzymes appearing in the plasma. A third alternative is that a portion of the amino acids could be absorbed as small polypeptides and thus does not appear as the free acid in the blood.

Further data on sheep fed at hourly intervals (*Table 4.5*) show that while most amino acids are absorbed into the blood, most of this net gut addition is removed by the liver. Other data on cattle and sheep fed a single meal are qualitatively similar (Hume, Jacobson and Mitchell, 1972; Baird, Symonds and Ash, 1975). The glucogenic amino acids, alanine, glycine and serine, were absorbed in the greatest amounts and even more were removed by the liver. Net movements of alanine and glycine from peripheral tissues thus surely must occur to supply this large liver uptake. The branched-chain amino acids, leucine, isoleucine and valine, were absorbed in the next greatest quantities and about one-third was removed by the liver. This also seems remarkable since, in non-ruminant species, muscle usually is considered to be the major site of branched-chain amino acid catabolism (Adibi, 1976; Goldberg and Chang, 1978). While small amounts of the three urea cycle amino acids were absorbed into the portal blood, the liver removed more arginine than was absorbed and released part of this back into the blood as citrulline and ornithine. Urea thus is produced by this process and, again, arginine must carry nitrogen from peripheral tissues for urea production in the liver.

Table 4.5. NET METABOLISM OF SOME PLASMA AMINO ACIDS BY PORTAL-DRAINED VISCERA AND LIVER OF FED AND FASTED SHEEP

| Amino acid | Portal viscera Fed (10) | | Liver Fed (17) | | Fasted 3 days (8) | |
	removed	added (mmol/h)	removed	added (mmol/h)	removed	added (mmol/h)
Alanine		2.3	2.6		3.7*	
Glutamine	1.5		2.0		1.8	
Glutamate	0.2			1.8		1.0*
Glycine		1.5	3.4		4.3	
Serine		1.1	0.9		0.9	
Arginine		0.3	0.8		1.1	
Citrulline		0.5		0.1		0.2
Ornithine		0.3		0.6		0.8
Leucine		1.1	0.3		0.6	
Isoleucine		0.7	0.3		0.4	
Valine		0.9	0.4		0.6	

* Significantly different from fed sheep ($P < 0.05$). Mean body weight of sheep was 50 kg and fed sheep received 800 g/day of lucerne nuts (20% protein) at hourly intervals. Summarized from Wolff, Bergman and Williams (1972) and Heitmann and Bergman (1980).

The metabolism of glutamine and glutamate differs from other amino acids. Glutamine was removed by both gut and liver in large quantities (*Table 4.5*) and again a net movement from peripheral tissues must occur. Conversely, glutamate was absorbed in only negligible amounts and was released by liver. The amount of glutamate released by liver apparently depends not only upon the amount of glutamine being deaminated but also by the amount of free ammonia being removed by the liver. Both free ammonia and the nitrogen removed from glutamine are used for urea production. As stated earlier, and even though no glutamate is absorbed, it definitely is present in the intestinal digesta in greater amounts than any other amino acid (*Table 4.4*; Clarke, Ellinger and Phillipson, 1966). Glutamate is thus metabolized by gut epithelium of the sheep and part of it is probably converted to alanine as is the case in monogastric animal species (Neame and Wiseman, 1958; Windmueller and Spaeth, 1977). The large removal of glutamine from the blood by gut tissues must also be related to this. Studies on the perfused rat intestine (Windmueller and Spaeth, 1977) have demonstrated that one-third of the nitrogen from glutamine can be transaminated for alanine release back into the circulation and one-half of the glutamine carbon can be used for gut CO_2 production. Glutamine thus is related to glutamate and is used for energy and alanine production by the gut but for nitrogen transport and urea and glucose production by the liver.

Fasting, of course, results in a virtual cessation of amino acid absorption and has a marked effect on their blood concentrations. It is also interesting that during fasting, hepatic removal of nearly all amino acids is maintained and even increased in the case of alanine and perhaps glycine. This can only be due to increased release of amino acids by peripheral tissues, especially the muscles. The increase in muscle release of amino acids has been witnessed in several mammalian species (Marliss *et al.*, 1971; Ruderman and Berger, 1974; Ballard, Filsell and Jarrett, 1976) and the liver must be their main site of utilization, especially for alanine, glutamine and glycine.

NET RENAL METABOLISM

Table 4.6 summarizes data on renal metabolism of the major amino acids in both fed and fasted plus acidotic sheep. Alanine was removed by the kidneys, as well as by liver, and probably is used for gluconeogenesis. Glutamine and glycine, however, were added to the blood and this would partially supply that removed by the liver. Further, renal glutamine and glutamate metabolism differed from that of other amino acids as it did in the gut and liver. In fed sheep, glutamine was added to the blood but this shifted to glutamine removal during fasting and acidosis. Glutamine is known to be the major source of urinary ammonia in most animal species and is needed to neutralize acids, especially during acidosis (Pitts, de Haas and Klein, 1963; Baruch *et al.*, 1975). The enzyme glutamine synthetase has been demonstrated in the kidneys of animals having a neutral or alkaline urine (e.g. rat and rabbit) but not in the dog whose urine is normally acid (Lyon and Pitts, 1969). Glutaminase enzyme activity, however, has been observed in all species tested. Since sheep kidneys show a glutamine release during feeding and a removal during fasting, renal glutamine metabolism in sheep bears more resemblance to the rat than to the dog. Clearly, glutamine serves as a nitrogen carrier between the kidneys and the gut plus liver, and may be of aid in avoiding ammonia toxicity as well as in urinary ammonia production for neutralization of acids.

Of further significance is that arginine was always released by the kidneys and this was partially balanced by a removal of citrulline (*Table 4.6*). Arginine synthetase activity thus has to exist in sheep kidneys which, again, is similar to

Table 4.6. NET RENAL METABOLISM OF SOME PLASMA AMINO ACIDS AND AMMONIA IN FED AND FASTED PLUS ACIDOTIC PREGNANT SHEEP

| Amino acid | Fed (5) | | Fasted + acidotic (5) | |
	removed (mmol/h)	added	removed (mmol/h)	added
Alanine	0.7		0.3*	
Glutamine		1.5	0.9*	
Glutamate	0.2			0.2*
Glycine		1.1		1.4*
Arginine		1.5		0.8*
Citrulline	1.4		0.5*	
Ammonia†		1.3		2.9*

* Significantly different from fed sheep.
† Added to blood plus urine. Mean body weight of sheep was 70 kg. Summarized from Bergman *et al.* (1974).

that of the rat but unlike that of the dog or human (Featherstone, Rogers and Freedland, 1973; Bergman *et al.*, 1974). The opposite reaction occurs in sheep liver where there is arginine removal but release of ornithine, citrulline and urea. Arginine thus serves as still another nitrogen carrier, in addition to glutamine and glycine, between the kidneys and liver of the ruminant.

NET METABOLISM BY THE HINDQUARTERS

Metabolism of amino acids by the hindquarters of sheep has been measured by multiplying concentration differences (between the lower vena cava and aorta) by rates of plasma flow. This mainly represents skeletal muscle metabolism, but metabolism by skin, adipose tissue and bone is also involved. *Table 4.7* shows

Table 4.7. NET METABOLISM OF SOME PLASMA AMINO ACIDS BY THE HIND-QUARTERS OF FED, FASTED AND ACIDOTIC SHEEP

Amino acid	Fed (8)		Fasted (7)		Acidotic (10)	
	removed	added (mmol/h)	removed	added (mmol/h)	removed	added (mmol/h)
Alanine		1.0		1.8*		1.2
Glutamine		1.1		1.7*		0.9
Glutamate	0.8		0.3*		0.5*	
Glycine		0.8		1.9*		0.9
Serine	0.2			0.3*	0.3	
Arginine		0.2		0.3		0.1
Citrulline	0.3		0.1		0.2	
Ornithine	0.4		0.3		0.4	
Leucine		0		0.2*	0.2	
Isoleucine		0		0.3*	0.2	
Valine		0.4		0.5		0.1

* Significantly different from fed sheep ($P < 0.05$). Body weight of sheep was 50–60 kg. Summarized from Bergman and Heitmann (1978) and Heitmann and Bergman (1980).

that, in fed sheep, different amino acids are either selectively removed from, or added to, the blood in net amounts. Ballard, Filsell and Jarrett (1976) have obtained qualitatively similar data on the basis of V−A differences alone.

While a number of amino acids were removed by the hindquarters in net amounts in fed sheep, there was a large release of alanine, glutamine and glycine together with small amounts of arginine. These fluxes, generally like that of the kidneys, were opposite to that of the liver. Acidosis had no effect on amino acid metabolism by the hindquarters, with the exception that glutamate removal may have decreased. After fasting, however, there was a larger overall release of amino acids again, especially alanine, glutamine and glycine. Similar results have been found in non-ruminant species, although alanine release by rat hind-limb and human forearm muscle may be greater than that of glutamine (Marliss *et al.*, 1971; Felig, Wahren and Raf, 1973; Ruderman and Berger, 1974). These facts thus point out the transport of carbon and amino groups away from muscle during period of net protein mobilization. Unlike the rat and human, however, some branched-chain amino acids can be released by sheep muscle rather than being totally catabolized.

As alluded to by others (Ruderman and Berger, 1974; Daniel, 1977; Bergman and Heitmann, 1978; Goldberg and Chang, 1978), the ability of muscle to continuously synthesize alanine, glutamine, glycine and small amounts of arginine from either the amino or carbon residues of other amino acids, affords several advantages to the body. Muscle cannot synthesize urea but it can still use amino acids as a fuel, or for deaminations, without the risk of releasing toxic quantities of ammonia into the intracellular or extracellular fluids. It thus provides the liver with easily utilized substrates so that urea, and also glucose, can be formed in that organ. Further, the kidneys are supplied with glutamine for neutralizing acids, and the gut also with glutamine for its energy requirements.

SUMMARY OF INTER-ORGAN TRANSPORT AND INTEGRATION OF METABOLISM

It is evident from the above that several major amino acids are involved in the transport of amino groups and carbon between various organs of the body. This surely is important for integration of metabolism in the whole body. *Figure 4.2*

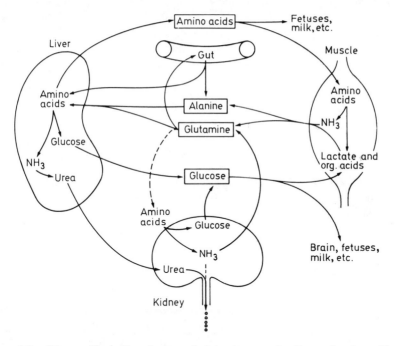

Figure 4.2   Diagram illustrating alanine and glutamine as major forms of amino acid and nitrogen transport between tissues of sheep. Dashed line indicates acidosis or fasting (From Bergman, 1973, courtesy of *Cornell Vet.*)

summarizes this point mainly for alanine and glutamine which seem to be the two most important amino acids involved. While nearly all amino acids are absorbed into the blood, large amounts are promptly removed by the liver. Variable amounts are transported to peripheral tissues for use in the turnover of body protein and any productive processes. Alanine is absorbed into the blood in larger amounts than any other amino acid, but glutamine is removed by gut tissues. Alanine and glutamine are always released by muscle and are taken up by the liver for urea and glucose synthesis. While animals at maintenance have a daily amino acid requirement, most amino acids are eventually converted to urea and glucose. The kidneys also participate in these movements of alanine and glutamine. While they always remove some alanine, the kidneys of fed sheep release glutamine and furnish about one-half of the glutamine removed by liver.

During fasting and acidosis, net qualitative as well as quantitative shifts occur in the movements of alanine and glutamine. In both conditions, the renal release of glutamine switches to that of removal (*Figure 4.2*), so that ammonia can be produced for neutralization of acids in the urine. Alanine and glutamine release by muscle is greatly increased during fasting to balance the even larger alanine removal by liver and the increased need for glutamine by the kidneys. If only acidosis is present, however, muscle does not appear to change significantly its amino acid metabolism.

In addition to alanine and glutamine, other amino acid cycles exist. In the sheep, but apparently not in the dog or human, the three urea cycle amino acids are involved in nitrogen transport between kidneys, muscle tissue and liver (*Figure 4.3*). The kidneys, and to a lesser extent muscle, remove citrulline and ornithine from the blood and add arginine. Conversely, the liver removes arginine and adds citrulline, ornithine and urea. This simple cycle of transport therefore seems to be another means of the body transporting amino groups between organs in order to allow peripheral tissues to oxidize amino acids

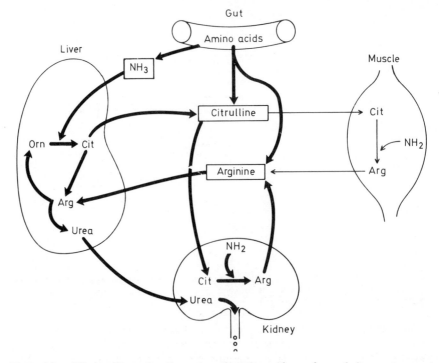

Figure 4.3   Diagram illustrating the urea cycle amino acids as a form of nitrogen transport between tissues of sheep. Heavy lines indicate the more major reactions. Arginine transports ammonia in a non-toxic form from kidneys and muscle so that the liver forms urea and releases ornithine and citrulline

without the risk of free ammonia being formed and causing highly toxic cellular effects.

Glycine transport also seems to be important and to differ from other amino acids. Less seems to be known of its real significance, however. In both fed and fasted sheep, it is removed by liver in larger molar quantities than any other amino acid. It is also continuously released by the kidneys and muscle with increased release occurring during fasting. The sheep, however, again seems to differ from the dog (Elwyn *et al.*, 1972) in this respect, since all of the glycine is probably not deaminated in the liver and it is not highly glucogenic (Wolff and Bergman, 1972). Instead, large amounts must be used for bile salt formation and also for hepatic detoxification of benzoic acid by way of hippuric acid formation (Wolff, Bergman and Williams, 1972).

Control and unidirectional metabolism of alanine and glutamine

Only limited studies have been made on the physiological regulation of amino acid metabolism, especially in ruminants. In all species, glucocorticoids are known to have long-term protein catabolic effects; they can even produce muscle wasting. They also increase the incorporation of total amino acid carbon into glucose (Reilly and Ford, 1974). Conversely, growth hormone is known to have long-term protein anabolic effects.

Since alanine and glutamine quantitatively seem to be the most important of the amino acids for transport between tissues, it is of special interest to study their specific tissue metabolisms and transport under varying and short-term conditions. In the preceding sections of this chapter, only the net fluxes or sum total of alanine and glutamine removal from (utilization), or addition to (production), the blood were described. Actual utilization and production, however, could be occurring simultaneously and in the same tissue. This section will discuss recent research on these unidirectional fluxes and control mechanisms of alanine and glutamine metabolism and how these can be influenced by hormones, fasting and acidosis. Utilization is considered to be the irreversible loss of labelled alanine or glutamine from the blood, and production the addition of new (unlabelled or *de novo*) alanine or glutamine to the blood.

SHORT-TERM HORMONAL EFFECTS

Brockman *et al.* (1975) have studied the effects of glucagon and insulin on alanine and glutamine metabolism in sheep. Glucagon was infused into alloxanized sheep maintained on a constant level of insulin, and when insulin was infused, sufficient glucose was given to prevent hypoglycaemia. In this way, a reflex release of either insulin or glucagon would not occur to counteract the effects of the other hormone being infused. Within 30 min, glucagon actually doubled the net hepatic extraction ratios and thus the hepatic removals of both alanine and glutamine. It also doubled the percentage conversion of $[^{14}C]$-alanine to glucose and increased the total hepatic glucose output. The arterial concentrations of both alanine and glutamine became reduced by about 30%. Insulin had little or no effect on net hepatic removal of any of the amino acids. It did, however, decrease the concentrations of several, including the branched-chain amino acids, indicating increased protein synthesis in muscle.

Table 4.8 shows unidirectional fluxes of alanine obtained before and during infusions of glucagon. Glucagon had no discernible effect on unidirectional alanine production in any tissue. It did, however, increase unidirectional alanine utilization by the liver and decrease its utilization by other tissues. It is interesting that the turnover of total plasma alanine did not change significantly, but its concentration in the blood did decrease by about 30%. The overall effect of

Table 4.8. UNIDIRECTIONAL RATES OF TISSUE ALANINE METABOLISM AND WHOLE-BODY PLASMA TURNOVER IN SHEEP

Fed sheep	*Art. conc.*	^{14}C *turnover*	*Utilized*			*Produced*		
			portal viscera	liver	periph. tissues	portal viscera	liver	periph. tissues
	(μM)	*(mmol/h)*	*(mmol/h)*			*(mmol/h)*		
Control (4)	98	10.4	2.3	4.7	3.5	4.3	1.9	4.2
Glucagon- infused (4)	67*	9.4	1.5*	6.6*	1.4*	3.8	1.0	4.6

* Significantly different from control ($P < 0.05$). Calculated from Brockman and Bergman (1975).

glucagon must therefore be that it quickly and specifically increases hepatic utilization and this, in turn, lowers the blood concentration so that utilization by muscle decreases. Insulin, however, seems to directly stimulate muscle to promote protein synthesis. Interestingly, both hormones seem to have no effect on *de novo* production *per se*, but instead affect only the irreversible disposal or utilization of alanine.

EFFECTS OF FASTING AND ACIDOSIS

Glutamine is like alanine, in that both are affected by glucagon and insulin and also that both are released by muscle and removed by liver even under feeding conditions. However, glutamine metabolism is different from alanine, in that it can be either released or removed in net amounts by the kidneys depending upon the acid-base or nutritional status of the animal. Further, muscle and liver glutamine metabolism must surely be altered to accommodate these shifts in renal metabolism.

Table 4.9 summarizes more detailed data on these aspects of glutamine metabolism in sheep during feeding, fasting and acidosis. The calculated unidirectional rates of metabolism show that glutamine, like alanine, is both utilized and produced in the liver, kidneys and muscles of the hindquarters. Both the glutaminase and glutamine synthetase enzyme systems are thus present in all three tissues. In the fed sheep, the muscles and kidneys produce more glutamine than is utilized, but in liver more is utilized than is produced. There is thus a net movement of glutamine from kidneys and muscle to the liver.

During both fasting and acidosis (*Table 4.9*), however, significant shifts occur. During fasting, renal glutamine utilization simultaneously increases as renal production decreases, showing that renal glutaminase activity is further switched on as glutamine synthetase is reduced. In compensation for this, muscle decreases

Table 4.9. UNIDIRECTIONAL RATES OF TISSUE GLUTAMINE METABOLISM AND WHOLE-BODY TURNOVER IN SHEEP

Sheep	Art. conc. (μM)	^{14}C turnover (mmol/h)	Utilized			Produced		
			liver	kidneys	hindq.	liver	kidneys	hindq.
				(mmol/h)			(mmol/h)	
Fed (5)	210	12.2	4.6	0.5	2.3	2.8	1.2	3.2
Fasted (4)	168*	11.5	4.2	1.2*	1.4*	2.2	0.9*	3.2
Acidotic (5)	178*	11.5	5.9*	1.6*	2.4	5.2*	0.3*	3.4

* Significantly different from fed sheep ($P < 0.05$). Summarized from Heitmann and Bergman (1978).

its glutamine utilization but liver metabolism is unchanged. During acidosis, renal glutamine metabolism is altered even more. Renal production of glutamine nearly ceases and renal utilization markedly increases. Muscle metabolism, however, does not compensate as it did during fasting; instead, liver glutamine metabolism is altered. Interestingly, both liver utilization and production of glutamine increased but with production increasing more than utilization. The overall effect of acidosis, therefore, was to decrease the liver's net removal of glutamine in response to the increased net renal removal. It is of further significance that the turnover of total plasma glutamine did not seem to be altered either by fasting or acidosis, even though its arterial concentration decreased. The whole-body production of glutamine must thus have remained nearly constant while the overall body utilization of glutamine increased.

The above findings in sheep are in overall agreement with net and veno-arterial data for rat hind-limb and human forearm muscle. In these experiments, net glutamine release from muscle increased during fasting and diabetes (Marliss *et al.*, 1971; Ruderman and Berger, 1974) and glutamine usually was removed by the splanchnic viscera. Glutamine, as well as alanine, must thus be an important nitrogen and carbon carrier from the muscles of the ruminant as well as other species, and is used by the liver for ureogenesis, by the kidneys for ammoniagenesis and by both organs for gluconeogenesis.

In conclusion, amino acid transport between organs is complex but it is undoubtedly important for the integration of whole-body metabolism. Some differences exist between ruminant and non-ruminant species, although the importance of these differences are not fully apparent. Alanine and glutamine are especially important in linking the nitrogen metabolism of muscle, kidneys, gut and liver but, also, the three urea cycle amino acids and glycine are involved to a significant extent.

References

ADIBI, S.A. (1976). *Metabolism,* **25,** 1287–1301

AOKI, T.T., BRENNAN, M.F., MULLER, W.A., SOELDNER, J.A., ALPERT, J.S., SALTZ, S.B., KAUFMAN, R.L., TAN, M.H. and CAHILL, G.F., Jr. (1976). *Am. J. Clin. Nutr.,* **29,** 340–350

BAIRD, G.D., HEITZMAN, R.J. and HIBBITT, K.G. (1972). *Biochem. J.,* **128,** 1311–1316

BAIRD, G.D., SYMONDS, H.W. and ASH, R. (1975). *J. Agric. Sci.,* **85,** 281–289

BALLARD, F.J., FILSELL, O.H. and JARRETT, I.G. (1976). *Metabolism,* **25**, 415–418

BARUCH, S.B., BURICH, R.L., EUN, C.K. and KING, V.F. (1975). *Med. Clins N. Am.,* **59**, 569– 583

BERGMAN, E.N. (1973). *Cornell Vet.,* **63**, 341–382

BERGMAN, E.N. and HEITMANN, R.N. (1978). *Fedn Proc.,* **37**, 1228–1232

BERGMAN, E.N., KAUFMAN, C.F., WOLFF, J.E. and WILLIAMS, H.H. (1974). *Am. J. Physiol.,* **226**, 833–837

BROCKMAN, R.P. and BERGMAN, E.N. (1975). *Am. J. Physiol.,* **228**, 1627–1633

BROCKMAN, R.P., BERGMAN, E.N., JOO, P.K. and MANNS, J.G. (1975). *Am. J. Physiol.,* **229**, 1344–1350

CHAISSON, J.L., LILJENQUIST, J.E., SINCLAIR-SMITH, B.C. and LACY, W.W. (1975). *Diabetes,* **24**, 574–584

CHAVEZ, E.R. and BAYLEY, H.S. (1976). *Brit. J. Nutr.,* **36**, 189–198

CLARKE, E.M.W., ELLINGER, G.M. and PHILLIPSON, A.T. (1966). *Proc. Roy. Soc., Ser. B.,* **166**, 63–79

DANIEL, P.M. (1977). *Lancet,* Aug. 27, 446–448

DREWES, L.R., CONWAY, W.P. and GILBOE, D.D. (1977). *Am. J. Physiol.,* **233**, E320–E325

ELWYN, D.H., LAUNDER, W.J., PARIKH, H.C. and WISE, E.M. (1972). *Am. J. Physiol.,* **222**, 1333–1342

FEATHERSTONE, ROGERS and FREEDLAND (1973) *Am. J. Physiol.,* **224**, 127–132

FELIG, P., WAHREN, J. and RAF, L. (1973). *Proc. Natn Acad. Sci. U.S.A.,* **70**, 1775–1779

GOLDBERG, A.L. and CHANG, T.W. (1978). *Fedn Proc.,* **37**, 2301–2307

HEITMANN, R.N. and BERGMAN, E.N. (1978). *Am. J. Physiol.,* **234**, E197–E203

HEITMANN, R.N. and BERGMAN, E.N. (1980) *Am. J. Physiol.* In the press

HUME, I.D., JACOBSON, D.R. and MITCHELL, G.E. (1972). *J. Nutr.,* **102**, 495–502

JOO, P.K., HEITMANN, R.N., REULEIN, S.S. and BERGMAN, E.N. (1976). *J. Dairy Sci.,* **59**, 309–314

KATZ, M.L. and BERGMAN, E.N. (1969). *Am. J. Physiol.,* **216**, 946–952

KAUFMAN, C.F. and BERGMAN, E.N. (1974). *Am. J. Physiol.,* **226**, 827–832

LINDSAY, D.B. and BUTTERY, P.J. (1980). These proceedings, pp. 125–146

LYON, M.L. and PITTS, R.F. (1969). *Am. J. Physiol.,* **216**, 117–122

MARLISS, E.B., AOKI, T.T., POZEFSKY, T., MOST, A.S. and CAHILL, G.F., Jr. (1971). *J. Clin. Invest.,* **50**, 814–817

NEAME, K.D. and WISEMAN, G. (1958). *J. Physiol.,* **140**, 148–155

PITTS, R.F., DE HAAS, J. and KLEIN, J. (1963). *Am. J. Physiol.,* **204**, 187–191

REILLY, P.E.B. and FORD, E.J.H.(1974). *J. Endocrinol.,* **60**, 455–472

RUDERMAN, N.R. and BERGER, M. (1974). *J. Biol. Chem.,* **249**, 5500–5506

TAGARI, H. and BERGMAN, E.N. (1978). *J. Nutr.,* **108**, 790–803

WINDMUELLER, H.G. and SPAETH, A.E. (1977). *Fedn Proc.,* **36**, 177–181

WOLFF, J.E. and BERGMAN, E.N. (1972). *Am. J. Physiol.,* **223**, 455–460

WOLFF, J.E., BERGMAN, E.N. and WILLIAMS, H.H. (1972). *Am. J. Physiol.,* **223**, 438–446

5

CONTROLLING FACTORS IN THE SYNTHESIS OF EGG PROTEINS

A.B. GILBERT
Agricultural Research Council, Poultry Research Centre, Edinburgh

Summary

The avian egg has been an object of interest and scientific study for a considerable length of time. Much is known of its structure and chemical composition, although knowledge of the chemistry of the egg has not always kept pace with the detailed information available of its physical form. Much is also known of the structure and physiology of the organs responsible for the production of the egg. Consequently it is now possible to give an accurate and reasonably comprehensible account of general reproduction processes in the bird.

However, one aspect remains for the most part largely unknown: by what means is each of the specific proteins deposited within the egg in the place and in the form in which they occur?

The chicken egg, typical of birds generally, contains about 12% by weight of protein. This is divided between the colloquial 'yolk', 'white' and 'shell' in the proportion of 12, 15 and 1, respectively. But none of these compartments is homogeneous and each consists of several proteins (up to 40 or so in the 'white') restricted to their respective position within the compartment. The proteins of 'yolk' are formed elsewhere than the ovary and are transported to it via the vascular system. Consequently these proteins have to be transported from the capillaries across several tissue layers and be incorporated within the oocyte while other plasma components are excluded. The proteins of the 'white' and 'shell' are almost exclusively produced within the oviduct and these have to be formed and released at the correct time and in the correct sequence.

The purpose of this chapter is to outline some of the intriguing mysteries which are still present and which still confront those attempting to understand how these specific substances are incorporated into the avian egg in the correct manner both in time and in space.

Introduction

With the growth of the poultry industry and the importance of the egg as a source of protein for human consumption, much work has been directed towards an understanding of general reproductive processes in domestic species, particularly the domestic fowl. It is now possible to give a clear picture of most aspects of avian reproduction; information can be obtained from the relevant chapters in many recent publications, e.g. Bell and Freeman (1971), Farner and King (1971–1975), Follett and Davis (1975) and King and McLelland (1979). Other more specific works include Romanoff (1960), Jordanov (1969), Hodges (1974), King (1975) and Murton and Westwood (1977).

Despite this general knowledge, certain aspects of the reproductive process still remain unclear. For example, the egg is the sole reproductive product and, as such, its function is to provide a viable chick: yet the biological significance of many of the various components of the egg are not entirely understood.

Furthermore, it is a strange paradox that the more information which becomes available from modern techniques about the structure and composition of the egg, the more difficult it becomes to reconcile these with our understanding of the mechanisms which must be involved in the deposition of material in the egg. It is not intended to give a general account of our present concept of how protein deposition occurs, but instead to outline some of the intriguing problems which still remain, because the precise manner in which most of the proteins are deposited in the egg remains an enigma.

The egg

The chicken egg is a commonplace object and few people will be unaware of its basic structure (*Figure 5.1*). A typical egg weighs about 60 g and consists, in colloquial terms, of a rigid but brittle 'shell' (6 g) overlying a watery, mucilagenous 'white' (34 g) which in turn surrounds the yellow 'yolk' (20 g). The egg consists of about 66% water, 12% protein, 11% lipid and 10% inorganic ions. However, none of the major compartments ('yolk', 'white' or 'shell') is homogeneous and each has a complex chemical and physical structure.

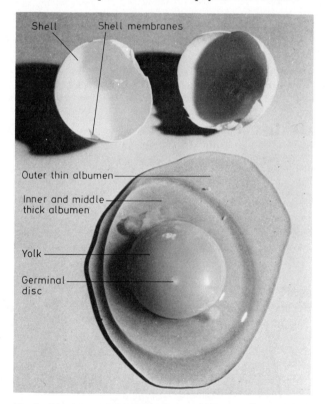

Figure 5.1   Egg broken out to show the various components. Encased within the middle thick albumen are the inner thin albumen and the chalaziferous layer

In the unfertilized egg the 'yolk' is surrounded by the composite 'yolk-membranes'. These consist of four layers; (a) the oocyte cytoplasmic membrane; (b) the perivitelline layer composed of a feltwork of proteinaceous fibres formed by the ovary; (c) a thin layer of granular material produced by the oviduct; and (d) an outer layer similar in composition to 'white' but forming a feltwork of fibres – it is also produced by the oviduct.

The main mass of the oocyte consists of the well-known deep-yellow yolk. A smaller fraction of paler 'white yolk' forms the central latebra, the neck and the nucleus of Pander. On the surface of the nucleus of Pander floats the germinal disc, the main cytoplasmic part of the original cell.

Immediately surrounding the yolk membranes, the thin chalaziferous layer is present as a dense gelatinous material which extends outwards towards both ends of the egg. Within this are the characteristic avian chalazae. Surrounding this layer, except at the two ends of the egg, is a more liquid albumen called the inner 'thin white' or 'liquid white'. Outside, and probably continuous with the chalaziferous layer, is a further coating of a viscous material (the middle dense albumen). These layers are covered by the outer thin (liquid) albumen, similar in consistency to the inner thin albumen. Although having a definite structure in the laid egg, the separation of the albumen into compartments occurs in the shell gland, and results from the addition of an amount of water equal to the mass of the protein material previously deposited. During the deposition of the albumen in the magnum only one layer is evident.

The shell consists of three distinct parts – the membranes, the calcified portion and the cuticle. It is generally accepted that two membranes exist, although Candish (1972) reasoned that they were probably formed as one. Even if this were so, the membrane does have two regions, an inner and an outer one, which can be distinguished easily by their morphology. The calcified part of the shell consists of a complex proteinaceous fibre network which acts as a framework for the formation of the calcareous crystals. The cuticle forms the extreme outer covering of the egg.

As far as has been ascertained, the structure and chemical composition of each of the various parts of the egg reflects its function (Gilbert, 1971e). The rigid shell protects and supports the internal soft structures. Together with the cuticle it prevents bacterial infection, while the cuticle also renders the shell relatively impervious to water so preventing desiccation of the embryo. The main function of the albumen appears to be to provide an aqueous environment in which the embryo can develop, although in most species of bird it provides some additional nutritional material towards the end of embryogenesis. Many of the proteins of albumen have bacteriocidal properties and consequently albumen will act as a further barrier to bacterial infection. Yolk forms the major nutritional material for embryonic growth, the yolk lipids probably providing the major part of the energy available to the embryo, at least in the early stages.

Protein synthesis in egg formation

The mechanisms for the accumulation of material in the oocyte and in the remainder of the egg differ in one important respect. Whereas the proteins of the albumen and shell are synthesized in the oviduct (see Gilbert, 1971b, for general discussion), the proteins for deposition within the oocyte are synthesized within the liver (McIndoe, 1971; Gruber, 1972; Tarlow *et al.*, 1977). Consequently the oviduct is a true secretory organ but the ovary, at least in this respect, is not.

The biochemical properties of many of the proteins of albumen have been known for many years (see, for example, Feeney and Allison, 1969). The earlier work of O'Malley (O'Malley and McGuire, 1967, 1968a, 1968b; O'Malley, McGuire and Middleton, 1968; O'Malley *et al.*, 1968; McGuire and O'Malley, 1968) firmly established the general outline of biosynthesis of avidin in the oviduct, and further work (e.g. Prasad and Peterkofsky, 1976; Palmitter *et al.*, 1977; Schutz *et al.*, 1977) has extended the available information. Nevertheless, few of the total number of proteins in albumen have been studied in any detail and little information of the proteins of the shell is available. Similarly little attention has been paid to the synthesis and biochemistry of yolk proteins, although interest in this aspect is growing (Clemens, 1974; Tarlow *et al.*, 1977).

For these reasons, it is not intended to deal with the biosynthesis of the proteins of the egg, since few conclusions of a general nature can be drawn. However, the information that is available suggests that synthesis follows the general pathways established in other tissues and in other animals; consequently reference should be made elsewhere in this volume, where such details are available.

Protein deposition in the ovary

The maturation and growth of the oocyte is a complex process (a general account may be found in Gilbert, 1979). The early stages are protracted and may last for many months (Marza and Marza, 1935), during which time the material deposited is pale in colour, hence the general term 'white yolk' for this material. When the oocyte reaches about 5 mm in diameter, the typical yellow yolk is deposited and this process continues for a period of about 8 days until ovulation. Because an ovary of a mature bird contains oocytes at all stages of maturation, the material being deposited in the oocyte must reflect the mechanisms resident either within the follicle or within the oocyte itself.

The follicle has a complex structure (*Figure 5.2*), most details of which have now been described; general accounts are available in Hodges (1974) and Gilbert (1971a, 1979). The oocyte is bordered by its own cytolemma which lies adjacent to the fibrous perivitelline layer. In the follicle of greater diameter than 5 mm, the granulosa is a single layer of cells but in smaller follicles it has a pseudostratified appearance; it surrounds the perivitelline layer. As an epithelial-cell tissue, the granulosa layer has a basal lamina separating it from the theca. Basically, the theca is composed of fibroblasts but it has extensive and complex vascular and nervous systems.

WHITE YOLK

Little is known of the chemistry and physical structure of white yolk (Bellairs, 1964): in comparison with yellow yolk it appears to have a higher protein to lipid ratio but it contains only 10–13% solid material.

It is not possible to give the source of white yolk with any degree of confidence. It could be derived from plasma components but generally the vascular system at this stage is poorly developed in comparison with later stages of follicular growth. If it is derived from the plasma some mechanisms must be responsible for its extraction from the capillary system, but nothing is known about the process. Also it would appear almost certainly to be transported

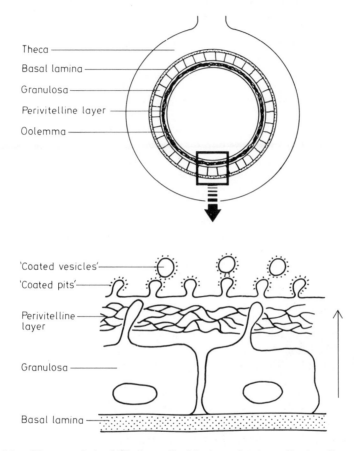

Figure 5.2   Diagram of the follicular wall with the region immediately adjacent to the oocyte shown in greater detail. The arrow shows the direction of flow of the yolk material

through the granulosa cells. On the other hand the granulosa may actually be involved in the synthesis of the yolk material. Hodges (1974), summarizing the information available (Press, 1964; Bellairs, 1965, 1967; Greenfield, 1966; Schjeide *et al.*, 1966, 1970; Paulson and Rosenberg, 1972), suggested there were five types of inclusion in the granulosa cell: (a) small (40 nm) smooth-surfaced vesicles; (b) large (500 nm) vesicles with granular inclusions; (c) vesicles containing electron-dense material; (d) neutral fat; and (e) large, apparently empty, vesicles forming interconnecting systems. However, nothing is known of the chemical nature of any of these: they could be yolk material in transport from the vascular system to the oocyte or they could be indicative of the *de novo* synthesis of yolk by the granulosa cells. Yet again they may have no relevance to yolk formation.

Similarly there is little evidence for the way in which white yolk is incorporated into the oocyte. There are numerous pit-like depressions in the oocyte cytolemma, at the bottom of which are small pinocytotic vesicles communicating with the perivitelline space. If they are involved in yolk uptake by the oocyte, as seems feasible, what is the material being transported?

In addition to the pinocytotic vesicles, there are some unusual interconnecting structures between the granulosa cells and the oocyte which have been called 'lining bodies' (Bellairs, 1964), 'transosomes' (Press, 1964) and 'premitochondria' (Schjeide and McCandless, 1962). In the region of these structures the oocyte cytolemma forms a deeper and wider pit than those associated with the pinocytotic vesicle, and lying in it is a long process from a granulosa cell. One or more lining bodies form between the granulosa process and the cytolemma, often also associated with the oocyte endoplasmic reticulum. These lining bodies are thought to represent some aspect of the mechanism for yolk transport since they appear to be incorporated into the oocyte.

YELLOW YOLK

In contrast to white yolk, the major protein components of yellow yolk are known to be produced by the liver (McIndoe, 1971; Gruber, 1972; Clemens, 1974).

Yellow yolk consists of approximately 50% water, 33% lipids and 17% protein. About 99% of the solid material of the yolk is proteinaceous, mainly in the form of lipoproteins and phospholipoproteins. These components have been relatively poorly studied in comparison to mammals, and usually they are classified (by centrifugation) as the low-density fraction (now divided into LDL and VLDL, very-low density lipoprotein), the water soluble fraction and the granular fraction. The LDL and VLDL are the most abundant, forming nearly 65% of the yolk solids, and they contain most of the lipid. Their chemistry has been reviewed by Cook (1968) and Cook and Martin (1969).

The water soluble fraction 'livetins' form 10% of yolk solids and the three major components (α-, β- and γ-livetin) are probably the plasma albumin, α_2-glycoprotein and γ-globulin, respectively (Williams, 1962). Trace amounts of other proteins (e.g. transferrin) are also present in this fraction. The granular fraction contains the remaining proteins divided into lipovitellin and phosvitin, although neither are homogeneous (Cook, 1968, Cook and Martin, 1969), and another fraction. Phosvitin consists of at least two phosphoproteins (Clark, 1970) and lipovitellin probably consists of several protein fractions.

After formation, the yolk proteins are carried through the vascular system to the follicle where they have to pass from the blood stream, cross the follicle wall and enter the oocyte. However, there are two major constraints which have to be placed on any explanation of this process: the majority (possibly all) of the lipoproteins in the plasma appears to be transported unchanged in form (Hillyard, White and Pangburn, 1972; Gornal and Kuksis, 1973; Holdsworth, Mitchell and Finean, 1974) and the relative composition of yolk differs in almost every respect from that of plasma (Everson and Souders, 1957; Draper, 1966; Cook, 1968; Shenstone, 1968; Gilbert, 1971d). Furthermore, there is a tendency for the larger molecules to be concentrated within the yolk and for the smaller ones to be excluded. For these reasons it seems *a priori* that the mechanism for the transfer of protein from the plasma to the oocyte is a special one. Very little experimental evidence is available on which to base an explanation for this transport, and the suggestions that have been put forward have been based mainly on ultrastructural studies.

In its journey from the plasma to the oocyte the protein has five successive potential barriers to cross: (a) the endothelial lining of the capillary; (b) the relatively thin tissue of the theca interna; (c) the basal lamina; (d) the granulosa layer; and (e) the oocyte cytolemma.

Because the usual capillary endothelium tends to prevent the passage of large molecules across it, since yolk is not a plasma filtrate and since the proteinaceous material of yolk probably passes into the oocyte largely unchanged, it must be reasoned that the vascular system in the follicle differs considerably from that in other organs. Certainly the vascular system is complex (Nalbandov and James, 1949) but recent evidence has been obtained showing that the capillaries may differ from those more commonly found.

A feature of the theca interna appears to be the common presence of erythrocytes outside the capillary network (Rothwell and Solomon, 1977; Perry, Gilbert and Evans, 1978a, 1978b); that this was not an artefact was confirmed by finding gaps in the endothelial lining of the capillaries (Perry, Gilbert and Evans, 1978a). What is more, many of the capillaries were shown to be without a basal lamina. It must be concluded that the capillaries within the follicle form a 'leaky' system unlike the usual terminal network of the vascular system. Since erythrocytes are able to escape in sufficient numbers to be common in sections viewed with the electron microscope, it must follow that the complex lipoprotein particles which form the major portion of yolk must also be able to escape freely from the confines of the vascular system.

Once outside the vascular system, it seems likely that the proteins would be able to pass between the loosely arranged cells of the theca interna: Gilbert (1971a) suggested that certain granular material seen in this region in some electron micrographs might be yolk material destined to pass into the oocyte. Although in retrospect this may seem to have been too ambitious a claim, particles thought to be VLDL (Evans, Perry and Gilbert, 1979) have been recently discovered between the thecal cells (Perry and Gilbert, 1980). Unlike the basal lamina elsewhere, that of the granulosa does not appear to prevent the passage of these particles (Perry, Gilbert and Evans, 1978b; Evans, Perry and Gilbert, 1979).

Grau and Wilson (1964) speculated that the yolk material would have to pass between the granulosa cells rather than through them, and this has been confirmed for the particles (Perry and Gilbert, 1979). Consequently, it is probable that the granulosa cells are not involved directly in the transport mechanisms of the VLDL. However, they may not be totally uninvolved in yolk formation since the long cellular processes passing into the intercellular spaces between the granulosa cells (Perry, Gilbert and Evans, 1978a) are similar to the cellular processes in the kidney tubule; it is therefore possible that they may have some unknown absorptive function.

The incorporation of the particles into the oocyte has been the subject of a recent study (Perry and Gilbert, 1979) and the following account is based mainly on this (*Figure 5.2*).

The particles appear to adhere preferentially to the outside surface of the oocyte cytolemma; no similar layer is present on the surface of the granulosa cells. In association with these particles, characteristic depressions (coated pits) occur in the oocyte cytolemma which gradually become withdrawn into the oocyte. These vesicles (200–350 nm diameter) have a general appearance similar to coated vesicles which have been described in other animals (Roth and Porter, 1964; Wallace and Dumont, 1968; Roth, Cutting and Atlas, 1976; Anderson, Brown and Goldstein, 1977), except that they are considerably larger. It is now accepted that the formation of coated vesicles is an important mechanism for the uptake of material by certain cells and it seems to apply also to the chicken oocyte. The actual mechanism for the formation of coated vesicles is not entirely known but it has been suggested that they form as a result of the specific properties of the proteins bound to the surface of the membrane. These proteins are thought to produce cross-linkages with each other, the linkages being smaller than those of the proteins of the membrane. In this way the membrane has to become progressively more curved inwards until a complete vesicle is formed (Ockleford, 1976). If this theory is correct then the apoprotein systems of the yolk lipoproteins and their special arrangement in the molecule are of considerable importance, although these aspects have received little attention (Rao and Mahadlevan, 1978).

Certainly the completed coated vesicles contain particles (Perry and Gilbert, 1979). Once the coated vesicles are withdrawn into the oocyte they appear to lose their coat and the naked but membrane-bound vesicles fuse together, a process which continues with the build up of larger masses of particles. Presumably this fusion process between larger and larger masses eventually leads to the formation of the yolk-spheres.

It is tempting to accept this description of the passage of material from the vascular system to the oocyte, since it does show how the lipoprotein material of the yolk could pass from the vascular system to the yolk without change. However, it must be stressed that at present it is only a framework on which a complete understanding of yolk accumulation could be built. When looked at critically there are many questions that remain unanswered. If, as the morphological studies indicate, the capillary system is 'leaky', it is difficult to envisage how it could function properly in its more usual role of conveying blood as part of a continuous system. Moreover, why does the blood not clot when it escapes from the system, because this is the normal response to capillary damage? Such a loss would elsewhere bring into operation well-known physiological mechanisms to prevent further loss. It is difficult to understand why similar ones are not brought into operation in the follicle.

Moreover, if plasma is released from the capillaries and if, as seems probable, there is an open pathway to the oocyte, why are certain of the plasma components less evident in the yolk? Perhaps the thecal and granulosa cells act in an absorptive capacity but, even so, there must be a limit to the quantity they can absorb. Moreover, where do they secrete the absorbed material or its metabolites, since a capillary system which can leak in this way seems unlikely to be an efficient one for the removal of these substances? An alternative would be a double capillary system, though no evidence for this has been produced.

The particles described by Perry and Gilbert (1979) and Evans, Perry and Gilbert (1979) have been tentatively identified as VLDL. But VLDL is not homogeneous and it is important that the chemical and physical properties of these particles be positively defined. With this information it may be possible to explain the apparent ease with which they pass through the basal lamina, a structure usually preventing the passage of such large molecules. Such information may help also to explain why these particles bind to the oocyte, but not to the granulosa cells. Explanations are also required for the binding sites in the oocyte cytolemma: is binding a property of the VLDL protein, of the oocyte membrane or both?

If the coated vesicles form as a result of the chemical and physical properties of either or both of the oocyte or particles, it is unlikely that metabolic energy is required. On the other hand, if energy is required, then it becomes difficult to see how this could be provided in the vicinity of the cytolemma by only the sparse cytoplasm and its inclusions which are present (Holdsworth, Mitchell and Finean, 1974; Perry, Gilbert and Evans, 1978a; Perry and Gilbert, 1979). Moreover, it seems unlikely that the oocyte nucleus could exert any influence over such metabolic processes, since it occupies a very small fraction of the oocyte during this phase of rapidly accumulating yolk.

From the number of coated vesicles which appear in electron micrographs (Perry and Gilbert, 1979) it is possible to obtain an estimate of the turnover of oocyte membrane, and this appears to be prohibitive unless there is some mechanism for the recycling of the membrane. Perhaps the loss of the coat soon after the vesicle enters the oocyte may indicate this is taking place, but what are the mechanisms responsible for the loss of the coat? Moreover, no evidence has yet been obtained for the reincorporation of membrane fragments into the cytolemma, and the recycling process, if it occurs, is unknown. The fusion of the membrane-bound vesicles thereafter may also be a necessary step to conserve membrane material, and the surplus membrane obtained from the fusion of two spheres would be available to be recycled and reincorporated into the cytolemma. Consequently it is possible that the structure of yolk, composed as it is of yolk-spheres, is the end result of the necessity to conserve the oocyte cytolemma.

So far, consideration has been given only to particles thought to be VLDL. Yet yolk contains many proteins other than VLDL. For the most part, the details of the uptake of these have not been studied; however, both immunoglobulins and phosvitin have specific receptor sites in the oocyte (Roth, Cutting and Atlas, 1976; Yukso and Roth, 1976) and these sites are likely to be different from those involved in transport of the VLDL. Moreover, it cannot be ruled out that some lipid material (lipoproteins?) may pass directly through the cytolemma without being incorporated into vesicles (Jordanov and Boyadjieva-Michailova, 1974).

Other aspects are equally intriguing. For example, Grau (1976) has shown that yolk from the laid egg has a concentric ring-like structure, but no complete explanation has been put forward for these rings. Nor is an explanation available for the increase in size of the yolk with the increase in age of the bird; this appears to involve a greater efficiency in the uptake of yolk (Gilbert, 1972). Clearly there are many components of yolk and many conceivable mechanisms for their transport into the oocyte: none has yet been adequately described.

Protein desposition in the oviduct

Among its many functions (Aitken, 1971; Gilbert, 1971b, 1971c, 1979; Hodges, 1974) the oviduct synthesizes and secretes four proteinaceous layers which cover the oocyte so forming the typical egg. These are: (a) the remaining parts of the composite yolk membranes; (b) the albumen; (c) the shell membrane; and (d) the calcified portion of the shell and cuticle. The oviduct is divided into five regions (*Figure 5.3*), but one – the vagina – plays no part in this activity. The structure of the oviduct is basically similar in each region, and the secretory cells lie either in the epithelium or in the composite tubular glands situated in the lamina propria (*Figure 5.3*). Consequently one of the puzzling aspects of

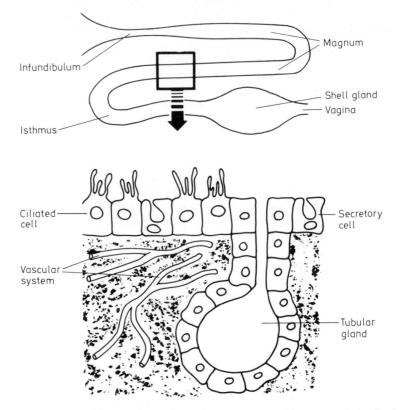

Figure 5.3   Diagram of the oviduct to show the major regions. The schematized wall of the oviduct (bottom) has been drawn at greater magnification to illustrate the glandular elements

oviduct function is why apparently homologous cells in the same organ have such different secretory products. The structure of the oviduct and that of the egg are well known. However, the advent of micro-chemical techniques, reliable histological methods and electron microscopy have produced a wealth of knowledge unavailable to earlier workers and this increased information has made the earlier, and of necessity, simpler concepts inadequate. Paradoxically, it is this increase in basic knowledge which has made it difficult precisely to relate form with function in the oviducal secretory processes.

THE COMPOSITE YOLK MEMBRANES

The oocyte boundary immediately after ovulation consists of only two proteinaceous structures. The cytolemma, contrary to earlier suggestions (Bellairs,

1965), persists (Perry, Gilbert and Evans, 1978a) and presumably has a composition similar to other cell membranes. The perivitelline layer (Bellairs, Harkness and Harkness, 1963; Wyburn, Aitken and Johnston, 1965) consists of a feltwork of long branching fibres, perhaps produced by the granulosa cells, although its formation remains unknown: it first appears in follicles of about 3 mm diameter (Hodges, 1974) as areas which are slightly more electron dense than the surrounding material in the perivitelline space. As the follicle grows these areas become more distinct and extensive, suggesting a condensation process rather than an active secretion of formed elements. The differences in the structure of the perivitelline layer between the germinal disc region and elsewhere (Perry, Gilbert and Evans, 1978a, 1978b), await an explanation.

Two further layers are added to the yolk membranes in the oviduct (Bellairs, Harkness and Harkness, 1963; Bellairs, 1964): both are apparently secreted as a fine granular material (Bain and Hall, 1969), although the outer layer subsequently coalesces into proteinaceous fibrous strands; the mechanism for this is unknown.

Little is known of the chemistry of the composite membrane but it appears to be mainly water and protein. Doran and Mueller (1961) suggested that the true perivitelline layer was similar to collagen but Bellairs, Harkness and Harkness (1963) regarded it as being non-collagenous. Of the two parts formed by the oviduct, the outer (fibrous) one has some affinities with albumen (Bellairs, Harkness and Harkness, 1963), and Shenstone (1968) suggested it was a mixture of lysozyme and conalbumin. The composition of the inner part is unknown.

Although not entirely certain, it seems probable that the first protein deposition on the oocyte occurs in the infundibulum (Bain and Hall, 1969); as well as the material for the remainder of the yolk membranes, this may include also some material which will contribute to the chalaziferous layer of the true albumen (Aitken, 1971). The major difficulty is to identify the cells responsible for the production of these proteins. It could be argued that the originally similar granular nature of the two outer yolk membranes indicates a common origin; the coalescence of the outer one into fibres may be thought of as some later (enzymatic) process about which information is lacking. Alternatively, this difference could be taken to indicate a separate secretory process for each layer and, therefore, different cells involved. Yet again the outer layer has a similar composition to albumen and it seems reasonable therefore to assume that the cells responsible for the formation of each would be similar; but, if so, why should the outer layer coalesce into fibres whereas albumen does not?

With such uncertainties, the identification of the cells responsible must remain speculative, particularly since at least four types of secretory cell have been identified (Aitken, 1971; Hodges, 1974). Moreover, Aitken (1971) thought that a further cell in the tubular glands was also secretory. The ciliated cell also contains granular material and hence may be secretory, although Aitken and Johnston (1963) thought not.

ALBUMEN

The egg takes about 2½ h to travel through the magnum (Warren and Scott, 1935a), the longest region of the oviduct, and during this time it accumulates about 20 g of a thick gelatinous material. This albumen is almost pure protein, with little water, since the main mass of water is added later in the shell gland. In the laid egg, albumen consists of 11% protein and 88% water. Its composition is known with considerable accuracy (Parkinson, 1966; Baker, 1968; Feeney and Allison, 1969; Gilbert, 1971b) and about 40 specific proteins have been identified: many exhibit genetic heterogeneity (Lush, 1961). There is remarkably little difference in the chemical composition of the various compartments in the laid egg, although the chalazae and chalaziferous layers tend to have more ovomucin than the others; this may be related to the relative abundance of epithelial 'goblet' cells in the infundibulum than elsewhere (Hodges, 1974) because these cells have been suggested to secrete ovomucin (Wyburn *et al.*, 1970; Sandoz, Ulrich and Brand, 1971). The further suggestion, that the material destined to become the inner (thin) albumen is formed in the cranial magnum and that of the middle (thick) albumen is formed in the caudal part, is also reasonable (Asmundson and Burmester, 1938; Scott and Burmester, 1939); it is disappointing not to be able to relate this more precisely with the known morphological differences between the two regions (Aitken, 1971; Hodges, 1974).

Of the many proteins in albumen, few have been positively linked with a specific cell type. Ovalbumen and lysozyme have been demonstrated in the tubular glands and avidin has been found in the non-ciliated epithelial cell (Kohler, Grimley and O'Malley, 1968; O'Malley *et al.*, 1969; Oka and Schimke, 1969; Tuohimaa, 1975). Schimke *et al.* (1977) produced evidence that ovotransferrin and ovomucoid are also likely products of the tubular glands, and Wyburn *et al.* (1970) reasoned that ovomucin was produced by the non-ciliated ('goblet') epithelial cells.

If this evidence is accepted, and the remainder of the proteins taken into account, it seems reasonable to assume that each secretory cell is able to produce more than one protein. This is even more likely when only four general types of secretory cell are known (Wyburn *et al.*, 1970; Sandoz, Ulrich and Brand, 1971). However, it cannot be entirely dismissed that there may indeed be many types of secretory cell, each producing its own protein, even though we are unable to distinguish them on morphological grounds (Wyburn *et al.*, 1970). It must also be questioned what the granular inclusions of the secretory cells actually represent. Certainly they are characteristic of each type of cell and in many cases they are depleted with the passage of the egg; consequently it must be accepted that they are proteinaceous. But are they formed of several proteins or just one? In electron micrographs the granules within one cell may be of different electron densities and the size is not completely uniform.

There is also no adequate explanation for the differences in relative abundance of the proteins in albumen; for example, ovalbumin forms about 54% of the total protein, ovomucoid and ovotransferrin between 11% and 13% each, ovomucin about 2% and avidin only 0.05%. To put this down to the needs of the embryo is simply to beg the question of the way in which it is brought about.

SHELL MEMBRANES

The shell membranes contribute a greater proportion of protein to the shell than either the calcified part or the cuticle (Candlish, 1972). They are secreted by the isthmus, a short region of the oviduct adjacent to the magnum, over a period of about 1½ h. They consist of a meshwork of fibres with the fibres of the inner layer being thinner and longer than those of the outer layer (Masshoff and Stolpmann, 1961; Simons and Wiertz, 1963, 1970; Fujii, Tamura and Okamoto, 1970). It is not certain that the fibres are arranged systematically (Simons and Wiertz, 1963) or at random (Tung, 1970; Fujii and Tamura, 1970). A third, more amorphous layer may be present.

The fibres of the membranes appear to consist of a central core of protein covered by a carbohydrate mantle (Masshoff and Stolpmann, 1961; Simons and Wiertz, 1963; Candlish, 1972). Candlish and Scougall (1969) suggested collagen may be a component, but there is still much of the chemistry of the shell membranes that requires elucidation.

The cells responsible for the secretions of the fibres have probably been correctly identified: the fibre core appears to be a product of the tubular glands (Richardson, 1935; Candlish, 1972) and the mantle a product of the glandular epithelial cells (Draper, Johnston and Wyburn, 1968; Candlish, 1972). However, the cellular granular inclusions can vary in size within a cell and between cells in different regions of the isthmus (Khairallah, 1966; Draper *et al.*, 1972). Candlish (1972) attempted to relate this to the different thicknesses of the fibres, but the argument is not entirely convincing. Also the ciliated cells contain a similar type of granule and, despite contradictory later observations (Wyburn, Johnston and Draper, 1970a; Aitken, 1971), Richardson (1935) did describe a secretory cycle in these cells: their function is not known. Similarly there is no known function for the peculiar tubular gland cells at the junction of the isthmus and shell gland, and the composition of their bizarre granules (Aitken, 1971) is unknown.

CALCIFIED SHELL AND CUTICLE

The structure of the calcified part of the shell and the cuticle is well known (for example, see Romanoff and Romanoff, 1949; Tyler, 1964, 1969; Simons, 1971; Simkiss and Taylor, 1971; Hodges, 1974; Gilbert, 1967, 1979). The organic matrix of the calcified part of the shell consists of the so-called 'spongy layer', composed of a regular array of large branching fibres (10 μm long and 10 nm thick) arranged parallel to the outer surface of the egg, and the mammillary layer (Robinson and King, 1968). The mammillary layer is composed of numerous conical-shaped knobs partly embedded in the outer part of the shell membrane (Terepka, 1963; Simons and Wiertz, 1963). In the cente of each mammilla is a small proteinaceous core.

Little is known of the chemical composition of the organic part, except that it is almost certainly a glycoprotein complex (Simkiss and Tyler, 1957; Tyler and Simkiss, 1959; Baker and Balch, 1962; Frank, Burger and Swanson, 1965). Moreover, it does not appear to be uniform throughout, since the carbohydrate components vary from region to region (Simkiss and Taylor, 1971).

The cuticle is a thin outer covering of the shell consisting usually of two parts, the inner one being more compact (Simons and Wiertz, 1963). Its proteins differ from those of the matrix, being richer in tyrosine, lysine and cystine (Simkiss and Taylor, 1971), but the specific proteins have not been identified.

The formation of the mammillary cores occurs probably in the 'red-region', a short section of the oviduct between the isthmus and the pouch of the shell gland (Wyburn *et al.*, 1973; Sternberger, Mueller and Leach, 1977). It is not known which cells are involved; however, the presence of curious 'bleb-cells' (Johnston, Aitken and Wyburn, 1963) only in this region of the oviduct may be significant.

The remainder of the protein of the shell is formed in the pouch of the shell gland. As yet the complex and varied secretory functions of this section of the oviduct have defied attempts to discover the mechanisms involved and the cells responsible for each (Gilbert, 1979). One consistent view is that the tubular glands are not involved in protein secretion, with the possible exception of those in one part claimed to be responsible for the cuticle (Fujii, 1963). For the remaining two types of secretory cell (apical and basal), conflicting viewpoints are held. Thus the formation of both the cuticle and the protein matrix have been ascribed to the apical cell (Richardson, 1935; Johnston, Aitken and Wyburn, 1963; Tamura and Fujii, 1966; Breen and de Bruyn, 1969), whereas others thought formation of the cuticle occurred in the basal cell (Tyler and Simkiss, 1959; Baker and Balch, 1962) and yet others thought that the carbohydrate component of the matrix was formed in the basal cell (Breen and de Bruyn, 1969). Pigment formation is thought to be a function of all the epithelial secretory cells (Aitken, 1971; Baird, Solomon and Tedstone, 1975). Although it is not impossible for one cell type to have more than one function, it makes it more difficult to understand how these various functions are separately controlled particularly when, during the major part of the time spent by the egg in the shell gland, matrix secretion is occurring whereas the cuticle is formed only during the last part. Moreover, pigment deposition may be superficial in some species and hence restricted to a short period of time after the cuticle has formed, but in others it may occur for most of the time that the egg remains in the shell gland.

SECRETORY PROCESSES OF THE CELLS OF THE OVIDUCT

The ciliated cells of the epithelium of the infundibulum have microvilli-like extensions at their apical border (Aitken, 1971), suggesting an apocrine type of secretory process. Similar structures occur in the ciliated cells in the shell gland but the release of the vesicular and granular material of these may occur also with disruption of the plasma membrane, hence resembling a holocrine type of secretion (Aitken, 1971). The ciliated cells of the isthmus and magnum are usually regarded as non-secretory although they do have granular inclusions. If they are secretory, the mechanism for release of their secretory product is not known.

The non-ciliated cells of the infundibulum and the shell-gland appear to have a mecrocrine type of secretion as well as an apocrine type; both have microvilli protruding from their apical borders but release of material may occur after these have been withdrawn (Johnston, Aitken and Wyburn, 1963; Breen and de Bruyn, 1969). It has not been established why such different processes should occur in one cell type and what secretory product each is related to; it is feasible that these different mechanisms may indicate a subtle difference in cell types usually classified as one. In contrast to the non-ciliated cell of the infundibulum and shell gland, those of the magnum appear to have only a mecrocine type of secretory mechanism: the cells become filled with secretory products, bulge into the lumen and discharge their secretions through rupture of the apical membrane

(Wyburn *et al.*, 1970; Aitken, 1971). The isthmian non-cilitated cell is considered to have an apocrine type of secretion (Draper, Johnston and Wyburn, 1968). Little is known of the 'bleb-cells' of the 'red-region', although presumably their structure indicates a mecrocrine type.

The microvilli of the cells of the tubular gland appear to be unconnected with their secretory function, although they are a characteristic feature in all regions. Within the magnum secretion is a mecrocrine (Wyburn *et al.*, 1970), despite the contrary suggestion of Hendler, Dalton and Glenner (1956), and the similarity of the tubular glands of the infundibulum to these (Aitken, 1971) suggests that their secretory mechanism is likely to be the same. The isthmian cells also appear to be mecrocrine-type cells, although Aitken (1971) described vesicles and granular material within the microvilli. Consequently it is possible that an apocrine type of secretion could occur, but this has not been observed.

STIMULATORY MECHANISMS FOR THE RELEASE OF THE SECRETORY PRODUCTS

In contrast to the knowledge of the way in which the secretory products are released from the secretory cells, there are many unsolved problems associated with the stimulus for their release.

Fundamentally the egg is formed as a series of concentric layers. As the egg descends the oviduct, only the outer portions of it are in direct contact with the luminal surface of the oviduct; consequently the secretions of each region of the oviduct are laid down on the structures previously produced. There are *a priori* four possibilities for the stimulus:

(1) Direct mechanical stimulation by the egg.
(2) A chemical activator from the egg.
(3) Hormonal coordination.
(4) Neural mechanisms.

There is strong evidence to suggest that mechanical stimulation plays a major part in the general response of the oviduct (Asmundson and Baker, 1940; Asmundson, Baker and Emlen, 1943) and foreign bodies and ovarian fragments are formed into 'eggs' even though they may be misshapen (Pearl and Surface, 1909; Pearl and Curtis, 1914; Hutt, 1939, 1946; Cole, 1946; Wentworth, 1960; Pratt, 1960; Gilbert, 1968; Tanaka, 1976). However, this general picture becomes less clear when each region of the oviduct is considered in detail.

Little is known of the response of the infundibulum; whatever causes release of the secretions it must act rapidly since the egg only remains there for about 15 min (Warren and Scott, 1935b). In the magnum the response is obvious, since the glandular cells discharge their products in the presence of the descending egg: glandular cells caudal to the egg are fully charged with secretory material whereas those cranial to the egg are for the most part discharged. However, not all of the glandular cells are discharged at one time (Aitken, 1971). It is therefore questionable that direct mechanical stimulation is the sole stimulus, although it could be argued that only cells fully charged with secretory products are able to respond to the mechanical effect of the egg and undischarged cells could be those not ready to respond. This would imply that ultimate control rests within the mechanisms for the synthesis of the protein secretions.

There is little doubt that the albumenous egg when it first enters the isthmus causes secretion of the shell membranes, and by careful selection of birds it is possible to recover from the isthmus partly coated eggs. Hence, it is tempting to accept that a straightforward movement through this region is all that is required to bring about completion of the membranes. However, movement through the isthmus is not regular (Richardson, 1935; Burmester, 1940) and, although the complex movements are likely to be related to secretory function, this aspect has not been considered. What is more, if mechanical stimulation were the only cause, as the egg progresses along the isthmus the leading surface of the egg would continually come into contact with undischarged glandular cells while the trailing surface would contact mainly discharged ones. Consequently the membrane should be thicker at the leading surface than the trailing surface, which is not noticeably the case.

It is not possible at present to reconcile the obvious mechanical effect of the egg as it enters the isthmus with the final form of the membrane-bound egg, although many suggestions are plausible; for example, it could be argued that only the naked albumen has the property of stimulating the release of secretions but this requires a chemical mediator. Candlish (1972) suggested that the cranial glandular elements produced the inner part of the membranes, while the outer part was formed in the caudal isthmus, and that the fibres were 'spun' by the rotation of the egg. This implies a continuous secretion from the gland cells which contrasts with the apparently 'all-or-none' response of the cells of the magnum.

Nothing is known of the mechanisms within the shell gland, although deposition is a continuous process lasting about 20 h.

Although mechanical stimulation is an important factor in causing release of the proteins from the secretory cells, by what means is the mechanical effect translated into physiological action? Many of the epithelial cells have microvilli (Aitken, 1971; Hodges, 1974) and it is reasonable to assume that they have a specific function, perhaps in the release mechanisms. The ciliated cells also usually contain granular inclusions (Aitken, 1971; Hodges, 1974) and so could be secretory; it is possible that these cells could be involved by producing a 'release factor' which may affect the cells actually secreting the proteins. It is not so easy to visualize how direct mechanical stimulation could affect the tubular glands buried in the tissue of the oviduct. The egg, being larger than the lumen of the oviduct, will tend to stretch the openings of the tubular glands, thereby allowing material in the glands to be voided (Wyburn *et al.*, 1970; Candlish, 1972). But this does not explain how the secretory cells release their products into the glandular lumen, unless this is a continuous process.

Clearly the concept of direct mechanical stimulation is unable to answer all the questions and other mechanisms have been suggested to be involved, although few have much credibility.

It has been proposed that chemical mediators from the yolk may be involved, since diffusable products obtained from the yolk will *in vitro* affect oviducal metabolism (Eiler, González and Horvath, 1970). However, it is difficult to conceive how these could be active because the oocyte is rapidly covered by a thick gelatinous layer of albumen. Moreover, if chemical transmitters are involved generally it is not easy to visualize how they could be retained on the surface of a developing egg which is constantly being coated with fresh secretory material.

Isolated loops of the oviduct have been reported to secrete when the intact portion is stimulated (Burmester and Card, 1939, 1941), which is indicative possibly of neural or hormonal mechanisms being involved. Certainly parts of the oviduct are well innervated (see Gilbert, 1967, 1979 and Hodges, 1974) but attempts made to determine a role for neural control have consistently failed (Sturkie and Weiss, 1950; Sturkie, Weiss and Ringer, 1954).

Excessive doses of steroid hormones have been shown to cause secretion of albumen by the magnum (Brant and Nalbandov, 1956). However, since steroid hormones are involved in the biosynthesis of some of the proteins of this region, such treatment may reflect an unphysiological overstimulation of the biosynthetic mechanisms. Nevertheless, some support for the view that steroid hormones are in some way involved in the release mechanisms has been obtained from ultrastructural studies (Kohler, Grimley and O'Malley, 1969). One difficulty with hormones is that, being carried in the blood, they are likely to affect all similar cells equally: with such a large region as the magnum, why then does secretion occur only in the presence of the egg? Perhaps the presence of the egg alters blood flow in its immediate area; certainly the capillary system in the oviduct is complex (Hodges, 1965, 1966; Gilbert, Reynolds and Lorenz, 1968).

There is growing evidence that prostaglandins play an important role in the transport of the egg along the oviduct (Talo and Kekäläinen, 1976; Verma, Prasad and Slaughter, 1976; Wechsung and Houvenaghel, 1976; Day and Nalbandov, 1977; Hertelendy and Biellier, 1978; Hammond, 1978) and it is possible that they could be involved also in secretory mechanisms.

Conclusions

It is clear that our knowledge of the structure of the egg and the organs responsible for its production (the ovary and the oviduct) is not sufficiently supported by biochemical or physiological information for it to be possible to give a comprehensive account of all aspects of the formation of the egg. In fact the details of the processes of protein deposition are almost totally unknown. The text has been deliberately speculative and more questions have been asked than have been answered. Nevertheless, it is hoped that this approach to the subject is of value, particularly at a time when there is a general belief that our knowledge of egg formation is complete, except for some minor details.

References

AITKEN, R.N.C. (1971). In *The Physiology and Biochemistry of the Domestic Fowl* (Bell, D.J. and Freeman, B.M., Eds), Academic Press, London

AITKEN, R.N.C. and JOHNSTON, H.S. (1963). *J. Anat.,* **97**, 87–99

ANDERSON, R.G.W., BROWN, M.S. and GOLDSTEIN, J.L. (1977). *Cell,* **10**, 351–364

ASMUNDSON, V.S. and BAKER, G.A. (1940). *Poult. Sci.,* **19**, 227–232

ASMUNDSON, V.S. and BURMESTER, B.R. (1938). *Poult. Sci.,* **17**, 126–130

ASMUNDSON, V.S., BAKER, G.A. and EMLEN, J.T. (1943). *Auk,* **60**, 34–44

BAIN, J.M. and HALL, J.M. (1969). *Aust. J. Biol. Sci.,* **22**, 653–656

BAIRD, T., SOLOMON, SARAH E. and TEDSTONE, D.R. (1975). *Br. Poult. Sci.,* **16**, 201–208

BAKER, C.M.A. (1968). In *Egg Quality : a Study of the Hen's Egg* (Carter, T.C., Ed.), Oliver and Boyd, Edinburgh

BAKER, J.R. and BALCH, D.A. (1962). *Biochem. J., 82*, 352–361

BELL, D.J. and FREEMAN, B.M. (1971). *The Physiology and Biochemistry of the Domestic Fowl,* Academic Press, London

BELLAIRS, R. (1964). In *Advances in Morphogenesis* (Abercrombie, M. and Brachet, J., Eds), Academic Press, London

BELLAIRS, R. (1965). *J. Embryol. Exp. Morph., 13*, 215–233

BELLAIRS, R. (1967). *J. Embryol. Exp. Morph., 17*, 267–281

BELLAIRS, R., HARKNESS, M. and HARKNESS, R.D. (1963). *J. Ultrastruct. Res., 8*, 339–359

BRANT, J.W.A. and NALBANDOV, A.V. (1956). *Poult. Sci., 35*, 692–700

BREEN, P.C. and DE BRUYN, P.P.H. (1969). *J. Morph., 128*, 35–66

BURMESTER, B.R. (1940). *J. Exp. Zool., 84*, 445–500

BURMESTER, B.R. and CARD, L.E. (1939). *Poult. Sci., 18*, 138–145

BURMESTER, B.R. and CARD, L.E. (1941). *Poult. Sci., 20*, 224–226

CANDLISH, J.K. (1972). In *Egg Formation and Production* (Freeman, B.M. and Lake, P.E., Eds), British Poultry Science, Edinburgh

CANDLISH, J.K. and SCOUGALL, R.K. (1969). *Int. J. Protein Res., 1*, 299–302

CLARK, R.C. (1970). *Biochem. J., 118*, 537–542

CLEMENS, M.J. (1974). *Prog. Biophys. Molec. Biol., 28*, 69–108

COLE, R.K. (1946). *Poult. Sci., 25*, 473–375

COOK, W.H. (1968). In *Egg Quality : a Study of the Hen's Egg* (Carter, T.C., Ed.), Oliver and Boyd, Edinburgh

COOK, W.H. and MARTIN, W.G. (1969). In *Structural and Functional Aspects of Lipoproteins in Living Systems* (Tria, E. and Scanu, A.M., Eds), Academic Press, London

DAY, S.L. and NALBANDOV, A.V. (1977). *Biol Reprod., 16*, 486–494

DORAN, M. and MUELLER, W.J. (1961). *Poult. Sci., 40*, 474–478

DRAPER, M.H. (1966). In *The Physiology of the Domestic Fowl* (Horton-Smith, C. and Amoroso, E.C., Eds), Oliver and Boyd, Edinburgh

DRAPER, M.H., JOHNSTON, H.S. and WYBURN, G.M. (1968). *J. Physiol., Lond., 196*, 7P–9P

DRAPER, M.H., DAVIDSON, MAIDA F., WYBURN, G.M. and JOHNSTON, H.S. (1972). *Q. Jl Exp. Physiol., 57*, 297–309

EILER, H., GONZÁLEZ, E. and HORVATH, A. (1970). *Biol. Reprod., 2*, 172–177

EVANS, A.J., PERRY, MARGARET M. and GILBERT, A.B. (1979). *Biochim. Biophys. Acta, 573*, 184–195

EVERSON, GLADYS J. and SOUDERS, HELEN J. (1957). *J. Am. diet. Ass., 33*, 1244–1254

FARNER, D.S. and KING, R.K. (1971–1975). *Avian Biology,* Vol 1–5, Academic Press, New York

FEENEY, R.E. and ALLISON, R.G. (1969). *Evolutionary Biochemistry of Proteins,* Wiley, New York

FOLLETT, B.K. and DAVIS, D.T. (1975). *Symp. Zool. Soc. Lond., 35*, 199–224

FRANK, F.R., BURGER, R.E. and SWANSON, M.H. (1965). *Poult. Sci., 44*, 63–69

FUJII, S. (1963). *Archvm Histol. Jap., 23*, 447–459

FUJII, S. and TAMURA, T. (1970). *J. Fac. Fish. Anim. Husb. Hiroshima Univ., 9*, 65–81

FUJII, S., TAMURA, T. and OKAMOTO, T. (1970). *J. Fac. Fish. Anim. Husb. Hiroshima Univ.,* **9**, 139–150

GILBERT, A.B. (1967). In *Advances in Reproductive Physiology* (McLaren, Anne, Ed.), Logos Press, London

GILBERT, A.B. (1968). *Br. Poult. Sci.,* **9**, 301–302

GILBERT, A.B. (1971a). In *The Physiology and Biochemistry of the Domestic Fowl,* Ch. 50 (Bell, D.J. and Freeman, B.M., Eds), Academic Press, London

GILBERT, A.B. (1971b). In *The Physiology and Biochemistry of the Domestic Fowl,* Ch. 54 (Bell, D.J. and Freeman, B.M., Eds), Academic Press, London

GILBERT, A.B. (1971c). In *The Physiology and Biochemistry of the Domestic Fowl,* Ch. 56 (Bell, D.J. and Freeman, B.M., Eds), Academic Press, London

GILBERT, A.B. (1971d). In *The Physiology and Biochemistry of the Domestic Fowl,* Ch. 58 (Bell, D.J. and Freeman, B.M., Eds), Academic Press, London

GILBERT, A.B. (1971e). In *The Physiology and Biochemistry of the Domestic Fowl,* Ch. 59 (Bell, D.J. and Freeman, B.M., Eds), Academic Press, London

GILBERT, A.B. (1972). In *Egg Formation and Production* (Freeman, B.M. and Lake, P.E., Eds), British Poultry Science, Edinburgh

GILBERT, A.B. (1979). In *Form and Function in Birds* (King, A.S. and McLelland, J., Eds), Academic Press, London

GILBERT, A.B., REYNOLDS, M.F. and LORENZ, F.W. (1968). *J. Reprod. Fert.,* **17**, 305–310

GORNALL, D.A. and KUKSIS, A. (1973). *J. Lipid Res.,* **14**, 197–205

GRAU, C.R. (1976). *Poult. Sci.,* **55**, 1418–1422

GRAU, C.R. and WILSON, B.W. (1964). *Experientia,* **20**, 26

GREENFIELD, M.L. (1966). *J. Embryol. Exp. Morph.,* **15**, 297–316

GRUBER, M. (1972). In *Egg Formation and Production* (Freeman, B.M. and Lake, P.E., Eds), British Poultry Science, Edinburgh

HAMMOND, R.W. (1978). *Poult. Sci.,* **57**, 1141

HENDLER, R.W., DALTON, A.J. and GLENNER, G.G. (1956). *J. Biophys. Biochem. Cytol.,* **3**, 325–337

HERTELENDY, F. and BIELLIER, H.V. (1978). *Biol. Reprod.,* **18**, 204–211

HILLYARD, L.A., WHITE, H.M. and PANGBURN, S.A. (1972). *Biochemistry, N.Y.,* **11**, 511–518

HODGES, R.D. (1965). *J. Anat.,* **99**, 485–506

HODGES, R.D. (1966). In *The Physiology of the Domestic Fowl* (Horton-Smith, C. and Amoroso, A.C., Eds), Oliver and Boyd, Edinburgh

HODGES, R.D. (1974). *The Histology of the Fowl,* Academic Press, London

HOLDSWORTH, G., MITCHELL, R.H. and FINEAN, J.B. (1974). *FEBS Letts, Amsterdam,* **39**, 275–277

HUTT, F.B. (1939). *Poult. Sci.,* **18**, 276–278

HUTT, F.B. (1946). *Auk,* **63**, 171–174

JOHNSTON, H.S., AITKEN, R.N.C. and WYBURN, G.M. (1963). *J. Anat.,* **97**, 333–344

JORDANOV, J.S. (1969). *On the Cytobiology of the Hen's Egg,* Publishing House of the Bulgarian Academy of Science, Sofia

JORDANOV, J.S. and BOYADJIEVA-MICHAILOVA, A. (1974). *Acta Anat.,* **89**, 616–632

KHAIRALLAH, L. (1966). *Ph.D. thesis,* Boston University

KING, A.S. (1975). In *Sisson and Grossman's The Anatomy of the Domestic Animals,* 5th edn (Getty, R., Ed.), Saunders, Philadelphia

KING, A.S. and McLELLAND, J. (1979). *Form and Function in Birds,* Academic Press, London

KOHLER, P.O., GRIMLEY, P.M. and O'MALLEY, B.W.(1968). *Science, N.Y.,* **160,** 86—87

KOHLER, P.O., GRIMLEY, P.M. and O'MALLEY, B.W.(1969). *J. Cell Biol.,* **40,** 8—27

LUSH, I.E. (1961). *Nature, Lond.,* **189,** 981—984

MARZA, V.D. and MARZA, R.V. (1935). *Q. Jl Microsc. Sci.,* **78,** 134—189

MASSHOFF, W. and STOLPMANN, H.(1961). *Z. Zellforsch. Mikrosk. Anat.,* **55,** 818—832

McGUIRE, W.L. and O'MALLEY, B.W. (1968). *Biochim. Biophys. Acta,* **153,** 187—194

McINDOE, W.M. (1971). In *The Biochemistry and Physiology of the Domestic Fowl* (Bell, D.J. and Freeman, B.M., Eds), Academic Press, London

MURTON, R.K. and WESTWOOD, N.J. (1977). *Avian Breeding Cycles,* Clarendon Press, Oxford

NALBANDOV, A.V. and JAMES, M.F. (1949). *Am. J. Anat.,* **85,** 347—378

OCKLEFORD, C.D. (1976). *J. Cell Sci.,* **21,** 83—91

OKA, T. and SCHIMKE, R.T. (1969). *Science, N.Y.,* **163,** 83—85

O'MALLEY, B.W. and McGUIRE, W.L. (1967). *J. Clin. Invest.,* **46,** 1101

O'MALLEY, B.W. and McGUIRE, W.L. (1968a). *J. Clin. Invest.,* **48,** 654—664

O'MALLEY, B.W. and McGUIRE, W.L. (1968b). *J. Clin. Invest.,* **47,** 75a

O'MALLEY, B.W., McGUIRE, W.L. and KORENMAN, S.G. (1967). *Biochim. Biophys. Acta,* **145,** 204—207

O'MALLEY, B.W., McGUIRE, W.L. and MIDDLETON, P.A. (1968). *Nature, Lond.,* **218,** 1249—1251

O'MALLEY, B.W., ARRONOW, A., PEACOCK, A.C. and DINGMAN, C.W.(1968). *Science N.Y.,* **162,** 567—568

O'MALLEY, B.W., McGUIRE, W.L., KOHLER, P.O. and KORENMAN, S.G.(1969). *Recent Prog. Horm. Res.,* **25,** 105—160

PALMITTER, R.D., MULUIHILL, E.A., McKNIGHT, G.S. and SENEAR, A.W. (1977). *Cold Spring Harb. Symp. Quant. Biol.,* **42,** 639—647

PARKINSON, T.L. (1966). *J. Sci. Fd Agric.,* **17,** 101—111

PAULSON, J. and ROSENBERG, M.D.(1972). *J. Ultrastruct. Res.,* **40,** 25—43

PEARL, R. and CURTIS, M.R. (1914). *J. Exp. Zool.,* **17,** 395—424

PEARL, R. and SURFACE, F.M. (1909). *Science, N.Y.,* **29,** 428—429

PERRY, MARGARET M. and GILBERT, A.B. (1979). *J. Cell Sci.,* **39,** 257—272

PERRY, M.M. and GILBERT, A.B. (1980). *Cell Biol. Int. Rpts.* (in press)

PERRY, M.M., GILBERT, A.B. and EVANS, A.J. (1978a). *J. Anat.,* **125,** 481—497

PERRY, M.M., GILBERT, A.B. and EVANS, A.J. (1978b). *J. Anat.,* **127,** 379—392

PRASAD, C. and PETERKOFSKY, A. (1976). *Arch. Biochem. Biophys.,* **175,** 730—736

PRATT, A. (1960). *Proc. S. Dakota Acad. Sci.,* **39,** 178

PRESS, N. (1964). *J. Ultrastruct. Res.,* **10,** 528—546

RAO, D.N., RAMAKRISHNA and MAHADLEVAN, S. (1978). *Poult. Sci.,* **57,** 1091—1093

RICHARDSON, K.C. (1935). *Phil. Trans. Roy. Soc. Lond., Ser. B,* **225,** 149—196

ROBINSON, D.S. and KING, N.R. (1968). *J. R. Microsc. Soc.,* **88,** 13—22

ROMANOFF, A.L. (1960). *The Avian Embryo,* MacMillan, New York

ROMANOFF, A.L. and ROMANOFF, ANASTASIA J. (1949). *The Avian Egg,* Wiley, New York

ROTH, T.F. and PORTER, K.R.(1964). *J. Cell. Biol.,* **20,** 313—332

ROTH, T.F., CUTTING, J.A. and ATLAS, S.B. (1976). *J. Supramolec. Struct.,* **4,** 522—548

ROTHWELL, B. and SOLOMON, SARAH E. (1977). *Br. Poult. Sci.*, **18**, 605—610
SANDOZ, D., ULRICH, E. and BRAND, E. (1971). *J. Microsc.*, **11**, 371—400
SCHIMKE, R.T., PENNEQUIN, P., ROBINS, DIANE and McKNIGHT, G.S. (1977). In *Hormones and Cell Regulation* (Dumont, J. and Nunez, J., Eds), North-Holland, Amsterdam
SCHJEIDE, O.A. and McCANDLESS, R.G. (1962). *Growth*, **26**, 309—321
SCHJEIDE, O.A., MUNN, R.J., McCANDLESS, R.G. and EDWARDS, R. (1966). *Growth*, **30**, 471—489
SCHJEIDE, O.A., GALEY, F., GRELLERT, E.A., I-SAN LIN, R.I., DE VELLIS, J. and MEAD, J.F. (1970). *Biol. Reprod.*, **2**, Suppl. 1, 14—43
SCHUTZ, G., NGUYEN-HUU, M.C., GIESECKE, K., HYNES, N.E., GRONER, B., WURTZ, T. and SIPPEL, A.E. (1977). *Cold Spring Harb. Symp. Quant. Biol.*, **42**, 617—624
SCOTT, H.M. and BURMESTER, B.R. (1939). *Proc. 7th World's Poult. Cong., Cleveland*, 102—106
SHENSTONE, F.S. (1968). In *Egg Quality : a Study of the Hen's Egg* (Carter, T.C., Ed.), Oliver and Boyd, Edinburgh
SIMKISS, K. and TAYLOR, T.G. (1971). In *The Physiology and Biochemistry of the Domestic Fowl* (Bell, D.J. and Freeman, B.M., Eds), Academic Press, London
SIMKISS, K. and TYLER, C. (1957). *Q. Jl Microsc. Sci.*, **98**, 19—28
SIMONS, P.C.M. (1971). *Ph.D. thesis*, Wageningen Centre for Agricultural Publishing and Documentation
SIMONS, P.C.M. and WIERTZ, G. (1963). *Z. Zellforsch. Mikrosk. Anat.*, **59**, 555—567
SIMONS, P.C.M. and WIERTZ, G. (1970). *Annls Biol. Anim. Biochim. Biophys.*, **10**, 31—49
STERNBERGER, BRIDGET H., MUELLER, W.J. and LEACH, R.M. Jnr. (1977). *Poult. Sci.* **56**, 537—543
STURKIE, P.D. and WEISS, H.S. (1950). *Poult. Sci.*, **29**, 781
STURKIE, P.D., WEISS, H.S. and RINGER, R.K. (1954). *Poult. Sci.*, **33**, 18—24
TALO, A. and KEKÄLÄINEN, R. (1976). *Biol. Reprod.*, **14**, 186—189
TAMURA, T. and FUJII, S. (1966). *J. Fac. Fish. Anim. Husb. Hiroshima Univ.*, **6**, 357—371
TANAKA, K. (1976). *Jap. Jl Zootech. Sci.*, **47**, 385—392
TARLOW, D.M., WATKINS, P.A., REED, R.E., MILLER, R.S., ZWERGEL, E.E. and LANE, M.D. (1977). *J. Cell. Biol.*, **73**, 332—353
TEREPKA, A.R. (1963). *Exp. Cell Res.*, **30**, 171—182
TUNG, M.A. (1970). *Ph.D. thesis*, University of British Columbia
TUOHIMAA, P. (1975). *Histochemie*, **44**, 95—101
TYLER, C. (1964). *Proc. Zool. Soc. Lond.*, **142**, 547—583
TYLER, C. (1969). *J. Zool. Lond.*, **158**, 395—412
TYLER, C. and SIMKISS, K. (1959). *J. Sci. Fd Agric.*, **10**, 611—615
VERMA, O.P., PRASAD, B.K. and SLAUGHTER, J. (1976). *Prostaglandins*, **12**, 217—227
WALLACE, R.A. and DUMONT, J.N. (1968). *J. Cell. Physiol.*, **82**, Suppl. 1, 75—89
WARREN, D.C. and SCOTT, H.M. (1935a). *Poult. Sci.*, **14**, 195—207
WARREN, D.C. and SCOTT, H.M. (1935b). *J. Agric. Res.*, **51**, 565—572
WECHSUNG, E. and HOUVENAGHEL, A. (1976). *Prostaglandins*, **12**, 599—608
WENTWORTH, B.C. (1960). *Poult. Sci.*, **34**, 782—784
WILLIAMS, J. (1962). *Biochem. J.*, **83**, 346—357

WYBURN, G.M., AITKEN, R.N.C. and JOHNSTON, H.S. (1965). *J. Anat.*, **99**, 469–484

WYBURN, G.M., JOHNSTON, H.S. and DRAPER, M.H. (1970). *Annls Biol. Anim., Biochim. Biophys.*, **10**, 53–55

WYBURN, G.M., JOHNSTON, H.S., DRAPER, M.H. and DAVIDSON, MAIDA F. (1970). *Q. Jl Exp. Physiol.*, **55**, 213–232

WYBURN, G.M., JOHNSTON, H.S., DRAPER, M.H. and DAVIDSON, MAIDA F. (1973). *Q. Jl Exp. Physiol.*, **58**, 143–151

YUKSO, S.C. and ROTH, T.F. (1976). *J. Supramolec. Struct.*, **4**, 89–97

6

METABOLISM OF THE FETUS

J.M. BASSETT
Nuffield Institute for Medical Research, University of Oxford

Summary

Utilizing chronic cannulation procedures, studies of the undisturbed fetus *in utero*, predominantly on sheep, have led to considerable advances in understanding of fetal metabolism and its regulation.

Determination of umbilical blood flow, together with venous-arterial concentration differences for oxygen and the principal metabolic substrates, has allowed definition of their possible contribution to oxidative metabolism in the fetus. However, while the high metabolic rate of placenta, predominantly a fetal tissue, is generally recognized neither its contribution to fetal metabolic requirements nor the overall metabolic requirements of the pregnant uterus have been so clearly defined.

Overall, amino acid nitrogen concentrations in fetal blood are higher than those in maternal blood and amino acids are evidently actively accumulated by fetal placental tissue from the maternal circulation. Uptake of individual amino acids by the pregnant uterus from the maternal circulation has not been well defined, so the extent of placental interconversions is unknown. However, umbilical A-V differences and blood flow indicate a net transfer to the fetus of most amino acids in amounts greater than required for protein synthesis. Fetal urea production determined from placental urea clearance and maternal-fetal arterial blood concentration differences, indicate that catabolism of amino acids could account for 25% oxygen consumption in the fetal lamb, although possibly less in the calf.

Release of insulin from the fetal pancreas is regulated by the maternal nutritional state and plays a major role in regulating fetal metabolism and growth. Glucagon's role is less certain, but its level relative to insulin is higher in starvation. While gluconeogenic enzymes are present in ruminant fetal liver, the extent of amino acid conversion to glucose in the fetus remains controversial.

Introduction

One impressive aspect of fetal growth and development is its rate. Successive cell divisions and multiplicative growth of the conceptus over a relatively few weeks lead to the development within the uterus of a young individual with most of the specialized organs and regulatory mechanisms of the adult already present. Postnatal life depends in large part on the success of this process. The substrate requirements for protein synthesis to achieve this rapid growth may be small relative to overall substrate turnover in the mother during most of gestation, although there is little quantitative information about this. Despite this and the recognition by Hammond (1944) that the fetus had a very high priority for maternal nutrient supplies, there is considerable evidence that growth during the latter part of gestation and size at birth may be markedly influenced by maternal nutrition and substrate supply. There is considerable interest in the way substrate supply to the fetus is regulated and in the alterations to maternal metabolic regulation which allow this to be optimized.

This chapter will consider the supply and utilization of substrates for protein synthesis by the fetus in relation to its overall metabolism and regulation by maternal nutritional status, particularly in later gestation. Development of pathways for amino acid metabolism in the fetus will only be mentioned where relevant to the theme.

Oxidative metabolism of the pregnant uterus and conceptus

Whereas the high metabolic rate of the pregnant uterus has been generally recognized for many years, a more detailed understanding of the way metabolic activity is partitioned within the various components in this complex system (*Figure 6.1*) has been limited by the difficulties of sampling blood supplying

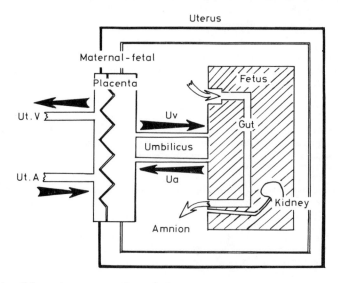

Figure 6.1   Schematic representation of the conceptus and its relation to the uterus, illustrating the major routes by which nutrients and waste products enter and leave

and leaving the particular compartments in the organ. The small size and difficult access to placental and umbilical vessels continues to present considerable problems for the separate *in vivo* assessment of metabolic activity in the uterine wall, placenta and fetus in small laboratory animal species such as rat, rabbit and guinea-pig. Knowledge in this field has been gained predominantly from studies on larger species such as sheep, cow and horse, where chronic cannulation and sampling of blood from uterine and fetal blood vessels is possible without serious interference to the normal function of the fetus or continuation of the pregnancy. However, even in these species measurement of placental metabolic activity separate from that of the uterus is difficult if not impossible and can only be estimated indirectly.

The relations of the fetus and its membranes with the uterus and maternal placenta are shown schematically in *Figure 6.1*. No attempt has been made to show separately the vascularization of the uterine wall and the maternal placenta, although it should be recognized that metabolically these behave very differently and it would be desirable for purposes of metabolic assessment to separate blood flowing from the placenta from that perfusing the myometrium.

Ideally one would wish to determine the metabolic activity of the conceptus, i.e. the fetus together with its surrounding membranes and placental attachments to the uterine wall, separately from that of the maternal endometrial and myometrial tissues. Although this is impracticable because of the intimate contact of maternal and fetal tissues in the placenta, it should be borne in mind that membranes and much of the placenta are fetal tissues and should not be ignored in assessment of fetal metabolism. For practical purposes fetal metabolism has usually been assessed by measurement of umbilical blood flow utilizing the Fick principle and metabolite concentration differences between umbilical artery and vein, measurements which take no account of metabolism in the extracorporeal fetal tissues or the flux of metabolites into and out of the amniotic fluid surrounding the fetus. Utilizing such techniques the available substrates and extent of oxidative metabolism in the fetus of several species has been delineated (Silver, 1976; Battaglia and Meschia, 1978).

In the sheep, the most widely studied species, estimates of fetal oxygen consumption lie in the range 6–8.5 ml/(kg min). Interestingly, despite wide variations in maternal body size and oxygen consumption rates, estimates for fetal oxygen consumption in most other species which have been studied lie within the same range (Battaglia and Meschia, 1978). Although fewer, estimates of uteroplacental oxygen uptake suggest that this is higher than that of the fetus with placental consumption, which may be as high as 30–35 ml/(kg min) in sheep, being the major contributor (Silver, 1976).

Even though glucose utilization by the pregnant uterus represents a major fraction of total body glucose utilization in sheep (Setchell *et al.*, 1972; Bergman, Brockman and Kaufman, 1974), determination of fetal utilization has shown that glucose cannot be the only substrate for oxidative metabolism

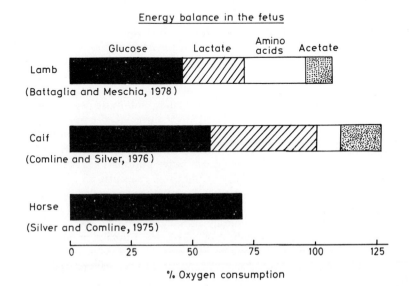

Figure 6.2 Estimated possible contributions to fetal oxidative metabolism, if totally oxidized, of the main metabolic substrates for which umbilical uptakes have been determined

(Tsoulos *et al.*, 1971). Subsequently, similar studies have determined the fetal uptake or output of other metabolites and, by relating these to oxygen uptake, their maximum possible contribution to fetal oxidative metabolism (*Figure 6.2*). The possible contributions of lactate produced in the placenta from maternal glucose and amino acids are particularly noteworthy. While such calculations indicate the potential contribution of each substrate to fetal oxidative metabolism, their actual contribution may be substantially less. For example, observations on glucose turnover in fetal lambs indicate that only some 50% or so of glucose appears as carbon dioxide (Setchell *et al.*, 1972). However, the large contributions of amino acids to oxidative metabolism in the fetal lamb could represent their actual contribution, as it is based on measurements of urea production by the fetus rather than on net uptake of amino acids.

Unlike measurements of fetal utilization of other metabolites, estimates for fetal urea production have been based on maternal-fetal differences in urea concentration and placental urea clearance because of the small umbilical vein-artery concentration difference (Gresham *et al.*, 1972). These studies showed clearly that there is a substantial net excretion of urea nitrogen by the fetal lamb *in utero* and that this excretion is increased markedly during maternal fasting (Simmons *et al.*, 1974).

Comparable maternal—fetal concentration differences in pregnant women (Gresham, Simons and Battaglia, 1971) suggest significant amino acid catabolism in human fetuses too, and although differences in the cow are smaller, suggesting less amino acid catabolism in this species (Comline and Silver, 1976), it is nevertheless evident that fetal amino acid catabolism may be more significant than formerly realized, and consistent with the high rates of amino acid turnover observed in fetal tissues (Chrystie *et al.*, 1977).

Uptake and utilization of amino acids by the conceptus

There is still a paucity of quantitative information about the uptake or utilization of amino acids by the gravid uterus or the conceptus. Christenson and Prior (1978) reported that uterine uptake of alpha amino acid nitrogen (AAN) was closely related to the amount of fetal tissue during the last third of gestation in sheep, and a similar conclusion is suggested by the results of Young and Widdowson (1975) for uptake of labelled amino isobutyric acid by normal or undernourished guinea-pig fetuses. Small A—V differences across the pregnant uterus have been found for some individual amino acids in the limited number of studies reported (Prenton and Young, 1969; Hopkins, McFadyen and Young, 1971), but there is still no adequate quantitative information about the uptake of individual amino acids by either the uterus or conceptus as a whole. (Some additional information on this point has recently been provided by the observations of Holzman *et al.* (1979), who have studied uterine uptake of individual amino acids in sheep with chronically cannulated fetuses. Such observations, however, still give no accurate indication of placental uptake or release separate from myometrial uptake and release.)

The observations by Young and Widdowson (1975) showed that placental uptake of the non-metabolized amino isobutyric acid was considerably greater than uptake by the maternal liver — a finding suggestive of rapid amino acid turnover in the placenta. Rates of mixed tissue protein turnover determined by continuous infusion of [^{14}C]-lysine or [^{14}C]-leucine (*Figure 6.3*) are also indicative of high rates of protein and amino acid turnover in the placenta and tissues of the fetus and newborn lamb, when compared with rates in maternal tissues (Soltez, Joyce and Young, 1973, Chrystie *et al.*, 1977; Young, 1979). Young (1979) has suggested that the very high rates of protein turnover and amino acid uptake of the placenta compared to maternal tissues may explain the high priority of the fetus for available supplies of nutrient amino acids.

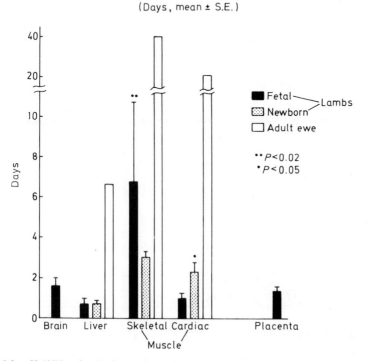

Figure 6.3   Half-life of mixed protein in the organs of fetal and newborn lambs and the adult ewe (From Young 1979, courtesy of Pitman Medical)

Consistent with the evidence for rapid turnover of placental and fetal tissue proteins, there is considerable evidence that total AAN and most individual amino acid concentrations in the placenta and fetal blood are substantially higher than those in maternal blood (Young and Prenton, 1969; Alexander *et al.*, 1970; Young and McFadyen, 1973; Girard *et al.*, 1977; Prior and Christenson, 1977; Phillipps *et al.*, 1978). There is also some information about the extent of net transfer of amino acids to the fetus via the umbilical circulation (Hopkins, McFadyen and Young 1971; Lemons *et al.*, 1976; Smith *et al.*, 1977). It is evident that the majority of amino acids must be actively accumulated from maternal blood by placental tissues against a considerable concentration gradient.

The magnitude of the concentration differences between maternal blood and placental tissue water varies considerably among the various amino acids and is undoubtedly related to their functions in the placenta and to the nature of the particular transport mechanisms involved in their uptake. These amino acid transport systems seem to correspond to those described for other tissues

(Young, 1976, 1979), but it is not entirely clear how their function relates to the rate of net transport from mother to fetus.

There must presumably be differences in permeability or in density of transport sites on maternal and fetal facing surfaces of placental cells but these differences are not necessarily the same for all amino acids. Indeed the amino acids concentrated to the greatest extent by placental tissue, glutamate, aspartate and taurine (Hill and Young, 1973; Phillipps *et al.*, 1978), appear not to be transferred to the fetus in significant quantity and may, in fact, be taken up by the placenta from the fetal circulation (Hopkins, McFadyen and Young, 1971; Lemons *et al.*, 1976; Smith *et al.*, 1977). However, umbilical vein–artery concentration differences indicate that most of the neutral and basic amino acids are transferred to the fetus in amounts commensurate with their incorporation into fetal tissues. Indeed umbilical uptakes of most amino acids, with the exception of the acidic acids, are substantially in excess of estimated carcass protein requirements of the ovine fetus (Lemons *et al.*, 1976) (*Figure 6.4*).

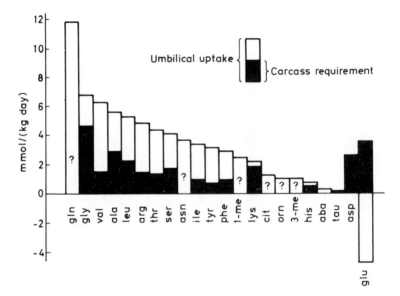

Figure 6.4   A comparison of the estimated carcass requirements of the fetal lamb with the mean umbilical uptake for each amino acid (From Lemons *et al.*, 1976, by permission of the American Society for Clinical Investigation)

Although it may be of considerable importance for the prematurely born human infant, questions about the essentiality of particular amino acids may have little relevance for the fetus *in utero*, as it appears to derive an adequate supply of virtually all essential and non-essential acids from the maternal circulation, under normal nutritional circumstances at least. However, it would be incorrect to regard the placenta solely as an organ for the transfer of amino acids and other nutrients from mother to fetus.

From varied evidence, it is apparent that in metabolic terms the placenta is an extremely active fetal organ, synthesizing many protein and steroid hormones for secretion into both maternal and fetal compartments as well as the wide range of enzymes needed to carry out its many metabolic and nutrient transporting activities. Its role in fetal amino acid metabolism is perhaps better understood if it is regarded as a fetal organ actively metabolizing amino acids and participating in the inter-organ amino acid transfer and interconversions now known to occur during postnatal life (Marliss *et al.*, 1971; Wolff and Bergman, 1972; Bergman and Heitmann, 1978; Heitmann and Bergman, 1978).

As already indicated, determination of umbilical vein–artery differences in fetal lambs shows there is no net transfer from the placenta to the fetus of the acidic amino acids, glutamate, aspartate or taurine, and clearly these must be synthesized within the conceptus. Indeed, Lemons *et al.* (1976) observed a large net transfer of glutamate from the fetus to the placenta in conscious undisturbed sheep, and the negative vein–artery differences in the acute studies of Hopkins, McFadyen and Young (1971) and Smith *et al.* (1977) are consistent with this. There is no evidence that the placenta takes up glutamate from the maternal circulation. Stegink *et al.* (1975) increased maternal glutamate concentration 10–20-fold in pregnant Rhesus monkeys by injecting radioactive glutamate, yet observed no transfer of radioactive glutamate to the fetus.

Pell, Jeacock and Shepherd (1979) also could find no evidence for uterine uptake of glutamate from the maternal circulation in studies of uterine A–V differences in conscious pregnant sheep. Lemons *et al.* (1976) concluded that the fetus excretes unwanted nitrogen by this route, particularly early in gestation. However, they also showed that there is a very large net transfer of glutamine from the placenta to the fetus. Pell, Jeacock and Shepherd (1979) found no significant arteriovenous difference across the uterus suggesting that glutamine is produced in the placenta, although placental uptake of glutamine produced in uterine musculature cannot be completely excluded by such studies. Pell, Jeacock and Shepherd (1979) and Young (1979) suggest that glutamate taken up by the placenta may be released back into the fetus as glutamine since glutamine synthetase has been demonstrated in ovine placenta, but Stegink *et al.* (1975) found litte conversion of radioactive glutamate to glutamine in fetal monkeys even though large amounts of glutamate were administered to the fetus.

It is well established that other peripheral tissues, particularly muscle, in postnatal life take up glutamate and release substantial amounts of glutamine (Marliss *et al.*, 1971; Ruderman and Berger, 1974; Bergman and Heitmann, 1978) derived from catabolism of a wide spectrum of amino acids in addition to glutamate. While the glutamine nitrogen could be derived from nucleotide catabolism it may be that, as in muscle, the production of glutamine by the placenta permits the oxidation of amino acids without release of toxic quantities of ammonia into the circulation, the glutamine produced then being available for use by other fetal tissues such as liver, kidneys and gastro-intestinal tract. However, although these tissues in sheep are known to use glutamine in postnatal life (Wolff, Bergman and Williams, 1972), its use by fetal tissues has not been quantified.

Smith *et al.* (1977) have drawn attention to the possible quantitative importance of amino acids including glutamine, aspartate and serine as precursors for nucleotide synthesis within the conceptus, but requirements for this purpose remain to be determined. The contribution of amino acids to oxidative metabolism within the placenta has also not been determined. Although there is some information about net flux of amino acids into the fetus, there is no comparable information about placental uptake from the maternal circulation. The high rate of protein turnover within the placenta (Young, 1979) suggests that transamination and catabolism could be considerable, but any conclusions about placental amino acid catabolism must await observations on metabolism of labelled amino acids within the placenta itself.

Despite the doubts about the details of amino acid metabolism within the placenta there can be little doubt that there is a net transfer to the fetus via the umbilical vein of most amino acids involved in carcass protein biosynthesis. The acute studies of Smith *et al.* (1977) show that the umbilical venous-arterial concentration differences of most amino acids correspond to their proportions in carcass proteins. The quantitative studies of umbilical uptake by Lemons *et al.* (1976; *Figure 6.4*) show a similar proportionality, but also show that in the sheep, at least, the amounts transferred are substantially in excess of the calculated requirements for fetal growth and are consistent with the earlier observations that urea synthesis in the fetal lamb is considerable (Gresham *et al.*, 1972).

Net fetal uptake figures give little indication about amino acid transactions within the conceptus. As already discussed, the fluxes of the acidic amino acids and glutamine within the conceptus suggest that amino acid interconversion and inter-organ transport may be considerable. Values for alanine and glycine turnover using radioactive tracers in fetal lambs *in utero* (Hatfield *et al.*, 1977; Prior and Christenson, 1977) are more than five times those for umbilical uptake and suggestive of considerable metabolic involvement. However, it is not clear that allowance has been made for transfer of radioactive tracer into the maternal circulation in these studies and there is still a dearth of validated quantitative information about amino acid metabolism within the fetus or about its relation to the rate of fetal growth.

One other aspect of amino acid transactions within the conceptus which also remains poorly understood is the rate at which protein and free amino acids within the amniotic fluid can recirculate through the fetus and contribute to fetal protein synthesis. It is well established that the fetus swallows large volumes of amniotic fluid throughout much of gestation. Furthermore, Pitkin and Reynolds (1975) showed that radioactively labelled protein added to the amniotic fluid in Rhesus monkeys was swallowed and digested in the fetal gut. Subsequently, radioactive amino acids were incorporated into fetal protein. They calculated that this route could possibly provide 10–15 per cent of fetal nitrogen requirements.

Studies in other laboratories (Heller *et al.*, 1975; Renaud *et al.*, 1975) have shown that free amino acids injected into the amniotic fluid disappear from the fluid very rapidly and are probably taken up by the fetus, since the rate of disappearance is negligible following fetal death *in utero* (Heller *et al.*, 1975). It seems likely that small molecules such as amino acids and sugars circulate between amniotic fluid and the fetal vascular compartment via routes other than, or additional to, absorption from the intestinal lumen. At present, however, the physiological and therapeutic importance of these observations remains to be assessed.

Nutritional and endocrine influences on amino acid metabolism in the fetus

With only limited information about amino acid uptake by the fetus, little is known about the way fetal amino acid metabolism responds to altered maternal nutrition, although there is good evidence that maternal undernutrition or fasting result in reduced fetal growth in sheep (Robinson, 1977) and rats (Girard *et al.*, 1977). In sheep, maternal fasting for 2 days does not result in significant changes in maternal or fetal plasma AAN concentrations (Bassett and Madill, 1974). In rats, however, maternal fasting for 4 days initially increased total free amino acid concentrations in both maternal and fetal plasma with the concentrations later returning towards pre-fasting values. In general, the changes in individual amino acids in fetuses reflected in a dampened manner the changes in maternal concentrations, the most striking being a marked decrease in alanine and serine, but a rise in the branched-chain acids, glycine, lysine and glutamine (Girard *et al.*, 1977).

Observations by Simmons *et al.* (1974) on pregnant ewes fasted for 8 days, show that fetal urea production is greatly increased during maternal fasting with the maximum increment occurring after 3—4 days of fasting (*Figure 6.5*). These

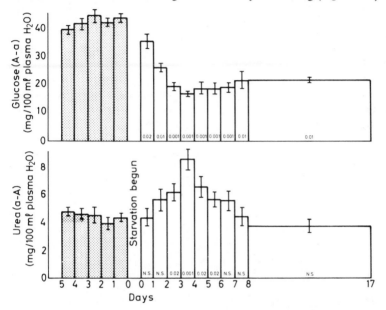

Figure 6.5   Maternal (A), fetal (a) plasma concentration differences for glucose and urea in ewes during 5 days on full feed followed by a 5-day fast. Glucose (A—a) differences are directly related to transplacental glucose uptake and urea (a—A) differences to urea production by the fetus. (From Simmons *et al.*, 1974, courtesy of International Pediatric Research Foundation)

authors calculated that amino acid catabolism at this time could account for most of the fetal oxygen consumption, although in more prolonged fasting other substrates appear to be utilized. Their observations indicate that fetal amino acid catabolism may be substantially increased when maternal nutrition is inadequate. Interestingly, Mellor and Matheson (1979) have recently reported a virtual cessation of fetal growth, as measured by change in fetal crown—rump length, with a day or two of starting severe maternal undernutrition — further evidence that amino acids are rapidly being diverted to catabolic pathways in this situation.

Fetal amino acid catabolism cannot of course be divorced from overall fetal oxidative metabolism and its endocrine control, and it is very likely that the net balance between fetal amino acid anabolism and catabolism reflects, like that of the postnatal animal, the prevailing endocrine environment brought about by the nutritional situation. While there is considerable evidence that fetal and placental endocrine mechanisms manipulate maternal metabolism to maintain nutrient supplies to the fetus, so cushioning it from the immediate effects of maternal nutritional fluctuations (Freinkel and Metzger, 1979), there can be little question that inadequate maternal nutritional status is quickly reflected in the fetus. This should cause little surprise when the magnitude of fetal energy requirements in late pregnancy are related to those of the mother.

Setchell *et al.* (1972) found that fetal glucose utilization might be as high as 70% of the total glucose utilization in the pregnant ewe, and Bergman, Brockman and Kaufman (1974) calculated from a comparison of glucose utilization

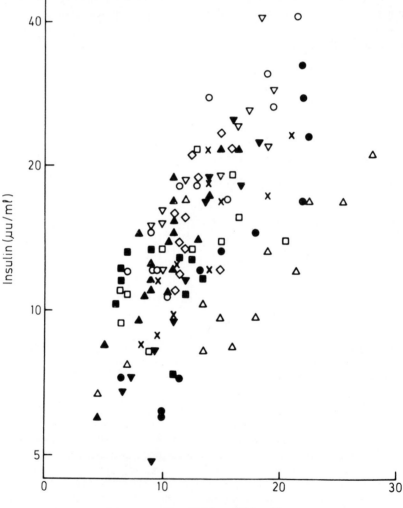

Figure 6.6   The relationship between the plasma concentrations of glucose and insulin in ten chronically cannulated fetal lambs. Samples were obtained at intervals over a period of 72 h. Their mothers were first fed *ad libitum* for 24 h and then fasted for 48 h. (Data from Bassett and Madill, 1974)

in pregnant and non-pregnant sheep that fetal metabolism must account for roughly 40% of the total rate during pregnancy and showed furthermore that glucose turnover progressively decreased during the development of maternal hypoglycaemia. As observed frequently in many species, not just in ruminants, maternal fasting in pregnancy results in a rapid fall in plasma glucose, presumably because of the continuing rapid transfer of glucose across the placenta to the fetus. Nevertheless, the fall in maternal concentration is quickly reflected in falling glucose concentrations in the fetus, leading as Tsoulos *et al.* (1971) have shown to a reduction in the possible contribution of glucose to fetal oxidative metabolism.

A similar situation probably prevails in other species too. Our own studies (Bassett and Madill, 1974) showed clearly that the adaptation to maternal fasting involved fetal endocrine adjustments. In particular, the declining plasma glucose concentration was closely associated with a decreased fetal insulin concentration (*Figure 6.6*), a finding which has recently been confirmed by Phillipps *et al.* (1978). Limited observations (Bassett, 1977) do not indicate, however, that maternal fasting results in any increase in fetal glucagon levels in sheep. However, Girard *et al.* (1977) observed a marked rise in glucagon as well as a fall in insulin in fetal rats when their mothers were starved during late pregnancy. Although the transfer of glucose across the placenta is determined primarily by the concentration in the maternal circulation there can be little question that its rate of utilization within the conceptus is regulated by endocrine mechanisms, particularly by insulin, and not solely by its circulating concentration. Colwill *et al.* (1970) and Shelley (1973, 1975) have shown clearly that insulin infused into fetal lambs will decrease their plasma glucose concentration relative to the maternal concentration, and more recently Simmons *et al.* (1978) have shown that fetal infusion of insulin at a rate of

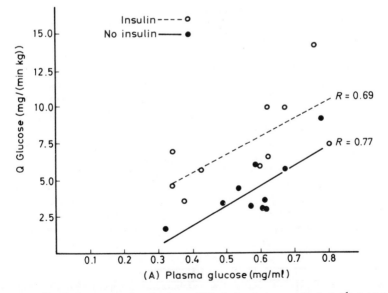

Figure 6.7   Fetal glucose uptake via the umbilical circulation, mg/(min kg⁻¹) fetal body weight), vs. maternal arterial (A) plasma glucose concentration. (From Simmons *et al.*, 1978, courtesy of International Pediatric Research Foundation)

0.24 u/(kg h) increased glucose uptake via the umbilical circulation compared to uninfused controls at all levels of maternal glucose concentration (*Figure 6.7*).

Insulin therefore plays an important role in directing the metabolism of glucose in the fetal lamb *in utero*, and it seems likely that this is true in other species too. However, insulin has a far wider role in integration of anabolic metabolism during postnatal life, acting as an important regulator of amino acid and fat metabolism as well as that of glucose. It seems likely that this is also the case in the fetus.

The role of glucagon remains more equivocal, although as reviewed elsewhere (Bassett, 1977) there is good evidence that it can increase plasma glucose concentration in the fetal lamb presumably by stimulating hepatic glycogenolysis. However, in postnatal life the balance between insulin and glucagon plays a very important role in regulating not only glycogenolysis but also hepatic amino acid catabolism and conversion to glucose via gluconeogenic pathways. Whether the fall in fetal insulin and glucose concentration in the fetus during maternal fasting also permits the fetus to synthesize glucose from amino acids under the stimulatory influence of glucagon, is a matter of considerable speculation. The marked increase in amino acid catabolism seen in fetal lambs during maternal fasting (Simmons *et al.*, 1974) (*Figure 6.5*) has led to speculation that gluconeogenesis may occur in the fetal liver.

There is little question that, in the long-gestation species such as the sheep (Warnes, Seamark and Ballard, 1977), cow (Prior and Scott, 1977) and guinea-pig (Jones and Ashton, 1976), the fetal liver possesses the necessary enzymatic machinery for gluconeogenesis. In short-gestation species, although maturation of the rate-limiting enzymes does not normally occur until birth, prolonged maternal fasting can lead to premature induction of them (Girard *et al.*, 1977). Despite this, direct evidence that fetal gluconeogenesis makes any significant contribution to the glucose homeostasis of the fetus remains hard to find. Girard *et al.* (1977), comparing the specific activity of fetal and maternal glucose during an infusion of $2[H^3]$-glucose to the mother, found a ratio of 1.0 in fed and 0.64 in fasted mothers.

In studies on sheep, Setchell *et al.* (1972) found that fetal glucose specific activity was only 70–80% and fructose specific activity 60% of the maternal value even after 5 h continuous infusion into the mother. Warnes, Seamark and Ballard (1977), however, could find no evidence for gluconeogenesis from lactate in fetal lambs even during stimulation by glucagon infusion, although they did not study undernourished ewes or ewes during prolonged maternal fasting. By contrast, Hodgson and Mellor (1977) have claimed substantial gluconeogenesis occurs in fetal lambs, and Prior and Christenson (1977) reported gluconeogenesis from alanine in fetal lambs *in utero*, although the actual incorporation of radioactivity into glucose was well within the limit for possible contamination of glucose derivatives by radioactive alanine in plasma reported by Hall *et al.* (1977). Clearly, far more studies of fetal gluconeogenesis in situations such as maternal fasting, where it is more likely to occur, are needed before any categorical statements can be made about whether fetal gluconeogenesis from amino acids or other substrates like lactate or glycerol makes a significant contribution to fetal glucose requirements.

Whereas the role of endocrine mechanisms in regulating fetal metabolic responses to maternal undernutrition remains somewhat controversial, there is a much greater unanimity of opinion about the positive role played by insulin in promoting fetal anabolism.

Although it is by no means clear that growth hormone has any major role in determining fetal growth and the exact relation between the various growth factors, such as somatomedins, and fetal growth remains to be defined (Brinsmead and Liggins, 1979), several recent studies and reviews have pointed to the clear positive association between fetal size and indices of fetal insulin secretion (Liggins, 1974; Hill, 1976; Girard *et al.*, 1976). Intrauterine growth retardation

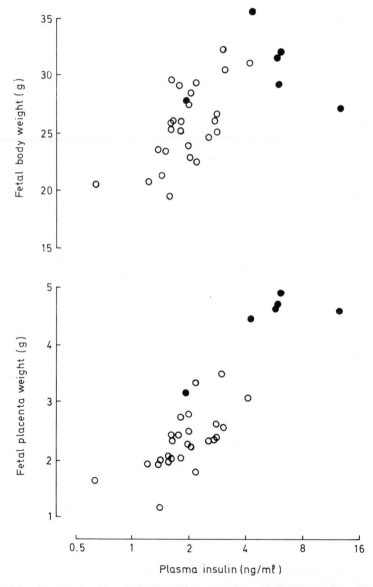

Figure 6.8   The relationship of plasma insulin (plotted on a logarithmic scale) with body weight and fetal placental weight in individual rabbit fetuses from normal litters (○) and litters where the number of fetuses was reduced to two at 9 days gestation (●). Samples were obtained from normal fetuses removed at 28 days gestation and litter-reduced fetuses at 27 or 28 days gestation. (J.M. Fletcher and J.M. Bassett, unpublished observations)

is clearly associated with low insulin and high glucagon concentrations (Lafeber, Jones and Rolfe, 1979) and fetal hypertrophy as in maternal diabetes (Beard and Oakley, 1976), or experimental litter reduction in rabbits (J.M. Fletcher, unpublished observations) with high insulin levels (*Figure 6.8*).

Interestingly, in the fetal rabbits body and placental weights of individual normal fetuses are closely related to their plasma insulin concentration at the time of removal from the uterus. The body weights and insulin concentrations in fetuses where litter size was reduced to only two at 9 days' gestation appear to fit on this relationship. These findings, together with the close relation between weight and plasma insulin in newly born rats (Girard *et al.*, 1976) give strong support to the view that changes in fetal insulin secretion brought about by altered substrate delivery from the mother, as observed in studies on sheep (Bassett and Madill, 1974; Phillipps *et al.*, 1978) play an important role in determining the rate of fetal anabolic metabolism.

Insulin binding sites have been identified in fetal tissues (Kelly *et al.*, 1974) and effects of insulin on amino acid uptake and incorporation into several fetal tissues *in vitro* have been described (Clark, 1971; Fricke and Clark, 1973; Asplund, 1975). Picon (1967) has reported that insulin given *in vivo* to fetal rats significantly increased their body weight and amount of nitrogen compared to control littermates. However, studies of protein turnover in fetal lambs infused with insulin *in utero* have so far failed to demonstrate a stimulatory effect of insulin on fractional protein synthesis rates as determined by radioactive lysine infusion (Young, Horn and Noakes, 1979). These studies did suggest that insulin decreased the rate of tissue protein catabolism, but a great deal more information is needed to establish clearly whether insulin directly regulates anabolic metabolism in the fetus *in utero* or acts indirectly through its effects on the rate of oxidative metabolism.

Obviously, insulin will interact with many other hormones in regulating anabolism in the fetus, and a complete picture of hormonal regulation of nitrogen metabolism in fetal tissues will only emerge when the actions of the newer growth regulatory hormones such as the somatomedins, their antagonists and their interrelations with other hormones have been clarified.

References

ALEXANDER, D.P., BRITTON, H.G., NIXON, D.A. and COX, B.D. (1970). *Biol. Neonate*, **15**, 304–308

ASPLUND, K. (1975). *Hormone Res.*, **6**, 12–19

BASSETT, J.M. (1977). *Ann. Rech. Veterin.*, **8**, 362–373

BASSETT, J.M. and MADILL, D. (1974). *J. Endocr.*, **61**, 465–477

BATTAGLIA, F.C. and MESCHIA, G. (1978). *Physiol. Rev.*, **58**, 499–527

BEARD, R.W. and OAKLEY, N.W. (1976). In *Fetal Physiology and Medicine*, pp. 137–157 (Beard, R.W. and Nathanielsz, P.W., Eds), Saunders, London

BERGMAN, E.N. and HEITMANN, R.N. (1978). *Fedn Proc.*, **37**, 1228–1232

BERGMAN, E.N., BROCKMAN, R.P. and KAUFMAN, C.F. (1974). *Fedn Proc.*, **33**, 1849–1854

BRINSMEAD, M.W. and LIGGINS, G.C. (1979). In *Reviews in Perinatal Medicine*, Vol. 3, pp. 208–242 (Scarpelli, E.M. and Cosmi, E.V., Eds), Raven Press, New York

CHRISTENSON, R.K. and PRIOR, R.L. (1978). *J. Anim. Sci.*, **46**, 189–200

CHRYSTIE, S., HORN, J., SLOAN, I., STERN, M., NOAKES, D. and YOUNG, M. (1977). *Proc. Nutr. Soc.*, **36**, 118A

CLARK, C.M. (1971). *Biol. Neonate*, **19**, 379–388

COLWILL, J.R., DAVIS, J.R., MESCHIA, G., MAKOWSKI, E.L., BECK, P. and BATTAGLIA, F.C. (1970). *Endocrinology*, **87**, 710–715

COMLINE, R.S. and SILVER, M. (1976). *J. Physiol.*, **260**, 571–586

FREINKEL, N. and METZGER, B.E. (1979). In *Pregnancy Metabolism, Diabetes and the Fetus*, pp. 3–23, Ciba Foundation Symposium 63, Exerpta Medica, Amsterdam

FRICKE, R. and CLARK, C.M. (1973). *Am. J. Physiol.*, **224**, 117–121

GIRARD, J., RIEUTORT, M., KERVRAN, A. and JOST, A.(1976). In *Perinatal Medicine*, pp. 197–202 (Rooth, G. and Bratteby, L.E., Eds), Almquist and Wiksell, Stockholm

GIRARD, J., FERRE, P., GILBERT, M., KERVRAN, A., ASSAN, R. and MARLISS, E.B. (1977). *Am. J. Physiol.*, **232**, E456–E463

GRESHAM, E.L., SIMONS, P.S. and BATTAGLIA, F.C.(1971). *J. Pediat.*, **79**, 809–811

GRESHAM, E.L., JAMES, E.J., RAYE, J.R., BATTAGLIA, F.C., MAKOWSKI, E.L. and MESCHIA, G. (1972). *Pediatrics, N.Y.*, **50**, 372–379

HALL, S.E., GOEBEL, R., BARNES, I., HETENYI, G. and BERMAN, M. (1977). *Fedn Proc.*, **36**, 239–244

HAMMOND, J. (1944). *Proc. Nutr. Soc.*, **2**, 8–12

HATFIELD, G.M., JEACOCK, M.K., JOYCE, J. and SHEPHERD, D.A.L.(1977). *Proc. Nutr. Soc.*, **36**, 19A

HEITMANN, R.N. and BERGMAN, E.N. (1978). *Am. J. Physiol.*, **234**, E197–E203

HELLER, L., FAHNENSTICH, E., GERNER, R., HALBERSTADT, E. and ROMER, E. (1975). *Infusiontherapie*, **2**, 57–62

HILL, D.E. (1976). In *Diabetes and other Endocrine Disorders During Pregnancy and in the Newborn*, Progress in Clinical and Biological Research, Vol. 10, pp. 127–139, (Alan, R., Ed.), Liss Inc., New York

HILL, P.M.M. and YOUNG, M. (1973). *J. Physiol.*, **235**, 409–422

HODGSON, J.C. and MELLOR, D.J. (1977). *Proc. Nutr. Soc.*, **36**, 33–40

HOLZMAN, I.R., LEMONS, J.A., MESCHIA, G. and BATTAGLIA, F.C. (1979). *J. Devl Physiol.*, **1**, 137–149

HOPKINS, L., McFADYEN, I.R. and YOUNG, M. (1971). *J. Physiol.*, **215**, 9P–10P

JONES, C.T. and ASHTON, I.K.(1976). *Archs Biochem. Biophys.*, **174**, 506–522

KELLY, P.A., POSNER, B.I., TSUSHIMA, T. and FRIESEN, H.G. (1974). *Endocrinology*, **95**, 532–539

LAFEBER, H.N., JONES, C.T. and ROLPH, T.P. (1979). In *Nutrition and Metabolism of the Fetus and Infant*, pp. 43–62, 5th Nutricia Symposium (Visser, H.K.A., Ed.), Martinus Nijhoff, The Hague

LEMONS, J.A., ADCOCK, E.W., JONES, M.D., NAUGHTON, M.A., MESCHIA, G. and BATTAGLIA, F.C. (1976). *J. Clin. Invest.*, **58**, 1428–1434

LIGGINS, G.C.(1974). In Size at Birth, pp. 165–183, Ciba Foundation Symposium 27, Associated Scientific Publishers, Amsterdam

MARLISS, E.B., AKOI, T.T., POZEFSKY, T., MOST, A.S. and CAHILL, G.F. (1971). *J. Clin. Invest.*, **50**, 814–817

MELLOR, D.J. and MATHESON, I.C. (1979). *Q. Jl Exp. Physiol.*, **64**, 119–131

PELL, J.M., JEACOCK, M.K. and SHEPHERD, D.A.L. (1979). *Proc. Nutr. Soc.*, **38**, 19A

PHILLIPPS, A.F., CARSON, B.S., MESCHIA, G. and BATTAGLIA, F.C. (1978). *Am. J. Physiol.,* **235**, E467—E474

PHILLIPPS, A.F., HOLZMAN, I.R., TENG, C. and BATTAGLIA, F.C. (1978). *Am. J. Obstet. Gynec.,* **131**, 881—887

PICON, L. (1967). *Endocrinology,* **81**, 1419—1421

PITKIN, R.M. and REYNOLDS, W.A. (1975). *Am. J. Osbstet. Gynec.,* **123**, 356—361

PRENTON, M.A. and YOUNG, M. (1969). *J. Obstet. Gynaec. Br. Commonw.,* **76**, 404—411

PRIOR, R.L. and CHRISTENSON, R.K. (1977). *Am. J. Physiol.,* **233**, E462—E468

PRIOR, R.L. and SCOTT, R.A. (1977). *Devl Biol.,* **58**, 384—393

RENAUD, R., KIRSTETTER, L., KOEHL, C., BOOG, G., BRETTES, J.P., SCHUMACHER, J.C., VINCENDON, G., WILLARD, D. and GANDAR, R. (1975). In *Therapy of Feto-placental Insufficiency,* pp. 265—291 (Salvadori, B., Ed.), Springer-Verlag, Berlin

ROBINSON, J.J. (1977). *Proc. Nutr. Soc.,* **36**, 9—16

RUDERMAN, N.B. and BERGER, M. (1974). *J. Biol. Chem.,* **249**, 5500—5506

SETCHELL, B.P., BASSETT, J.M., HINKS, N.T. and GRAHAM, N.McC. (1972). *Q. Jl Exp. Physiol.,* **57**, 257—266

SHELLEY, H.J. (1973). In *Foetal and Neonatal Physiology,* pp. 360—381 (Comline, K.S., Cross, K.W., Dawes, G.S. and Nathanielsz, P.W., Eds), Cambridge University Press

SHELLEY, H.J. (1975). *J. Physiol.,* **252**, 66—67P

SILVER, M. (1976). In *Foetal Physiology and Medicine,* pp. 173—193 (Beard, R.W. and Nathanielsz, P.W., Eds), Saunders, London

SILVER, M. and COMLINE, R.S. (1975). *J. Reprod. Fert.,* Suppl. **23**, 589—594

SIMMONS, M.A., MESCHIA, G., MAKOWSKI, E.L. and BATTAGLIA, F.C. (1974). *Pediat. Res.,* **8**, 830—836

SIMMONS, M.A., JONES, M.D., BATTAGLIA, F.C. and MESCHIA, G. (1978). *Pediat. Res.,* **12**, 90—92

SMITH, R.M., JARRETT, I.G., KING, R.A. and RUSSELL, G.R. (1977). *Biology Neonate,* **31**, 305—310

SOLTEZ, G., JOYCE, J. and YOUNG, M. (1973). *Biol. Neonate,* **23**, 139—148

STEGINK, L.D., PITKIN, R.M., REYNOLDS, W.A., FILER, L.J., BOAZ, D.P. and BRUMMEL, M.C. (1975). *Am. J. Obstet. Gynec.,* **122**, 70—78

TSOULOS, N.G., COLWILL, J.R., BATTAGLIA, F.C., MAKOWSKI, E.L. and MESCHIA, G. (1971). *Am. J. Physiol.,* **221**, 234—237

WARNES, D.M., SEAMARK, R.F. and BALLARD, F.J. (1977). *Biochem. J.,* **162**, 617—634

WOLFF, J.E. and BERGMAN, E.N. (1972). *Am. J. Physiol.,* **223**, 447—454

WOLFF, J.E., BERGMAN, E.N. and WILLIAMS, H.H. (1972). *Am. J. Physiol.,* **223**, 438—446

YOUNG, M. (1976). In *Fetal Physiology and Medicine,* pp. 59—79 (Beard, R.W. and Nathanielsz, P.W., Eds), Saunders, London

YOUNG, M. (1979). In *Placental Transfer,* pp. 142—158 (Chamberlain, G. and Wilkinson, A., Eds), Pitman Medical, London

YOUNG, M. and McFADYEN, I.R. (1973). *J. Perinat. Med.,* **1**, 174—182

YOUNG, M. and PRENTON, M.A. (1969). *J. Obstet. Gynaec. Br. Commonw.,* **76**, 333—344

YOUNG, M. and WIDDOWSON, E.M. (1975). *Biol. Neonate,* **27**, 184—191

YOUNG, M., HORN, J. and NOAKES, D.L. (1979). In *Nutrition and Metabolism of the Fetus and Infant,* pp. 19–27, 5th Nutricia Symposium (Visser, H.K.A., Ed.), Martinus Nijhoff, The Hague

7

METABOLISM IN MUSCLE

D.B. LINDSAY
Department of Biochemistry, Agricultural Research Council, Institute of Animal Physiology, Cambridge

and

P.J. BUTTERY
Department of Applied Biochemistry and Nutrition, University of Nottingham School of Agriculture

Summary

Skeletal muscle is a major reserve of protein in large animals as much as in laboratory animals. Even at nitrogenous equilibrium there is a large turnover of skeletal muscle protein. Although this is not accompanied by net changes in uptake or output of plasma amino acids by muscle, there is evidence of substantial exchange between plasma and muscle free amino acids. In depletion of muscle protein a net output of amino acids is measurable. Most of the amino-N output is in the form of alanine, glutamine and, less consistently, glycine. Both glucose and amino acids have been proposed as the carbon source of alanine and glutamine, and evidence supporting both these propositions will be discussed. On the whole the evidence favours glucose-C as the major source of alanine and some amino acids as the source of glutamine-C. Only a limited number of amino acids are oxidized in muscle. In rats, leucine is the major amino acid oxidized in muscle, but in ruminants the evidence in support of this concept is more equivocal.

Introduction

The function of muscle with which we are most familiar is its capacity to contract — to fulfil both locomotory and postural needs. In this context, the term 'metabolism in muscle' is concerned with the energy sources used for contraction either for maximum energy output or for more moderate but sustained energy production. Muscle has, however, a further function. It contains the largest amount of protein in an animal and, perhaps adventitiously, because animals appear to have no specific protein reserves it constitutes the major site of mobilizable protein.

Mobilization of muscle protein

As an example of the mobilization of protein in large animals, we may consider a study by Butterfield (1966). Four groups of steers were slaughtered, in order to determine their body composition. A control group was killed at about 20 months of age, and for comparison with this, a second group was killed about a year later, when they had gained an additional 104 kg; a third group was killed following underfeeding for about three months, during which time they lost 57 kg; and a fourth group was also underfed for a similar period, but this was followed by a period of liberal refeeding when 'catch-up' growth occurred. Of the 104 kg gained by the second group compared to control group, about 60% was in the carcass, and nearly one-third of the gain in this fraction was in muscle (*Table 7.1*). Of the 57 kg of weight loss in underfeeding (third group compared

Table 7.1. CHANGES IN SOME BODY COMPONENTS OF STEERS SUBJECTED TO VARYING NUTRITIONAL TREATMENTS (RESULTS FROM BUTTERFIELD, 1966)

| | *Changes in weight* (kg) | | regrowth |
	normal fed	underfed	(after underfeeding)
Body weight	+104	−57	+125
Carcass weight*	+ 61	−42	+ 52
Muscle	+ 17	−11	+ 19

* 'Carcass' is not a well-defined term, but here probably means the flayed headless body with viscera and intestinal fat removed.

to control group), more than 70% was loss from the carcass, of which again more than one-third was of muscle; and finally during regrowth (fourth group compared to second group) although less than half the gain was in the carcass, again about one-third of this gain was in muscle.

Although these results show clearly that a substantial amount of muscle protein can be mobilized, they do not show how large this is in relation to the total amounts of protein that may be mobilized. This is shown more clearly in a study of Lodge and Heaney (1973). The body composition of a group of pregnant ewes (19) at term was compared with that of a group of non-pregnant ewes which had been of roughly similar weight at the beginning of pregnancy and were given a similar diet. As *Table 7.2* shows, both groups lost weight on this diet, although this was much greater for the non-pregnant than the pregnant group. Of much greater interest in the present context is the indication that

Table 7.2. COMPARISON OF WEIGHT CHANGES IN PREGNANT AND NON-PREGNANT SHEEP GIVEN SIMILAR FEED (8.5–10.5 MJ APPARENTLY DIGESTIBLE ENERGY PER DAY). (RESULTS FROM LODGE AND HEANEY, 1973)

	Pregnant	*Non-pregnant*	*Difference*
Initial body wt (kg)	63.6	63.0	−0.6
Final body wt (kg)	61.0	54.9	−6.1
	Protein content of tissues (g)		
Carcass	3882	4566	− 684
Skin + viscera	1245	1358	− 113
Liver	130	112	+ 18
Blood	369	392	− 23
Fleece	639	818	− 179
Udder	218	19	+ 199
Uterus	130	11	+ 119
Fetus + placenta	*c.*1500	−	−

there was a substantial redistribution of tissue proteins in the pregnant group. Of the tissue protein that decreased in amount, about 70% was carcass protein; most of this may be assumed to be in muscle. A similar finding in sheep subjected to a smaller degree of underfeeding has been reported by Robinson *et al.* (1978). In a group of sheep receiving a basal (i.e. without taking into account the additional requirements for fetal development) energy intake of about 70% that of a control group, there was an apparent loss of protein from a number of tissues of about 500 g, over the whole period of pregnancy. About 50—70% of this protein mobilized was of carcass protein.

Although we are not aware of comparable measurements in lactation, there is some evidence that in cattle in the early stage of lactation, muscle protein is mobilized (Roberts *et al.*, 1979) and since the liver and the gastro-intestinal tract probably increase in size in early lactation, it is likely that in this case also most of the reserve protein is derived from muscle.

Protein synthesis

In mature non-productive animals in nitrogen equilibrium, although there is no net loss of muscle protein, we know that there is a substantial rate of protein degradation, this being matched by a corresponding rate of protein synthesis (*Table 7.3*). Although most measurements of muscle protein synthesis have been made in laboratory animals, there are now several estimates of rates of synthesis (and thus of degradation) in a number of farm animals.

Several qualifications need to be made. First, no attempt has been made to discriminate between results using different experimental techniques, although

Table 7.3. RATES OF PROTEIN SYNTHESIS IN MUSCLE OF VARIOUS SPECIES. MUSCLE IS ASSUMED TO BE 20% PROTEIN AND REPRESENT 40% BODY WEIGHT IN NON-RUMINANTS, 30% in RUMINANTS. VALUES ARE CALCULATED FROM MILLWARD *et al.* (1975), LAURENT *et al.* (1978), GARLICK, BURK AND SWICK (1976), EDMUNDS, BUTTERY AND FISHER (1978), BUTTERY *et al.* (1975), SOLTESZ, JOYCE AND YOUNG (1973); LOBLEY, REEDS AND PENNIE (1978)

	Body weight (kg)	Rate of protein synthesis* (g protein/day per kg muscle$^{0.75}$)
Rat	0.04	20
	0.12	11
	0.5	7
Chicken	2	4—32
Pig	20	20
	70	19
Sheep	4	52
	50	7
Cow	250	12
	630	7

* Rates of protein synthesis have been expressed per kg muscle$^{0.75}$ because this has been found the most suitable way of comparing rates of energy-yielding processes between species of widely different body weight.

there is evidence that differences can be obtained depending on the assumptions of a particular technique. Secondly, although the results have been shown as related to body weight or, more precisely, to metabolic muscle weight $(M)^{0.75}$, a more important variable is the age of the animals. This is known to be of importance at least in laboratory rats. In some of the studies, the age of the animals was not given, although for a particular species the larger animals were always older. Thus in large animals, as in laboratory rats, the rate of synthesis of muscle protein is substantially greater in younger animals even after allowing for differences in 'metabolic' body weight.

Exchange of amino acids across muscle

At least in part, the synthesis of protein in muscle is known to involve the plasma amino acids. Since, however, the rates of synthesis and of degradation are matched, we might not perhaps expect to be able to measure any significant uptake of blood amino acids by muscle. This was in fact observed by Lindsay, Steel and Buttery (1977a, 1977b) who were unable to demonstrate any significant uptake or release of plasma amino acids in studies in sheep fed hourly and in which arteriovenous differences for amino acids were measured, the venous blood being taken from a site draining the hind-limb muscles (biceps femoris and semimembranosis). However, when [^{14}C]-alanine was infused, we were able to demonstrate a marked extraction (about 25%) of alanine. It is thus clear that there is in fact a marked uptake, matched by a corresponding output of amino acid. More recently, we have observed a similar change with respect to [^{14}C]-threonine (Morton, Lindsay and Buttery, unpublished observations).

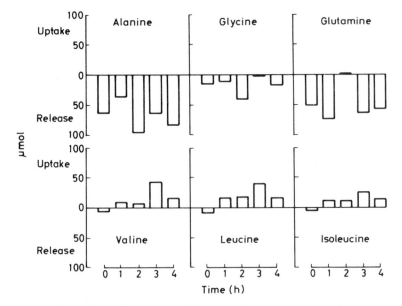

Figure 7.1   Whole-blood arteriovenous differences for some amino acids across human forearm muscle (deep vein) a meal of 200 g protein was taken at zero time. (Values taken from Aoki *et al.*, 1976)

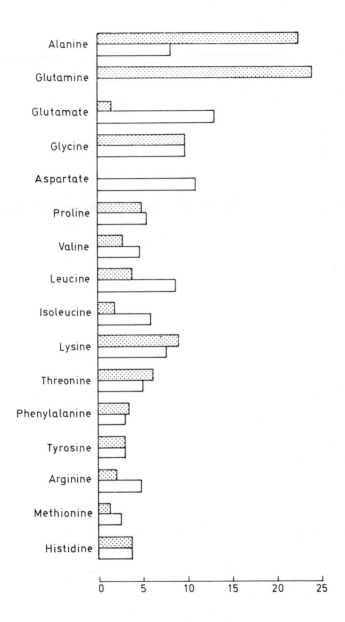

Figure 7.2   Comparison of the percentage output of individual amino acids (mol/100 mol total output) in perfusate during perfusion of the isolated rat hindquarters (results of Ruderman and Berger, 1974) with the amino acid compositions (mmol/100 mmol) of rat muscle (diaphragm) protein. These latter values are the average of values by Low (cited by Odessey, Khairallah and Goldberg, 1974) and Change and Goldberg (1978b). The values for glutamic and aspartic acids, in the muscle protein composition, include any amide present. Stippled bars, muscle output; unshaded bars, protein composition

In non-ruminant animals, food intake and the absorption of nutrients is normally intermittent. In rats, it has been shown that protein synthesis varies over 24 hours, in relation to meals (Garlick, Millward and James, 1973). Thus, at nitrogenous equilibrium a relatively greater rate of protein synthesis compared to degradation after a meal is matched by a greater rate of degradation, relative to synthesis in the post-absorptive period. In these circumstances one might expect to be able to detect a net movement of amino acids following a meal.

In *Figure 7.1*, which is compiled from results obtained by Aoki *et al.* (1976), the arteriovenous difference across human forearm muscle is shown in subjects who were given a 200 g protein meal which was eaten within about 10 min following an overnight fast. There is an appreciable, and statistically significant, uptake of all the branched-chain amino acids. For other indispensable amino acids, the result was less clear — there was no appreciable uptake of phenylalanine, but probably some uptake of lysine. In striking contrast, there is a marked and sustained output of alanine, and to a lesser degree glutamine and glycine. In the post-absorptive state, or following a moderate fast, there is in all species studied (rat, man, sheep, steer), a readily measurable outflow of amino-N from muscle. The outflow of amino acids, however, is not in the proportions that they occur in muscle protein. Ruderman and Berger (1974) investigated the appearance of amino acids in the perfusate of the isolated perfused rat hind-limb, the rats being previously starved about 48 h (*Figure 7.2*). The output for each amino acid is expressed as a percentage of the total output. The amino acid composition of muscle protein is expressed in a comparable way.

The proportions differ in several important respects. There is a much larger output in the perfusate of alanine and glutamine than occurs in muscle protein, whereas there is a much smaller output of glutamic and aspartic acid than might be expected on the same basis. The major differences in the essential amino acids are the much reduced outputs of the branched-chain amino acids.

Oxidation of amino acids

Several factors might explain the reduced output of the branched-chain amino acids, aspartic and glutamic acids. One possibility is that they are oxidized in muscle.

A study of the probable sites of oxidation of most of the amino acids was made by Miller (1962). He used rat liver perfusions and also eviscerated rats, in some of which the livers had also been removed, to which was added a number of ^{14}C-labelled amino acids, and determined the recovery of $^{14}CO_2$. The results were not quite unambiguous, since to some amino acids he added variable amounts of 'carrier' amino acid, and sometimes this was the DL-form of the amino acid. Nevertheless, his results (*Table 7.4*) showed that for purposes of oxidation the amino acids fell into three classes. Most of the essential amino acids were readily oxidized in the liver, but in eviscerate rats in which the liver had been removed there was very little oxidation. However, the branched-chain amino acids were only oxidized to a rather small extent by the perfused liver, but in contrast were readily oxidized by eviscerate rats whether or not the liver was present. Finally, the non-essential amino acids (glutamic, aspartic, glycine, proline) were apparently oxidized equally readily both in the perfused liver and in the eviscerate animal, whether or not the liver was present.

Table 7.4 SITES OF $^{14}CO_2$ PRODUCTION FROM ^{14}C AMINO ACIDS IN THE RAT. VALUES ARE FOR (a) PERCENTAGE DOSE ADDED, APPEARING AS CO_2; (b)–(e) PERCENTAGE DOSE TAKEN UP BY TISSUE WHICH APPEARS AS CO_2. RESULTS FROM (a) MILLER (1962); (b) MANCHESTER (1965); (c) GOLDBERG AND ODESSEY (1972); (d) BEATTY *et al.* (1974); (e) CHANG AND GOLDBERG (1978a)

^{14}C amino acid	Eviscerate rat (a)		Isolated perfused liver (a)	Isolated diaphragm muscle		Isolated red + white skeletal muscle	
	liver present	liver absent		(b)	(c)	(d)	(e)
Leucine	14	13	7	25–50	38	50	29
Isoleucine	18	28	4	33–58	36	–	–
Valine	31	19	–	18–31	23	–	22
Threonine	–	1.2	66	–	0	–	0
Lysine	–	0.4	25	–	0	–	0
Phenylalanine	28	0.1	26	<0.5	0	–	0
Tyrosine	–	–	–	–	0	–	0
Histidine	–	0.4	37	–	0	–	–
Methionine	22	2	48	–	0	–	–
Arginine	–	3	16	–	0	–	–
Tryptophan	21	1.6	17	–	0.6	–	–
Glutamic	63	40	48	45–50	47	29–32	–
Aspartic	63	53	–	–	52	–	7–8
Alanine	–	–	–	41–60	57	15–17	–
Proline	–	17	–	1–2	2–3	–	–
Glycine	19	8	48	2–3	3	0.8	0
Serine	–	–	–	–	4	–	–

It thus appeared that whereas most amino acids could be oxidized in the liver, the branched-chain and non-essential amino acids could be oxidized at least by some peripheral tissues. Evidence supporting these findings has been obtained from studies with isolated rat muscle by Manchester (1965), Goldberg and Odessy (1972), Beatty *et al.* (1974), and Chang and Goldberg (1978a) (*Table 7.4*). Whereas there was ready production of $^{14}CO_2$ from some ^{14}C-amino acids (leucine, isoleucine, valine, alanine, aspartic and glutamic acids), there was no measurable production from phenylalanine, tyrosine, threonine, lysine, histidine, methionine or serine. Only for glycine, serine and proline are the results at all equivocal. If they are oxidized in muscle, the rate must be quite low. It is thus clear that muscle could constitute a large part of the 'peripheral tissue' that was observed by Miller to have the capacity to oxidize a number of amino acids.

Fate of amino groups produced in catabolism of amino acids in muscle

Thus the smaller output from the muscle of fasting animals, of aspartic glutamic and the branched-chain relative to the other amino acids could be due, at least in part, to their oxidation within muscle. The fate of the nitrogen has, however, still to be explained. This can be accounted for by the mechanism shown in *Figure 7.3*. Branched-chain transaminases transfer an amino group from the branched-chain acids to 2-oxo-glutarate; the glutamate formed then reacts with pyruvate, through the action of alanine aminotransferase to produce alanine, re-forming the 2-oxoglutarate. Free ammonia, which can be formed in muscle during contraction, or through the 'purine nucleotide' cycle (Lowenstein, 1972), would also remove glutamate in part, by the formation of glutamine. Thus this cycle can account for the large release of alanine by muscle, the amino group being derived from the branched-chain amino acids which were oxidized, while the carbon is derived from glucose. Chang and Goldberg (1978a) have carried out balance studies using the isolated rat diaphragm, as shown in *Table 7.5*. The

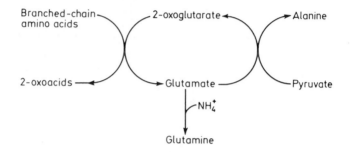

Figure 7.3   Scheme for the production of alanine and glutamine in skeletal muscle as a consequence of the catabolism of the branched-chain amino acids (leucine, isoleucine and valine). The corresponding 2-oxoacids are 2-oxoisocaproic, 2-oxo-3-methyl valeric and 2-oxo-isovaleric

Table 7.5. COMPARISON OF THE EXTENT OF CATABOLISM OF BRANCHED-CHAIN AMINO ACIDS AND THE INCREASE IN MEDIUM OF ALANINE, GLUTAMINE AND GLUTAMATE. RAT DIAPHRAGMS WERE INCUBATED WITH 0.5 mM BRANCHED-CHAIN ACID. VALUES ARE IN nmol/mg TISSUE PER 2 h. (RESULTS FROM CHANG AND GOLDBERG, 1978a)

| ^{14}C-amino acid | $^{14}CO_2$ recovered | ^{14}C in 2 oxoacid | Sum | Increase in medium | | | Sum |
				alanine	glutamine	glutamate	
Leucine	5.2	0.6	5.8	1.2	3.2	0.6	5.0
Isoleucine	4.7	1.2	5.9	1.1	3.8	1.0	5.9
Valine	2.7	1.2	3.9	0.8	2.5	0.3	3.6

diaphragms were incubated with ^{14}C-labelled branched-chain amino acids. In the catabolism of these acids, the first product is a 2-oxoacid, and this is then decarboxylated, releasing $^{14}CO_2$. Thus, the sum of these two products will give the extent of total transamination. For each amino acid there is reasonable agreement with the sum of the gains in the medium during incubation, of alanine, glutamic acid and glutamine.

Although this scheme is now fairly generally accepted, a number of workers have added further extensions to the scheme. In particular, they have suggested that a large part of the pyruvate used in the formation of alanine is derived from amino acid carbon — the branched-chain and other amino acids (Goldstein and Newsholme, 1976; Garber, Karl and Kipnis, 1976a, 1976b; Snell and Duff, 1977). This extension has been strongly disputed by Chang and Goldberg (1978a, 1978b).

The carbon source of the alanine and glutamine released from muscle

The main arguments relating to the carbon source of alanine (and glutamine) may be conveniently listed under the following headings.

EFFECT OF GLUCOSE UPTAKE

Garber, Karl and Kipnis (1976a) emphasized that in their studies with the isolated rat epitrochlaris muscle, while alanine and glutamine output declined over the period of incubation, the uptake of glucose and output of pyruvate and lactate remained constant. They also observed that while insulin increased glucose uptake four-fold, there was no effect on alanine and glutamine release. Grubb (1976) also observed that in the perfused rat hind-limb there was no correlation between glucose uptake and alanine output, whereas when insulin was added there was a small but significant decrease in alanine output, in addition to the expected increase in glucose uptake (*Figure 7.4*). Pozefsky and Tancredi

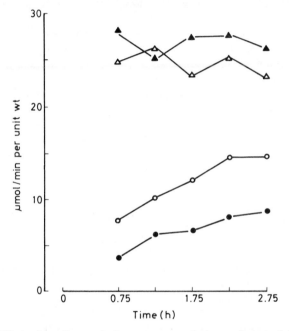

Figure 7.4   Effect of insulin on alanine appearance during perfusion of the isolated rat hindquarters. ▲ ● are the values in absence, and △ ○ the values in presence, of insulin (about 7 mU/ml). ▲ △ represent total alanine output as estimated enzymatically; ● ○ represent alanine derived from glucose (as estimated from ^{14}C appearing in alanine when [^{14}C]-glucose is present in the medium). (Results of Grubb, 1976)

(1972), however, found that intra-arterial infusion of pyruvate into human fore-arm resulted in a significant (about 7%) increase in alanine output. Moreover, Chang and Goldberg (1978a) were able to show that, in experiments with the rat diaphragm, glucose did increase alanine output, the effect being greatest when leucine or a mixture of amino acids (at plasma concentrations) was also added.

It is perhaps understandable that rather varied effects should be obtained. Glucose (or pyruvate) is only likely to be an effective stimulus when the supply of NH_2 groups is not limiting. Moreover, it is quite possible that glycogen may be a more immediate source of pyruvate, for transamination to alanine. Thus a lack of correlation between glucose or pyruvate uptake and alanine output cannot be interpreted as implying glucose has no precursor. Grubb (1974) (see later) estimated that only about 3% of glucose taken up by the hind-limb was actually converted to alanine.

EFFECT OF ADDITION OF AMINO ACIDS

Odessey, Khairallah and Goldberg (1974) found that addition of the branched-chain, but not other, amino acids (at even up to five times their plasma concentrations) increased the output of alanine by the rat diaphragm. This effect of branched-chain acids was confirmed by Chang and Goldberg (1978a) (see *Table 7.5*). In contrast to these results, Garber, Karl and Kipnis (1976b) found that alanine output by epitrochlaris muscle was stimulated by a number of amino acids (aspartate, cysteine, methionine, threonine, serine and glycine) as well as the branched-chain amino acids. Aspartate, cysteine, methionine, tyrosine, lysine and the branched-chain amino acids were found to stimulate glutamine release. Only ornithine, arginine and tryptophan had no effect on release of either alanine or glutamine. The concentration of the amino acids used for these experiments was high (about 10 mM). It was claimed that much lower concentrations (below 1 mM) were effective also, but only aspartate, leucine and methionine were used in these studies. It should be noted that even at 1 mM, the concentration of methionine used is between 20 and 50 times the normal plasma concentration.

Ruderman and Berger (1974) have also demonstrated, using the isolated perfused rat hindquarters, that 10 mM leucine, valine, lysine, glycine, histidine or proline increase the rate of output of alanine and glutamine. It may be argued that it is the supply of amino groups that is commonly limiting for alanine output, and thus the amino acid addition simply makes more amino groups available. However, several amino acids (e.g. threonine and lysine) do not participate in transamination reactions and thus it is unlikely that they at least are effective by stimulating transamination. Moreover, the use of amino acid concentrations which are at least ten-fold above normal plasma concentrations may well stimulate pathways which are normally only of minor importance. It should also be borne in mind that amino acids at high concentration may well affect the breakdown of muscle glycogen.

Goldstein and Newsholme (1976), who also demonstrated the stimulating effect of isoleucine and glutamate on alanine output in the isolated rat diaphragm, argued that the increased alanine could not be accounted for in terms of glycogen breakdown. The additional alanine output, however, in their experiments was only about 1 μmol/g in the period of incubation. In fact the mean glycogen content was about 1 μmol/g less in glutamate treated compared with control diaphragms. Because the standard error was about $\pm$ 1 μmol/g, this change was not significant, but one cannot assume that the change did not therefore occur. Finally, it is to be noted that leucine is as effective as any amino acid in stimulating alanine production, yet there is no known mechanism by which such a ketogenic amino acid could yield a net synthesis of alanine. Garber, Karl and Kipnis (1976b) suggested some unknown pathway might be involved, but it is difficult to accept this without much stronger evidence than has yet been presented.

In a study by Snell and Duff (1977), in which rat diaphragms were incubated with 3 mM leucine, or valine, the increased alanine output (compared with a control) with both amino acids was almost balanced with decreased production of pyruvate and lactate. Only for glutamate was there an appreciable net increase in 'alanine + pyruvate + lactate'.

EFFECTS OF INHIBITORS

The effects of inhibitors are often difficult to interpret because their specificity is rarely absolute. Thus, iodoacetic acid has been used by Garber, Karl and Kipnis (1976b) in studies with isolated epitrochlaris, to estimate the role of glucose in determining alanine output. Glycolysis was certainly inhibited (lactate production fell to about 20—36% of control value) while alanine and, to a lesser extent, glutamine and glutamic output was increased (by 25—45%). Garber, Karl and Kipnis concluded that the alanine carbon could not be derived from glycolysis. In contrast, Chang and Goldberg (1978a), who found that iodoacetate *decreased* alanine output by 20%, also observed that iodoacetate inhibited the transamination of branched-chain amino acids, and increased proteolysis (by about 70%), and concluded, therefore, the effects of iodoacetate were uninterpretable. The different action of iodoacetate on alanine output may have been due to differences in concentration of the inhibitor used (4—10 times greater by Chang and Goldberg) but their general point seems a valid one. Sodium fluoride has similar multiple effects to that of iodoacetate, and for this reason Chang and Goldberg preferred to use 2-deoxyglucose as a glycolytic inhibitor. This, at 1 mM concentration, produced a 50% inhibition of alanine output yet had no effect on branched-chain acid transamination; it did decrease lactate formation to 28% of control values, while proteolysis (estimated from the tyrosine output) was increased by about 16%. This finding would thus be consistent with the view that a large fraction of alanine-C was derived by glycolysis, with an increased availability of amino acids from proteolysis being ineffective in stimulating alanine production.

Gerber, Karl and Kipnis (1976b) also studied the effect of amino-oxyacetate, a transaminase inhibitor, on alanine output. The additional output of alanine induced by leucine was decreased to about 35%. This suggests that transamination is involved in the production of much of the alanine released, but offers no information as to the source of the alanine carbon. Ruderman and Berger (1974) found that cycloserine (which inhibits alanine amino transferase) decreased the leucine-induced stimulation of alanine output in the isolated perfused rat hindlimb to about 40% of control. The inhibitor had much less effect on the basal (unstimulated) alanine output, and it may be that an appreciable fraction of this is derived from alanine produced by proteolysis. Although some alanine must be derived *de novo*, it is sometimes forgotten that perhaps 40% of alanine could be derived from proteolysis.

The only inhibitor affecting amino acid metabolism which offers direct information on the possible contribution of amino acid carbon to alanine is that studied by Snell and Duff (1977). These authors observed that 3-mercaptopicolinate (1 mM), which is an inhibitor of phosphoenolpyruvate carboxykinase, decreased alanine output from the isolated rat diaphragm. The authors claimed that the effect was seen with valine-stimulated but not leucine-stimulated alanine output, and since valine but not leucine is glycogenic, this suggested that the effect was related to the production of pyruvate for alanine formation. Their results may, however, be interpreted differently. In the *absence* of a branched-chain amino acid, the inhibitor significantly decreased alanine output. Thus, the effect may be due to a non-specific effect of the inhibitor in reducing proteo-lysis. If the difference in control values is ignored, then the effect of leucine-stimulated alanine output is very similar to that of valine-stimulated output, and the effect cannot be attributed to a 'glycogenic' effect of valine.

STUDIES WITH [^{14}C]-GLUCOSE

Odessey, Khairallah and Goldberg (1974) determined directly the amount of alanine that might be derived from glucose in the isolated rat diaphragm, by incubating with [^{14}C]-glucose. ^{14}C from glucose was recovered only in alanine, glutamate, glutamine and asparagine of the amino acids. In fasted rats, it was calculated that about 60% of the alanine released was derived from glucose, whereas when alanine output was stimulated by the addition of the branched-chain amino acids (leucine, isoleucine and valine, each at 0.5 mM) the proportion of alanine derived from glucose was actually increased – to about 69%. It was thus calculated that the additional alanine produced in response to the branched-chain amino acids was nearly 80% derived from glucose. The proportions were rather smaller in comparable studies using diaphragms derived from fed rats. The basal alanine output was calculated to be about 29% derived from glucose, that in the presence of the branched-chain amino acids about 46% derived from glucose, and thus the additional alanine generated by the added amino acids was about 60% derived from glucose.

Grubb (1976) has also studied the origin of alanine carbon. In her study, she used the isolated perfused rat hind-limb, to which [^{14}C]-glucose was added. Incorporation of [^{14}C]-glucose from glucose into alanine increased steadily during the perfusion, being almost double after 3 h that observed after 1 h. This rather suggests that incorporation into alanine, at least in part, occurs via some intermediate that is turning over relatively slowly – muscle glycogen would seem a very likely possibility.

As *Figure 7.4* illustrates, the total output of alanine was only slightly (but significantly) reduced by the addition of insulin. The amount derived from glucose (as indicated by ^{14}C incorporation), was, however, greatly increased by insulin. It may be estimated from these results that the amount of alanine derived from glucose was about 33% in the absence, and about 51% in the presence, of insluin (at plasma concentration about 7 μU/ml). The rats used for these experiments were not fasted up to the surgery; thus these results are in good quantitative agreement with those of Odessey, Khairallah and Goldberg (1974), obtained with the rat diaphragm.

EXPERIMENTS WITH ^{14}C-LABELLED AMINO ACIDS

Studies with ^{14}C-labelled amino acids comparable to those made with [^{14}C]-glucose present the following difficulties:

(1) Since it is being postulated that *endogenous* amino acids are alanine/gluta-mine precursors, it is necessary to determine the specific activity of the relevant amino acid in the precursor pool. The specific activity of amino acid

in the medium will certainly not be satisfactory and even the mean intra-cellular specific activity may not be adequate.

(2) Since several amino acids are supposed to be potential precursors, the contribution of a particular amino acid may at best be quite small.

(3) Quantitative comparison is further complicated by the reactions of the tri-carboxylic cycle. This point merits a more extensive consideration.

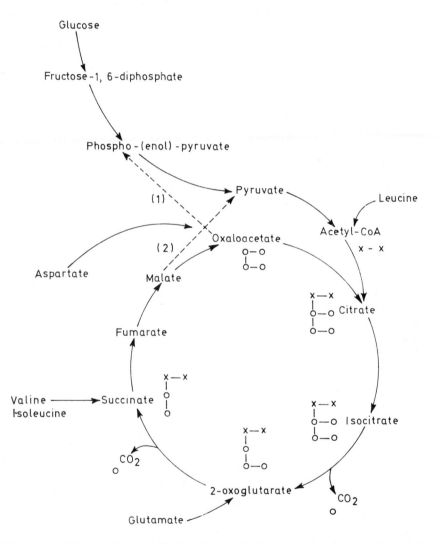

Figure 7.5   Pathways for the oxidation of valine, leucine, isoleucine, aspartic and glutamic acids in skeletal muscle. The reactions (1) and (2) represent those catalysed by phosphoenol-pyruvate carboxykinase and 'malic' (NADP:malate dehydrogenase) enzymes, respectively. The presence of these enzymes in skeletal muscle has been demonstrated by Nolte, Brdiczka and Pette (1972). – X – represents carbons originating in acetyl-CoA, and – O – represents carbons originating in succinic, malic or oxaloacetic acids. CO_2 is derived from these 4-carbon compounds and the carbon of acetyl-CoA is converted to CO_2 only after one or more turns of the cycle

It is generally accepted that alanine synthesized *de novo* will be derived from pyruvate. Thus, estimation of the contribution of a particular amino acid is obtained from the ratio of the specific activities of intracellular pyruvate and the amino acid precursor. However, the conversion of amino acids such as aspartate, glutamate, isoleucine or valine to pyruvate, through known pathways, requires that they first be converted to tricarboxylic acid cycle intermediates (*Figure 7.5*). As a consequence of the exchange of carbons, as shown in the illustration, the specific activity of these intermediates will be decreased due to dilution of carbon from acetyl-CoA and the contribution of the amino acids will be under-estimated. This problem of 'exchange' or 'crossover' is exactly comparable to that complicating studies of gluconeogenesis in the liver (Weinman, Strisower and Chaikoff, 1958).

There are some additional features of amino acid catabolism illustrated in *Figure 7.5*. Since $^{14}CO_2$ produced in the cycle is derived from the 4-carbon inter-mediates, production of $^{14}CO_2$ from amino acids, although necessary, is not a sufficient condition to demonstrate that there is oxidation of an amino acid. Only acetyl-CoA suffers a net oxidation, and thus for an amino acid to be oxi-dized through the tricarboxylic acid cycle it is necessary that a pathway to acetyl-CoA be demonstrated. For leucine and two carbons of isoleucine this occurs via acetoacetyl-CoA, but for the remaining carbons of isoleucine, and for valine, aspartate and glutamate, this appears to involve prior conversion to pyru-vate, which is known to be in part converted to acetyl-CoA in muscle and then oxidized to CO_2. One consequence would appear to be that *if* amino acid carbon is transferred to pyruvate then some of the carbon will pass to acetyl CoA and must appear as CO_2. As a corollary, only those amino acids whose labelled forms give rise to $^{14}CO_2$ may contribute carbon to pyruvate, and thus to alanine.

Thus from the earlier studies on the oxidation of amino acids in muscle, it would seem that only the branched-chain (except leucine) and some dispensable (glutamic, aspartic and possible glycine) amino acids can possibly give rise to pyruvate. This argument would hold even for amino acids whose possible conver-sion to pyruvate does not necessarily involve passage through reactions of the tricarboxylic acid cycle (e.g. glycine, serine, and possibly threonine), since in so far as they are converted to pyruvate they must be to some degree oxidized. For amino acids for which the labelled form does give rise to CO_2, the question then arises: how much of the amino acid is in fact converted to CO_2 or oxidized?

Chang and Goldberg (1978b) have attempted to make some estimate for aspartate and valine, using the isolated rat diaphragm (*Table 7.6*). After incubation with ^{14}C-labelled amino acids, they measured recovery of label in glutamine/

Table 7.6. RECOVERY OF RADIOACTIVITY IN SOME METABOLIC PRODUCTS OF THE CATABOLISM OF VALINE AND ISOLEUCINE FOLLOWING INCUBATION WITH ISOLATED RAT DIAPHRAGMS. VALUES ARE IN nmol/mg TISSUE PER 2 h, AND ARE EXPRESSED FOR EACH AMINO ACID RELATIVE TO THE RECOVERY IN GLUTAMINE PLUS GLUTAMIC ACID TAKEN AS UNITY. (RESULTS FROM CHANG AND GOLDBERG, 1978b)

Amino acid	CO_2	Glutamine + glutamate	Lactate + pyruvate	Alanine
Aspartate	9.5	1.0	0.58	<0.02
Valine	12.6	1.0	0.54	<0.02

glutamate, in lactate/pyruvate, in CO_2 and in alanine. Taking recovery in glutamate/glutamine as unity, while ten times as much activity was found in CO_2, recovery in lactate and pyruvate was about 50—60% of that in glutamine/glutamate. Recovery in alanine, however, was only about 4% of this. In separate experiments they estimated, using 3,4-^{14}C-labelled glucose, that the flow of glucose carbon to lactate and pyruvate was about 1.5-fold greater than the flow to CO_2. Thus the amount of ^{14}C-aspartate or ^{14}C-valine which represented true oxidation (via pyruvate and acetyl CoA) was only about 40% (1/1.5) of that in (pyruvate + lactate). Most of the [14]CO_2 appearing must arise by exchange reactions, through repeated recycling in the tricarboxylic acid cycle. The contribution to alanine-C is small, roughly 0.02/(0.54—0.58), or 4% of that of glucose. On the other hand, it is likely that valine and aspartate make a significant contribution to glutamine carbon, the repeated recycling ensuring that carbon from amino acids feeding into the cycle will be equilibrated with glutamic acid and glutamine.

It is to be noted that the argument above overestimates the contribution of glucose to CO_2, since the presence of phospho-enol-pyruvate carboxy kinase, and malic enzyme, both of which have been shown to be present in muscle, means that some glucose or pyruvate carbon reaches the cycle intermediates other than through acetyl-CoA. Thus not all the [14]CO_2 derived from glucose represents 'true' oxidation. The estimate of the amount of valine and aspartate oxidized, and the amount of their carbon likely to appear in alanine, is therefore further reduced. In contrast, if pyruvate formed from amino acids was compartmentally distinguished from that formed from glucose — it might for example be largely intra-mitochondrial, that from glucose largely cytosolic — then the amount of amino acid oxidized might be larger than indicated by the amount of glucose oxidized.

CONCLUSION : PATTERN OF AMINO ACID CATABOLISM IN RAT MUSCLE

The picture that emerges is one in which most of the amino acids produced in the catabolism of muscle protein are either used in the re-synthesis of muscle protein, or are released from muscle largely unchanged. The branched-chain amino acids, however, are catabolized to a significant degree. Leucine carbon, after transamination (Chang and Goldberg estimate about 50% of leucine produced is transaminated), is fully oxidized via acetyl-CoA (and is estimated to account for about 5% of muscle CO_2). Only small amounts of isoleucine, valine, aspartate and glutamate are fully oxidized (in sum, rather less than 5% of muscle CO_2); a small part is released as alanine (for which most of the carbon is derived from glucose) but the greater part must be released as glutamine. The remaining surplus nitrogen appears as alanine.

Generality of the findings between muscles of the same and different species

Whether this general picture that has been established for the rat diaphragm is also true for skeletal muscle, remains to be established. It seems plausible, however, since the pattern of release of amino acid from diaphragm muscle and that from any skeletal muscle is broadly similar (*Table 7.7*). In this table, the pattern

Table 7.7. OUTPUT OF INDIVIDUAL AMINO ACIDS AS PERCENTAGE OF TOTAL OUTPUT FROM MUSCLES OF VARIOUS SPECIES. RESULTS FROM (a) CHANG AND GOLDBERG (1978b); (b) FELIG *et al.* (1970); (c) POZEFSKY *et al.* (1976); (d) BELL *et al.* (1975); (e) LINDSAY, STEEL AND BUTTERY (1977a); (f) BALLARD, FILSELL AND JARRETT (1976)

Species:	Rat (a)	Human (b)	Human (c)	Cattle (d)	Sheep (e)	Sheep (f)
Muscle:	diaphragm	forearm	forearm	hind-limb	hind-limb	hind-limb
Length of fast:	(48 h)	(overnight)	(60 h)	(20 h)	(48–96 h)	(144 h)
Amino acid						
Ala	11	30	26	24	21	20
Gln	25	9.9	27	27	10	24
Gly	9	10	8	10	24	5
Ser	6	2	3	–	4	4
Pro	5	8	3	–	7	–
Glu	6	–	–	uptake	0.5	uptake
Asp	0.8	–	–	0	uptake	–
AsN	–	–	–	–	uptake	2.4
Val	2.3	4.1	2.9	8.1	7.1	10.9
Leu	4.5	3.7	3.3	6.4	5.9	10.9
Ileu	3.4	2.9	1.9	6.4	1.9	5.0
Thr	5.3	7.4	6.4	–	8.3	6.9
Lys	5.6	9.9	8.0	uptake	5.1	uptake
His	3.5	3.8	2.6	0.5	2.2	2.8
Met	3.2	1.5	1.3	5.8	0.9	–
Phe	3.4	2.1	1.9	3.7	2.4	3.8
Tyr	2.7	2.4	1.9	0.5	1.0	3.8
Arg	3.9	6.2	3.5	2.7	4.0	–
Try	1.1	–	–	–	–	–

of release of amino acid from the isolated rat diaphragm (Chang and Goldberg, 1978b) is compared with that from human forearm muscle (Felig *et al.*, 1970; Pozefsky *et al.*, 1976) from the muscles of the sheep hind-limb (Lindsay, Steel and Buttery, 1977a), from the sheep hindquarters (Ballard, Filsell and Jarrett, 1976) and cattle hindquarters (Bell *et al.*, 1975). For most of the amino acids, their fractional output is very similar. Some differences are seen, however, for alanine, glutamine and glycine and for the branched-chain amino acids.

Whereas alanine output is always large, the output from the rat diaphragm seems to be somewhat lower than that for the skeletal muscles. Whether this reflects a real difference between diagphragm and skeletal muscle is uncertain, since this is the only study made *in vitro*. There is also some variation in the output of glutamine. Here too it is difficult to decide if there are real differences in glutamine output. There are differences in length of fasting between studies, and it is particularly difficult to compare ruminants and non-ruminants in this respect because of the buffering effect exerted by the microbial population of the rumen in determining the absorption of nutrients in ruminants. In addition, in some studies glutamine was not separated from asparagine, and in one study (Ballard, Filsell and Jarrett, 1976) whole blood was used for measurement. Felig, Wahren and Räf (1973) have shown that this may result in quantitative differences in estimates of the transfer of amino acids. Similar reservations may apply in assessing the significance of the larger output to glycine from sheep muscle observed by the authors (Lindsay, Steel and Buttery, 1977a). It may also be pertinent to observe that this was found to vary greatly between sheep, much more so than for other amino acids.

The one change that may reflect an important difference between species is in the output of the branched-chain amino acids. In all the studies with ruminants there is an appreciably greater proportional output of valine, leucine and isoleucine than is seen in studies of man and the rat. Indeed, the fractional output of these acids in ruminants is of the same order as the proportions in which they occur in sheep and cattle muscle (leucine 7—8%; isoleucine 4—5% and valine 5—6%) when expressed on a comparable basis (mmol/100 mmol amino acid).

These observations thus suggest that the catabolism of the branched-chain amino acids occur to a much smaller extent in ruminant muscle, and there are other findings that tend to support this conclusion. Thus, Coward (1979) has found with the perfused sheep diaphragm that addition of leucine (3 mM) to the perfusing medium resulted in only a small or no stimulation of alanine production; that leucine infusion into the sheep hind-limb muscle *in vivo* does not stimulate alanine output has also been demonstrated by Mackenzie and Lindsay (unpublished). Coward and Buttery (1979) have also studied the catabolism of the branched-chain amino acids by the isolated perfused sheep diaphragm. When ^{14}C-labelled amino acids were added to the perfusate, the production of $^{14}CO_2$ was substantially less than that reported in comparable studies with rat muscle. There was probably a small amount of catabolism, since some activity appeared in the volatile acids in the perfusate. This activity was found in the higher acids (probably isobutyric, isovaleric and 2-methyl butyric) which are believed to arise from side reactions in the catabolism of the branched-chain acids (Coward, 1979). Nevertheless the VFA released, relative to the amount of branched-chain acid released, was small, consistent with there being only a small catabolism of the acids by muscle (*Table 7.8*).

Table 7.8. COMPARISON OF THE CAPACITY FOR OXIDATION OF THE BRANCHED-CHAIN AMINO ACIDS BY THE PERFUSED SHEEP DIAPHRAGM (COWARD AND BUTTERY, 1979) AND THE ISOLATED RAT DIAPHRAGM (GOLDBERG AND ODESSEY, 1972). ALL VALUES IN μmol/h PER 30 g WET WEIGHT

| | | ^{14}C-amino acid | | |
		valine	isoleucine	leucine
Sheep	$^{14}CO_2$	<0.07	0.2—0.3	0.2
	^{14}C-VFA	0.2—0.3	0.2	0.1
	amino acid output	2—6	−3—+3	3—4
Rat	$^{14}CO_2$	1.4	2.3	3.3
	amino acid output	3—4	3—4	3—4

The limited catabolism of the branched-chain acids is, however, by no means invariable. The authors recently measured the output of amino acids by the hind-limb muscles *in vivo* in diabetic sheep. Diabetes was initiated by alloxan, controlled by insulin, and the diabetic state reproduced by withholding insulin and food for 36 h. Despite these attempts to produce a uniform degree of diabetes, there was appreciable variation in its severity, as judged by the degree of ketonaemia, and perhaps because of this the output of amino acids was more variable than is usually observed. Nevertheless there was a clear indication of a change in pattern relative to that of the normal sheep. The total output was increased only moderately — perhaps about 30%; but there was a much increased output of glutamine, a slightly increased output of alanine and a considerably reduced output of glycine. What was most striking, however, was the marked fall in the output of the branched-chain amino acids. Leucine and isoleucine, which in normal animals contribute about 8% of the total output, in diabetic animals provided less than 1%; while valine was actually taken up in substantial amounts.

These findings strongly suggest that in diabetic sheep there is a marked increase in the catabolism of the branched-chain amino acids in muscle. The oxidation of leucine in the rat diaphragm is known to be increased in diabetes (Buse, Herlong and Weigand, 1976) but since there is substantial oxidation of leucine by normal rat diaphragm, the effect is much less dramatic than is seen in ruminant muscle.

At present we can only speculate as to the reason for the differences in pattern of amino acid catabolism in ruminants. It is possible that the slight catabolism of the branched-chain acids in normal ruminants is a consequence of the significant utilization of 3-hydroxybutyrate by ruminant muscle, which in contrast to non-ruminants occurs even in normal fed animals. This is a result of the appreciable concentration (0.2–0.3 mM) of 3-hydroxybutyrate (derived from butyrate produced by ruminal fermentation) in the blood of fed ruminants. According to Landaas (1977), 3-hydroxybutyrate inhibits the further catabolism of the ketoacids derived from the branched-chain acids. This effect is at least partly dependent on elevation of the NADH/NAD ratio in the muscle mitochondria, and thus requires that the ratio of 3-hydroxybutyrate/acetoacetate be high. In the blood of normal ruminants, although the 3-hydroxybutyrate is elevated, that of acetoacetate is very low and thus the full inhibitory effect of 3-hydroxybutyrate would be exerted. In diabetes, although there is a further increase in the concentration of 3-hydroxybutyrate, there is a much larger proportional rise in the concentration of acetoacetate, so that the 3-hydroxybutyrate/acetoacetate ratio actually falls. Thus there may be a fall in the mitochondrial 3-hydroxybutyrate/acetoacetate ratio and this should attenuate the inhibitory effect of 3-hydroxybutyrate, with an increase in the oxidation of leucine (and other branched-chain amino acids) as apparently occurs.

The apparent lack of catabolism of the branched-chain acids in normal ruminant muscle, the small increase in alanine output when there is a large increase in branched-chain amino acid metabolism in diabetic muscle, and the lack of increase in alanine output in response to an increase in leucine concentration, all suggest that alanine output is not linked to the catabolism of the branched-chain amino acids as in non-ruminant muscle. This leaves uncertain the origin of the amino group of the alanine put out by ruminant muscle. As *Table 7.7* indicates, alanine constitutes about the same proportion of the amino-N output and more than double the proportion found in protein from muscle in ruminants as in non-ruminants, and thus 50% or more of the alanine-N must be derived from some other source than the alanine of muscle protein. It is possible that there may be rather more oxidation of aspartate and/or glutamate in ruminants than occurs in the rat diaphragm. This does fit in well with the apparently lower output of glutamine which the present authors (although not others) observed (*Table 7.7*) in normal, but not in diabetic sheep. On the other hand, we have seen that Chang and Goldberg (1978b) argued that since glutamate and aspartate could undergo oxidation after passing through the pyruvate and acetyl-CoA pool, this would imply that aspartate and glutamate oxidation must be less than that of pyruvate oxidation in muscle. Although there is no direct evidence as to the extent of pyruvate oxidation in ruminant muscle, we have observed that glucose oxidation is variable, and rather small. This apparent discrepancy could be due to incomplete equilibration between mitochondrial and cytosolic pyruvate, the amino acids being in equilibrium with the former and glucose with the latter. However, in the absence of any direct evidence on this, we have to say that at present the origin of the alanine-N in ruminants is not known.

Much the same limitation applies with respect to alanine-C. About half the alanine-C may be derived from alanine in protein. Lindsay, Steel and Buttery (1977b) have found that enough carbohydrate (glucose uptake corrected for lactate and pyruvate output) was available for the alanine-C released. However, Coward (1979), in studies with the perfused sheep diaphragm to which [14C]-glucose was added, estimated that only about 5% of alanine-C came from glucose. Chandrasena (1977) has estimated that about 13% of alanine is derived from glucose in wether sheep fasted 48 h. His results were based on experiments in which [14C]-glucose was infused intravenously, and the specific activity of plasma glucose and alanine compared. Thus the site at which alanine is formed from glucose was not established. In current similar experiments with pregnant sheep, the present authors have obtained as great or larger values for incorporation of [14C]-glucose carbon into alanine. Studies across muscle, however, have given inconclusive results as to the transfer of ^{14}C from glucose to alanine at this site. Large rates of infusion of isotope would be needed to give estimates of any precision for the amount of alanine-C derived from glucose in muscle *in vivo*. It is in any case likely, as we have seen, that a significant amount of alanine-C is derived from muscle glycogen. Thus, although some of the carbon of alanine produced by sheep muscle is derived from glucose, it is not possible to give a quantitative estimate.

It was seen earlier (*Table 7.7*) that the present authors (although not others) have observed a rather larger (but variable) output of glycine from sheep, compared with non-ruminant, muscle. This might be a consequence of the minimal oxidation of the branched-chain amino acid catabolism, since O'Brien (1978) has shown that the 2-oxo acids produced by transamination from the branched-chain amino acids can both inhibit glycine oxidation and stimulate its synthesis. If this were a factor in the large output of glycine observed, then one would except that in the diabetic state, where the authors have suggested the oxidation of the branched-chain acids may be more extensive, the output of glycine should fall; and this is what is observed.

If these speculations turn out to have substance, the differences in the catabolism of amino acids in the muscle of ruminants may reflect not any fundamental difference from the metabolism of non-ruminants, but stem rather as a consequence of other metabolic changes, produced ultimately as a consequence of the dietary fermentation that is so characteristic of ruminants.

Interaction between the rate of degradation and rate of protein synthesis in muscle

The rate of net protein deposition in muscle is obviously determined as the difference between the rate of protein synthesis and the rate of degradation.

One question, as yet unanswered, is whether the rates of synthesis and degradation are controlled quite independently. In the liver, at least in the perfused organ, the rate of protein synthesis is markedly affected by an increase in the supply of amino acids above that normally present in plasma (Jefferson and Korner, 1969). In the rat heart, this effect is also seen but its magnitude is much less; moreover, the effect can be reproduced simply by increasing the supply of branched-chain amino acids (Rannels, Hjalmarson and Morgan, 1974). In the rat diaphragm the effect can be further localized, since while addition of the branched-chain acids stimulates protein synthesis, the addition of all the other amino acids is ineffective (Fulks, Li and Goldberg, 1975). These authors further showed that the stimulating effect could be reproduced by addition of leucine alone, or of isoleucine plus valine. The effect of leucine alone has been confirmed by Buse and Reid (1975) and Buse and Weigand (1977). These authors were, however, unable to confirm the effectiveness of isoleucine or valine, alone or in combination. It also appears that the branched-chain acids or leucine alone are effective in stimulating protein synthesis in rat skeletal muscle (Li and Jefferson, 1978; Buse, Atwell and Mancusi, 1979). Since heart (Buse *et al.*, 1972) as well as diaphragm and skeletal muscle (*Table 7.4*) can readily oxidize the branched-chain acids, the possibility has to be considered that a catabolic product of the branched-chain acids might be responsible for the stimulating effect on protein synthesis (Rannels, Hjalmarson and Morgan, 1974). In the rat diaphragm at least this seems improbable, because 2-oxo-isocaproate — the first product of leucine catabolism — although it is oxidized at least as readily as leucine, was generally ineffective in stimulating protein synthesis (Buse and Weigand, 1977).

In several conditions where the rate of protein synthesis is decreased — in fasting (Garlick *et al.*, 1975), in diabetes (Buse and Weigand, 1977) or under the influence of haemorrhagic shock or corticosteroids (Ryan *et al.*, 1974) — it would appear that there is increased oxidation of the branched-chain amino acids, particularly of leucine. It is therefore tempting to suppose that a major factor in increasing protein synthesis is a decrease in the oxidation of leucine. However, at present it is still possible to suppose that some third factor initiates protein synthesis or catabolism, with catabolism of the branched-chain acids passively following. Although there is obviously some control over muscle metabolism by the branched-chain amino acids (see, for example, Li and Jefferson, 1978; Buse, Atwell and Mancusi, 1979) the exact nature and mechanism of this control is still unknown. The major control of protein synthesis in muscle *in vivo* appears to be hormonal, and this is dealt with in Chapter 9 of this book.

References

AOKI, T.T. BRENNAN, M.F., MULLER, W.A., SOELDNER, J.S., ALPERT, J.S., SALTZ, S.B., KAUFMANN, R.L., TAN, M.H., CAHILL, G.F. Jr. (1976). *Am. J. Clin. Nutr.,* **29**, 340–350

BALLARD, F.J., FILSELL, O.H. and JARRETT, I.G. (1976). *Metabolism,* **25**, 415–418

BEATTY, C.H., CURTIS, S., YOUNG, M.K. and BOCEK, R.M. (1974). *Am. J. Physiol.,* **227**, 268–272

BELL, A.W., GARDNER, J.W., MANSON, W. and THOMPSON, G.E. (1975). *Br. J. Nutr.,* **33**, 207–217

BUSE, M.G. and REID, S.S. (1975). *J. Clin. Invest.,* **56**, 1250–1261

BUSE, M.G. and WEIGAND, D.A. (1977). *Biochim. Biophys. Acta,* **475**, 81–89

BUSE, M.G., ATWELL, R. and MANCUSI, V. (1979). *Horm. Metab. Res.,* **11**, 289–292

BUSE, M.G., HERLONG, F. and WEIGAND, D.A. (1976). *Endocrinol.,* **98**, 1166–1175

BUSE, M.G., BIGGERS, J.F., FREDERICI, K.H. and BUSE, J.F. (1972). *J. Biol. Chem.,* **247**, 8085–8089

BUTTERFIELD, R.M. (1966). *Res. Vet. Sci.,* **7**, 168–179

BUTTERY, P.J., BECKERTON, A., MITCHELL, R.M., DAVIES, K. and ANNISON, E.F. (1975). *Proc. Nutr. Soc.,* **34**, 91A

CHANDRASENA, L.G. (1977). 'Aspects of lactate metabolism in sheep, *Ph.D. thesis,* University of Liverpool

CHANG, T.W. and GOLDBERG, A.L. (1978a). *J. Biol. Chem.,* **253**, 3677–3684

CHANG, T.W. and GOLDBERG, A.L. (1978b). *J. Biol. Chem.,* **253**, 3685–3695

COWARD, B.J. (1979). 'Ruminant muscle metabolism', *Ph.D. thesis,* University of Nottingham

COWARD, B.J. and BUTTERY, P.J. (1979). *Proc. Nutr. Soc.,* **38**, 139A

EDMUNDS, B.K., BUTTERY, P.J. and FISHER, C. (1978). *Proc. Nutr. Soc.,* **37**, 32A

FELIG, P., WAHREN, J., RÄF, L. (1973). *Proc. Natn. Acad. Sci. U.S.A.,* **70**, 1775–1779

FELIG, P., POZEFSKY, T., MARLISS, E. and CAHILL, G.E. (1970). *Science, N.Y.,* **167**, 1003–1004

FULKS, R.M., LI, J.B. and GOLDBERG, A.L. (1975). *J. Biol. Chem.,* **250**, 290–298

GARBER, A.J., KARL, I.E. and KIPNIS, D.M. (1976a). *J. Biol. Chem.,* **251**, 826–835

GARBER, A.J., KARL, I.E. and KIPNIS, D.M. (1976b). *J. Biol. Chem.,* **251**, 836–843

GARLICK, P.J., BURK, T.L. and SWICK, R.W. (1976). *Am. J. Physiol.,* **230**, 1108–1112

GARLICK, P.J., MILLWARD, D.J. and JAMES, W.P.T. (1973). *Biochem. J.,* **136**, 935–945

GARLICK, P.J., MILLWARD, D.J., JAMES, W.P.T. and WATERLOW, J.C. (1975). *Biochim. Biophys. Acta,* **44**, 71–84

GOLDBERG, A.L. and ODESSEY, R. (1972). *Am. J. Physiol.,* **223**, 1384–1391

GOLDSTEIN, L. and NEWSHOLME, E.A. (1976). *Biochem. J.,* **154**, 555–558

GRUBB, B. (1976). *Am. J. Physiol.,* **230**, 1379–1384

JEFFERSON, L.S. and KORNER, A. (1969). *Biochem. J.,* **111**, 703–712

LANDAAS, S. (1977). *Scand. J. Clin. Lab. Invest.,* **37**, 411–418

LAURENT, G.L., SPARROW, M.P., BATES, P.C. and MILLWARD, D.J. (1978). *Biochem. J.,* **176**, 393–405

LI, J.B. and JEFFERSON, L.S. (1978). *Biochim. Biophys. Acta,* **544**, 351–359

LINDSAY, D.B., STEEL, J.W. and BUTTERY, P.J. (1977a). *Proc. Nutr. Soc.,* **36**, 33A

LINDSAY, D.B., STEEL, J.W. and BARKER, P.J. (1977b). *Proc. Nutr. Soc.,* **36**, 80A

LOBLEY, G.E., REEDS, P.J. and PENNIE, K. (1978). *Proc. Nutr. Soc.,* **37**, 96A

LODGE, G.A. and HEANEY, D.P. (1973). *Can. J. Anim. Sci.,* **53**, 479–489

LOWENSTEIN, J.M. (1972). *Physiol. Rev.,* **52**, 382–414

MANCHESTER, K.L. (1965). *Biochim. Biophys. Acta,* **100**, 295–298

MILLER, L.L. (1962). In *Amino Acid Pools,* pp. 708–721 (Holden, J.T., Ed.), Elsevier, New York

MILLWARD, D.J., GARLICK, P.J., STEWART, R.J.C., NNANYELUGO, D.O. and WATERLOW, J.C. (1975). *Biochem. J.,* **150**, 235–243

NOLTE, J., BRDICZKA, D. and PETTE, D. (1972). *Biochim. Biophys. Acta,* **284,** 497–507

O'BRIEN, W.E. (1978). *Archs Biochem. Biophys.,* **189,** 291–297

ODESSEY, R., KHAIRALLAH, E.A. and GOLDBERG, A.L. (1974). *J. Biol. Chem.,* **240,** 7623–7629

POZEFSKY, T. and TANCREDI, R.G. (1972). *J. Clin. Invest.,* **51,** 2359–2369

POZEFSKY, T., TANCREDI, R.G., MOXLEY, R.T., DUPRE, J. and TOBIN, J.D. (1976). *J. Clin. Invest.,* **57,** 444–449

RANNELS, D.E., HJALMARSON, A.C. and MORGAN, H.E. (1974). *Am. J. Physiol.,* **226,** 528–539

ROBERTS, C.J., REID, I.M., PIKE, B.V. and TURFREY, B.R. (1979). *Proc. Nutr. Soc.,* **38,** 68A

ROBINSON, J.J., McDONALD, I., McHATTIE, I. and PENNIE, K. (1978). *J. Agric. Sci., Camb.,* **91,** 291–304

RUDERMAN, N.B. and BERGER, M.(1974). *J. Biol. Chem.,* **249,** 5500–5506

RYAN, N.T., GEORGE, B.C., ODESSEY, R. and EGDAHL, R.H. (1974). *Metabolism,* **23,** 901–904

SNELL, K. and DUFF, D.A. (1977). *Biochem. J.,* **162,** 399–403

SOLTESZ, G., JOYCE, J. and YOUNG, M. (1973). *Biol. Neonate,* **23,** 139–148

WEINMAN, E.O., STRISOWER, E.H., CHAIKOFF, I.L. (1957). *Physiol. Rev.,* **37,** 252–272

8

DIETARY CONSTRAINTS ON NITROGEN RETENTION

K.N. BOORMAN
Department of Applied Biochemistry and Nutrition, University of Nottingham School of Agriculture

Summary

Three dietary constraints on nitrogen retention are considered: nitrogen intake, protein quality and dietary energy.

The relation between nitrogen retention and nitrogen intake is well defined and an animal can exist at several states of nitrogen retention without impairment of function. The concept of protein 'storage' arose from such observations. In growing or producing animals, growth or production will be limited by states other than 'maximal' nitrogen retention. Nitrogen stores may have use in such animals in sustaining production in times of insufficiency of dietary nitrogen supply.

The quality of the dietary protein is an important determinant of the efficiency of deposition of dietary nitrogen within the body. The pattern of essential amino acids comprising 'good quality' will depend largely on the amino acid composition of the product (e.g. muscle). Other needs may influence the pattern, but only marginally in the growing animal. Also, other needs may comprise or include 'maintenance' needs, depending on the definition of the latter adopted.

If the pattern of amino acids is to be used to define dietary requirements, the input of one amino acid in the pattern must be defined exactly in quantitative terms. Existing information shows that rates of input for given rates of growth of less than maximal, for individual amino acids, found under different conditions are remarkably similar within a species (chick), and even between species (rat, chick and turkey). For a much larger species (pig), there is an inevitable decrease in gross efficiency of utilization of the amino acid.

Departures from ideal amino acid balance have been reviewed thoroughly elsewhere and generally result in decreases in nitrogen retention. This largely reflects a decrease in nitrogen intake in the *ad libitum* fed animal, due to the decrease in food intake which accompanies such departures, rather than a decrease in the efficiency of utilization of the limiting amino acid. The latter view of the cause of much of the adverse effect of imbalance, once accepted, is now not generally defensible.

Energy and protein interact because dietary protein is a source of dietary energy, because dietary energy is needed for protein turnover and deposition and because deposited protein represents part of the body's energy store. The complexity of this interaction is such as to defy adequate description so far. In simple terms inadequate dietary energy will curtail nitrogen retention, but the form of the relationship between nitrogen retention and dietary energy has only rarely been described. Increments in nitrogen retention with increasing dietary energy have been observed at most dietary energy contents studied. The situation is, however, confused because many observations reputed to demonstrate the interaction are explicable in terms of indirect effects on the intake or efficiency of utilization of a limiting amino acid, mediated through changes in food intake or carcass composition

Introduction

If optimum nitrogen retention is encouraged in an animal by optimizing dietary and environmental conditions, rates of nitrogen retention in animals of different ages or weights can be compared. It would be expected that the rate per unit body weight would be maximal in the young animal and tend to zero in the mature, non-producing animal. Carr, Boorman and Cole (1977) correlated the rates of nitrogen retention of pigs with body weight (from 2.5 to 170 kg) using published observations which were adjudged to include no hindrance to optimal nitrogen retention. These data (*Figure 8.1*) were not segregated for breed or sex.

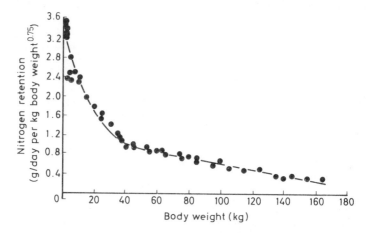

Figure 8.1   The relationship between nitrogen retention (g/day per kg body weight $^{0.75}$) and body weight (*W*, kg) in the pig. Line is fitted: nitrogen retention = 3.324 ~ 0.098*W* + 0.001*Z*, where *Z*=*W²* for values of *W*≤ 45 kg and 45 (2*W* − 45) for values of *W*≥ 45 kg. (From Carr, Boorman and Cole, 1977, 'Nitrogen retention in the pig', sources of data given in the original. Reproduced with permission from *British Journal of Nutrition,* published by Cambridge University Press)

Similar curves must underlie optimal nitrogen retention in other animals but are difficult to demonstrate because results are not available for a sufficient range of body weights (rats, poultry), there are uncertainties about the values to ascribe to nitrogen inputs and hence optimal nitrogen retention is difficult to define (ruminant animals), or because of apparently large inherent variability in bodily nitrogen retention (human). The parameters of these curves will depend on growth rates of animals in relation to body size and the proportion of tissue accretion attributable to protein. For each species the curve represents the upper limit to nitrogen retention at each body weight.

Within a species this upper limit will vary with sex and breed in so far as there are differences in growth rate and body composition due to such variables. These differences are, however, relatively small compared with the effect of body weight or of diet. Three aspects of diet will be considered below: protein (nitrogen) intake, protein quality and energy supply.

Protein intake

The dose—response relationship between nitrogen input, whether ingested or absorbed, and nitrogen retention has been defined for several species and is clearly explained by Allison (1959, 1964). An idealized version of this relationship is shown in *Figure 8.2*. The negative intercept on the ordinate axis (nitrogen retention) represents the obligatory nitrogen losses of the animal receiving no dietary nitrogen ('endogenous loss'). Carr *et al.* (1977) found a remarkable consistency in this value (0.16 g N/day per kg $W^{0.75}$) in the data from different sources for the pig, which was reflected in the maintenance nitrogen requirement they found from the intercepts on the abscissa axis (nitrogen intake). These constancies agree with the consistency originally found between species by Smuts (1935).

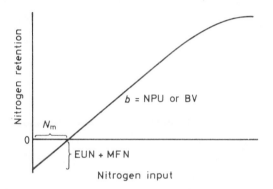

Figure 8.2 Theoretical relationship between nitrogen retention (mass/unit time) and nitrogen input either as nitrogen ingested or nitrogen absorbed. In the former case the slope (*b*) represents net protein utilization (NPU), in the latter biological value (BV). The intercepts represent obligatory nitrogen losses (endogenous urinary nitrogen (EUN) and metabolic faecal nitrogen (MFN) (ordinate) and ingested or absorbed nitrogen required for 'maintenance' (N_m) (abscissa)

The positive response in nitrogen retention to nitrogen inputs of less than maximal is often represented as rectilinear (*Figure 8.2*); however, data may support this (Allison, Anderson and Seeley, 1946) or be more suggestive of a diminishing-return curve (Allison, 1957). Since, under controlled conditions, this slope represents the quality of the protein fed (biological value, net protein utilization or nitrogen balance index), the assumption that the relationship is linear has tended to become fixed. Most data are linear to a good approximation, but as the data examined by Carr *et al.* (1977) show, a curvilinear response is probably a more exact interpretation.

Carr *et al.* (1977) cursorily examined the effect of body weight on the slope of this relationship in conditions of 'good' protein quality and could find no real evidence that there was such an effect. The commented that 'there is no conclusive published information for any species concerning the effect of age or weight on the classical indices of protein quality . . . .'. It should be noted, however, that if the pattern of amino acids needed by an animal changes with age, the capacity of an individual dietary protein to satisfy those needs will alter and this should be reflected in a changed slope of the nitrogen retention response curve. Some evidence of such changes in amino acids needs will be produced below.

The form of this response curve at adequate or near-adequate protein intakes is usually represented as asymptotic (Allison, 1959, 1964; Rufeger, 1966). Experimental data are rarely sufficiently free from variation to allow resolution of the true nature of the curve in this region. Fisher *et al.* (1964) fed young chicks on diets containing adequate (220 g/kg) or more than adequate (280 g/kg) protein and found small but consistent increases in total (including feathers) carcass nitrogen. These increases are still apparent if the values are corrected for the small decreases in body weight that occurred when the diet of higher protein content was fed. This indicates that increases in nitrogen retention are possible at more than adequate nitrogen intakes.

That much of the nitrogen retention represented by this curve is depletable and repletable has been commented on and described as a capacity for protein storage (Holt, Halac and Kajdi, 1962; Allison and Wannemacker, 1965; Fisher, 1967; Swick and Benevenga, 1977; Lindsay and Buttery, 1980). This store largely resides in skeletal muscle and skin and its use to the animal has been assessed by Fisher (1967), Swick and Benevenga (1977) and Lindsay and Buttery (1980). Swick and Benevenga (1977) encapsulated the nature of protein stores in the statement: 'Thus the concept of protein reserve appears to be valid in that body composition is not fixed but is variable within wide limits which are still compatible with life and that it represents a facet of normal, dynamic protein metabolism.'

Thus the mature animal can exist in different 'normal' states of nitrogen retention and the dietary nitrogen required to maintain the *status quo* will reflect its current nitrogen retention. This concept is accepted in the human and accounts for some of the difficulty encountered in attempting to define a protein 'requirement' for this species. In animals, for which conventional ideas relate to 'maximal' or 'optimal' nitrogen retention, the full significance of the dynamic status of body protein content is only now being appreciated. In the growing animal the situation is confused, since it is impossible to demonstrate a depletable—repletable capacity without influencing growth rate, except possibly at or near the nitrogen requirement for maximal retention (see discussion of Fisher *et al.*, 1964, above). However, to the extent that catch-up growth is exhibited by animals (McCance, 1977), some degree of depletion is reversible.

Much of the discussion of the usefulness of stored protein to the animal has been concerned with maintenance of essential functions during protein or amino acid depletion (Fisher, 1967). The usefulness of such stores for supplementing inadequate dietary supplies of protein to maintain productive processes has been much less considered and rarely quantified. In such processes the size of the animal considered will be critical. The protein produced in the product will increase in size relative to total body protein (and therefore protein store) as the animal decreases in size. Thus the capacity of stores to supplement dietary protein in a large animal is likely to be greater than that of a small animal, provided stored protein represents a similar proportion of body protein in large and small animals. The latter is arguable; the value of one-fifth to one-quarter of total nitrogen retention originally suggested by Allison (1964) for the dog being accepted as realistic for the cow by Paquay, Debaere and Lousse (1972), while Wessels and Fisher (1965) found a value of 70 g stored N per kg body N for the mature cockerel.

Empirical observations support the general idea of a greater capacity to supplement dietary protein with body protein as body size increases. The dairy cow seems able to maintain high rates of output in periods of low nitrogen intake (van Es and Boekholt, 1976), while any interference with nitrogen input in the laying hen is very rapidly reflected in egg output. The latter observation should not, however, give rise to the idea that protein stores are unimportant in the laying hen. Owing to the probable phasic nature of egg white synthesis, the demand for protein at certain times of the day may be greater than dietary supply. Withdrawal of protein from a store and replacement later in the day may be an important mechanism in maintaining the regular sequence of egg production. Hurwitz and Bornstein (1973) have discussed this and constructed a model describing amino acid needs for egg production which incorporates this process (see also Smith, 1978).

Protein quality

As already stated, when other variables are controlled, the quality of the protein fed is reflected in the slope of the curve relating nitrogen retention to nitrogen input. The quality of a dietary protein is the extent to which its essential amino acid pattern coincides with the needs of the animal. Theoretically the poorer a protein is, the more of it that is needed to satisfy the needs for a particular nitrogen retention. Therefore, maximum nitrogen retention is achievable with a poorer dietary protein, but more of the protein will be needed and the rate of increase in nitrogen retention per unit intake of the protein will be less than that for a better protein source. In practice, maximum retention may not be achieved with poor dietary proteins. Wethli, Morris and Shresta (1975) showed that young chicks did not achieve maximum weight gain when receiving proteins of poor quality, even if sufficient of such proteins were included in the diet of the chick to satisfy the requirement for the limiting amino acid expressed as a proportion of the diet. Although nitrogen retention was not directly measured in this study, it is fair to assume that growth rate is a good indicator of nitrogen retention. The failure of practice to match theory is accounted for by the deleterious effects of the excesses of amino acids which must be supplied in order to meet the requirements for the most deficient in the protein. These deleterious effects are evident in a decreased food intake (see Boorman, 1979) and deranged amino acid metabolism (Harper, Benevenga and Wohlhueter, 1970). Some effects of amino acid imbalance are discussed below.

A mature animal may be less sensitive to such effects than a growing animal, the former having relatively greater protein reserves with which to buffer effects of poor dietary amino acid patterns. If this is so, maximal nitrogen retention in products (milk, eggs) from mature animals may be possible from high inputs of poor amino acid patterns.

OPTIMAL AMINO ACID BALANCE

Much of the investigation of optimal amino acid patterns and amino acid requirements has been empirical and this has led to much duplication. For the young growing animal the chief determinant of the pattern of amino acids required will be the pattern deposited in body-weight gain, largely muscle. Within the limits of the accuracies of determinations, therefore, there should be good correlations between the patterns of amino acids in muscles and those required in the diet. Since muscle composition does not differ greatly between animals, these patterns should be similar for different animals. *Figure 8.3* shows these correlations for

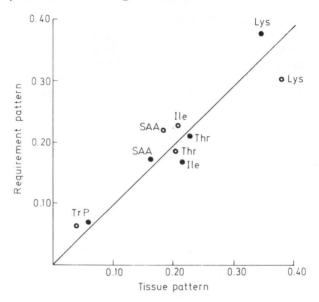

Figure 8.3   Relationship between pattern of requirements found by feeding experiments for lysine (Lys), methionine plus cystine (SAA), isoleucine (Ile), threonine (Thr) and tryptophan (Trp), and pattern of the same amino acids in tissue (muscle) of growing pigs (●) and chick (○). Each amino acid is represented as a fraction of all five in each pattern. The line represents coincidence of requirement and tissue patterns. Requirement patterns from Cole (1978) for pig, and Lewis (1978) for chick. Tissue patterns from Lawrie (1974) for pig, and Scott, Nesheim and Young (1976) for chick.

the pig and the chick, for five amino acids, using recent information on requirements. In view of the inexactness of measures of amino acid requirements it would be unwise to attempt a detailed analysis on this basis, but the values suggest that isoleucine requirement in the pig may be anomalously low, while lysine is possibly high. The deviations of the lysine and sulphur amino acid requirements from muscle composition in the chick may reflect feather growth (see below).

This approach to developing an optimal pattern for dietary amino acids takes no account of differences between amino acids in their fates in the body other than for protein synthesis and retention. Measurements of amino acid retentions in the body have shown differences in gross efficiencies of utilization between different amino acids (Bunce and King, 1969a, 1969b; Aguilar, Harper and Benevenga, 1972; Scott, Nesheim and Young, 1976). These other fates may be conceptualized as 'maintenance' and will therefore be dependent upon a function of body weight. Using conventional values for endogenous nitrogen loss it is possible to compare the proportion of utilized dietary protein that would be diverted to maintenance in a 350-g broiler chick with that in an 80-kg pig. These approximate calculations indicate that only a small proportion of utilized dietary protein is diverted in the chick (about 10%), while the proportion is approximately doubled in the pig, although still relatively small. These calculations suggest that, unless one amino acid is used very much more for maintenance purposes than others, if maintenance is seen as the 'basal rate of amino acid oxidation' (Millward *et al.*, 1976) this process will not greatly influence the pattern of requirements.

If the approach to maintenance includes 'a reasonable allowance for the nitrogen contained in the growth of tissues that continue to grow throughout life' (Mitchell, 1962), it is possible to calculate the effect of this process on the pattern of amino acids required. *Table 8.1* shows the effect of feather growth

Table 8.1. EFFECT OF FEATHER GROWTH ON THE PATTERN OF AMINO ACIDS REQUIRED BY THE YOUNG CHICK, CALCULATED FROM SOFT TISSUE GAIN*

| | *Pattern* *(each amino acid as fraction of whole)* | |
	without feathers	with feathers
Arginine	0.139	0.139
Histidine	0.041	0.033
Isoleucine	0.085	0.093
Leucine	0.137	0.140
Lysine	0.155	0.121
Methionine + cystine	0.072	0.098
Phenylalanine + tyrosine	0.133	0.134
Threonine	0.083	0.082
Tryptophan	0.017	0.015
Valine	0.139	0.144

* Assumptions: 350-g chick gaining 30 g/day (28 g soft tissue + 2 g feathers); soft tisssue = 180 g protein /kg (amino acid composition from Scott, Nesheim and Young, 1976), feathers = 820 g protein/kg (amino acid composition from Block and Bolling, 1951)

in the chick on the pattern of amino acids that would be predicted as the requirement from muscle composition alone. It is evident that the composition of feather protein could influence the pattern required by the young chick, especially in the cases of the sulphur amino acids, which represent a much larger proportion of feather protein than muscle, and lysine and histidine of which the converse is true. These adjustments which feathers make to the required pattern would improve the correlations between predicted and observed requirements for lysine and sulphur amino acids shown in *Figure 8.3*.

Since the relative rates of soft tissue and feather growth will differ at different ages, the pattern required might change with age. Chung, Griminger and Fisher (1973) determined the responses of chicks to lysine and methionine at 1–3 and 5–7 weeks of age. The birds were very slow-growing, which would tend to emphasize differences due to maintenance effects, and those fed on the methionine-limiting diets grew more rapidly than those fed on the lysine-limiting diets, which produces an anomalously small lysine : methionine requirement ratio when these are expressed as rates of intake (mg/day). In addition, the responses to methionine were not very satisfactory. However, the results (*Figure 8.4*) allow comparison of the ratios between these requirements under similar conditions at different ages. Using values chosen by inspection it is evident that the ratio decreases markedly with age. This presumably reflects the changing relationship between soft tissue gain and feather growth with increasing body weight.

In animals such as the pig, in which coat growth is much less significant, such effects will have less effect on the amino acid pattern required. The subject of the effect of keratins on amino acid requirements has been considered at length by Mitchell (1959, 1962).

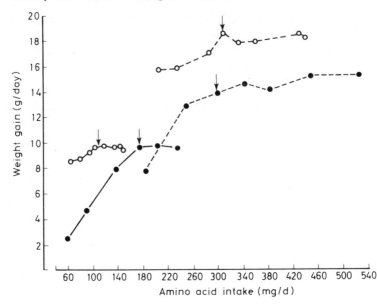

Figure 8.4   Response in body-weight gain (g/day) of chicks to dietary methionine (○) or lysine (●) (mg/day) at 1–3 weeks (continuous lines) or 5–7 weeks (broken lines). Arrow represent requirements for maximum growth chosen by inspection by present reviewer. Ratios of lysine : methionine requirements at 1–3 weeks = 1.58 (174 : 110 mg/day) and at 5–7 weeks = 0.97 (300 : 310 mg/day). (Recalculated from Chung, Griminger and Fisher, 1973)

These 'maintenance' processes do not entirely describe the alternative fates of amino acids in the body. Pathways other than protein deposition will presumably be stimulated by growth and production. The use of methionine, for example, in acting as a carrier and supplier of methyl groups, will itself be divisible into 'maintenance' and 'production' compartments. However, the quantitative effects of these processes on the pattern of amino acids required in the fast growing animal seem small in comparison with depositions in tissues, from evidence such as that presented in *Figure 8.3* and *Table 8.1*.

Other productive processes, such as milk secretion and egg laying, must also impose their amino acid patterns on dietary requirements. *Table 8.2* shows the pattern of the experimentally determined requirements for lysine, tryptophan

Table 8.2. PATTERN OF REQUIREMENTS FOR LYSINE, TRYPTOPHAN AND METHIONINE IN THE LAYING HEN, AND PATTERN OF THE SAME AMINO ACIDS IN EGG PROTEIN

	Pattern	
	(each amino acid as fraction of all three)	
	requirement*	egg protein†
Lysine	0.591	0.595
Tryptophan	0.134	0.124
Methionine	0.276	0.281

* From Agricultural Research Council (1975).
† From Scott, Nesheim and Young (1976).

and methionine for the laying hen in comparison with that of the same amino acids in egg protein. The influence of the pattern in egg protein on the requirement pattern is evident.

It is possible to do much more complex analyses of the responses of animals to amino acids and partition requirements into maintenance and production components. This has been done for the laying hen (Fisher, 1976) and the effect of the amino acids in egg on the partial requirement for production is even more marked than that on overall requirement (Morris, 1974). A similar approach can be adopted with the growing bird (Freeman, 1979). These approaches pertain to the quantitation of response discussed below and by Fisher (1980). The foregoing in this review indicates that, in the absence of more detailed information, the pattern of essential amino acids in the product can be used as a basis for the pattern in the diet.

CONVERTING OPTIMAL BALANCE TO QUANTITATIVE INPUTS

If a satisfactory pattern of amino acids can be defined for inclusion in the diet, and this approach has been adopted for poultry (Agricultural Research Council, 1975) and suggested for pigs (Cole, 1978), it is necessary to define the quantitative input of one amino acid to allow all the others to be converted to quantitative terms. There is a need, therefore, to define the dose—response relationship for a reference amino acid. It is often assumed that this relationship will need to be defined for different production circumstances, different diets (e.g. energy

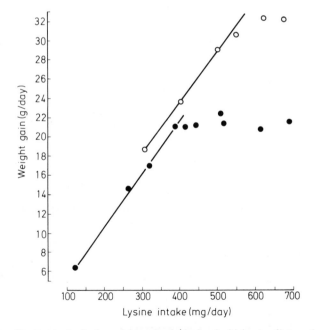

Figure 8.5   Responses in body-weight gain (g/day) of chicks to dietary lysine intake (mg/day) in experiments of Griminger and Scott (1959) (●) and Lewis 1978 (○). Lines are fitted to means of data for linear ranges (Griminger and Scott : $Y = 0.055X - 0.405$; Lewis : $Y = 0.052X - 2.619$)

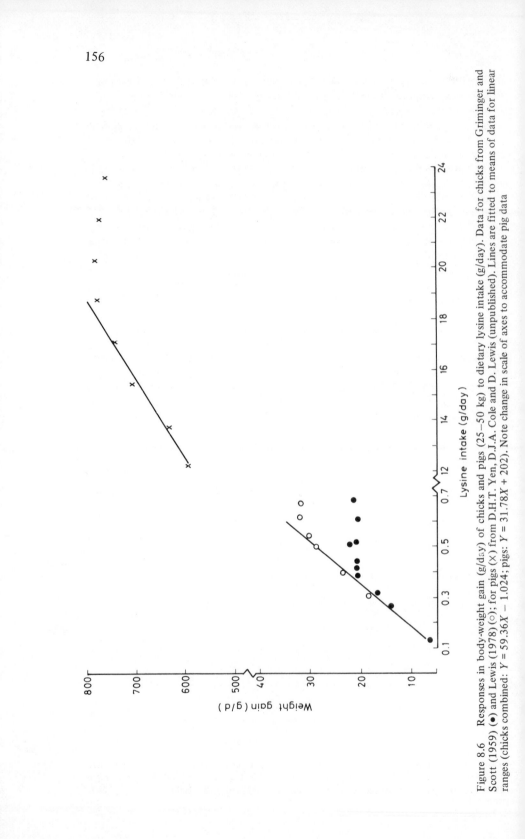

Figure 8.6 Responses in body-weight gain (g/day) of chicks and pigs (25–50 kg) to dietary lysine intake (g/day). Data for chicks from Griminger and Scott (1959) (●) and Lewis (1978) (○); for pigs (×) from D.H.T. Yen, D.J.A. Cole and D. Lewis (unpublished). Lines are fitted to means of data for linear ranges (chicks combined: $Y = 59.36X − 1.024$; pigs: $Y = 31.78X + 202$). Note change in scale of axes to accommodate pig data

contents) and different stock characteristics (i.e. breed, sex). However, it is possible that by choosing the appropriate variables for dose and response many of these differences can be rationalized and a single relationship might be possible for a species, or even for several species.

D'Mello (1978) rationalized several of the differences in the requirements of growing poultry for sulphur amino acids by plotting output (weight gain) as a function of intake of amino acid. It is probable that further variation would have been removed by using lean gain as the output. The latter approach would be of benefit in extending the idea to pigs and larger animals.

The similarity between animals of the same species is exemplified in *Figure 8.5*, which shows the response of young chicks to lysine in two experiments separated by 20 years. During that period maximum rate of gain increased by 50% and the heavier birds require more lysine; however, the rate of response to lysine is very similar in both cases and the lysine need of the heavier birds could have been predicted quite accurately from a knowledge of their growth rate and the rate of response found by Griminger and Scott (1959).

D'Mello (1978) also pointed out that the response rate to methionine was similar in the young rat and turkey to that in the young chick. These animals are all of similar body weight when young, but whether the response rate is similar for much larger animals is questionable. Ideally, if maintenance requirement is related to body weight and entirely independent of production the effect

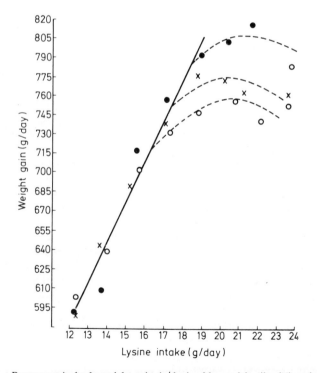

Figure 8.7   Responses in body-weight gain (g/day) of boars (●), gilts (X) and castrates (○) to dietary lysine intake (g/day). Continuous line is fitted to means of data from linear ranges for each sex, pooled ($Y = 31.75X + 202$). Broken lines are indications of limit for each sex. (Data of D.H.T. Yen, D.J.A. Cole and D. Lewis, unpublished)

of the larger body size should be on the intercept of such response curves rather than the slope. *Figure 8.6* shows the response of the 25–50 kg pig to lysine in comparison with that of the chick. The response rate of the pig is evidently less, indicating presumably that with such large changes in body weight not all of the effects of body weight are accounted for by changes in the intercept of the response curve. This is further evidence of the unsatisfactory nature of the term 'maintenance'. It should also be noted that there may be poorer utilization of free lysine in the pig under meal feeding conditions (Batterham, 1979).

In the pig alone, much of the differences due to sex can be rationalized simply by using lysine intake as the input (*Figure 8.7*). As with the chick, the requirement is essentially determined by growth rate, there being little difference between sexes in the rate of response. On the basis of these data resort to lean gain to express output might be unnecessary, although fuller analysis might reveal differences in response rates between boars and castrates.

CONSEQUENCES OF DEPARTURE FROM IDEAL BALANCE

This subject has been reviewed extensively elsewhere (Harper, Benevenga and Wohlhueter, 1970; Rogers, 1976) and only some points pertaining to nitrogen retention will be discussed here.

Classical amino acid imbalance is caused by the addition of an incomplete mixture of essential amino acids to a balanced lower-protein diet, which serves as the control diet. The distinction between this condition and deficiency, where the control diet is a higher-protein balanced diet has been explained recently by Boorman (1979). The effect of amino acid imbalance on absolute nitrogen retention, compared with control animals is unpredictable. Amino acid imbalance will generally cause a decreased food intake which produces a decrease in growth rate. This may be reflected in a decrease in nitrogen retention; however, body composition may change and partially offset this decrease. The net effect of these trends will depend on the severity of the imbalance imposed, its duration, the animal's previous protein status and probably other factors. The efficiency of utilization of dietary nitrogen for protein deposition will decrease in severe imbalance, as the diet is providing a large supplement of un-utilizable amino acids. These trends will be illustrated more specifically below.

The response in food intake to amino acid imbalance is now seen as a direct and primary phenomenon (see Harper, Benevenga and Wohlhueter, 1970; Boorman, 1979) and is assumed to be largely responsible for the effect on growth rate. (Fisher *et al.*, 1960; Kumta and Harper, 1961; Klain, Vaughan and Vaughan, 1962). Since addition of the amino acid lacking in the imbalancing mixture (limiting amino acid) allows restoration of food intake and growth rate, it was originally assumed that imbalancing mixtures impaired the utilization of the limiting amino acid. Fisher *et al.* (1960) were the first to question this, and since then a deeper understanding of amino acid metabolism during imbalance has illustrated that poor utilization of the limiting amino acid is unlikely and that the converse may be the case.

The feeding of an imbalanced amino acid supply stimulates the uptake of the limiting amino acid into some proteins, especially in the liver (Harper, Benevenga and Wohlhueter, 1970). The supply of the limiting amino acid for these syntheses is replenished ultimately from the muscles via the plasma free amino acid pool. It is possible, therefore, after an imbalanced diet has been fed for some days to a young animal for there to be an enhanced anabolic state in liver proteins while there is a net catabolic state in the muscles (Sidransky and Verney, 1970). No increase in the overall catabolic rate of the limiting amino acid can be demonstrated (Harper, Benevenga and Wohlueter, 1970). Thus poorer utilization of the limiting amino acid seems unlikely.

The conflict between the former view of impaired utilization of the limiting amino acid and subsequent understanding of imbalance is a conflict of semantics, there being a poorer growth rate due to the decreased food intake but not a poorer utilization of the limiting amino acid within the body.

It is clear therefore that studies on nitrogen retention in imbalance should make comparisons between animals eating equal or similar amounts of imbalanced and control diets. The study of Kumta, Harper and Elvehjem (1958) showed no impairment of the proportion of dietary nitrogen retained, which implies no decrease in absolute nitrogen retention, in rats receiving amounts of relatively mildly imbalanced diet equal to those of a control diet fed to other rats. Fisher and Shapiro (1961) equalized intake of imbalanced and control diets fed to chicks by adjusting energy concentrations and measured carcass nitrogen. They found no decrease in birds fed on the imbalanced diet. In fact, in both these studies there appeared to be some evidence of increased efficiency of nitrogen retention. This supports the more general statement of Harper, Benevenga and Wohlhueter (1970) that 'in some . . . studies it appeared as if an imbalance might actually increase efficiency of amino acid utilization'.

Tobin, Boorman and Lewis (1973) studied the effects of increasing severities of amino acid imbalance on chicks, including some very mild imbalances. Such mild imbalances do not cause depressions in food intake for several days and under these conditions a small transitory increase in growth rate in response to imbalance occurs which is reflected in an increased food intake (Boorman, 1979). Some data for histidine imbalance are shown in *Table 8.3*, in which the transitory increases are sufficient to be reflected in trends over 10 days in young chicks. Although absolute nitrogen retentions were not measured, final carcass

Table 8.3. EFFECT OF INCREASING SEVERITY OF HISTIDINE IMBALANCE ON THE YOUNG CHICK (AFTER TOBIN, 1974)

Imbalancing addition to diet (g/kg)*	Food intake (g/bird per day)	Weight gain (g/bird per day)	Carcass nitrogen (g/kg)	Nitrogen retained‡ (mg/bird per day)	(mg/mg N ingested)
0†	20.4	6.00	26.8	132	0.403
25	22.0	8.08††	26.8	187	0.426
50	18.8	7.35**	28.1	189	0.418
75	15.4††	5.55	29.8††	163	0.379
100	12.1††	4.30††	30.2††	131	0.338

* Imbalancing mixture of essential amino acids lacking histidine.
† Control diet containing 16 g N/kg as balanced protein.
‡ Assuming initial carcass nitrogen content of 30 g/kg.
** Values are means of six replicates of four birds each, differences from control significant at $P < 0.05$.
†† Values are means of six replicates of four birds each, differences from control significant at $P < 0.01$.

nitrogen contents were, and from initial weights and an assumed initial carcass nitrogen content typical of these birds, relative trends in nitrogen retention can be computed. As observed by Fisher and Shapiro (1961), carcass nitrogen content increases with dietary amino acid imbalance, although with severe imbalance this does not represent a net increase in nitrogen gain because of the small body-weight gain. At less severe intensities of imbalance, food intake is not so severely affected, or may be increased, body-weight gain is therefore maintained and the increased carcass nitrogen content is reflected in an increased absolute nitrogen retention. Whether or not this represents a more efficient utilization of dietary nitrogen for nitrogen deposition will depend on the size of the increase in nitrogen retention and the size of the imbalancing mixture. Large imbalancing mixtures represent increases in un-utilizable protein in the diet and will inevitably decrease the efficiency of utilization of dietary nitrogen.

Thus it seems very likely that amino acid imbalance increases the efficiency of utilization of the limiting amino acid. This increased efficiency allows utilization of some of the imbalancing mixture for protein deposition. Increases in absolute nitrogen retention and efficiency of utilization of dietary nitrogen are therefore possible. These will only be detectable, however, if the decrease in food intake and the size of the imbalancing mixture, respectively, do not mask them.

Dietary energy and nitrogen utilization

Dietary energy and protein interact because protein provides part of the dietary energy, because energy is needed for protein turnover and deposition, and because body protein is part of the stored energy of the body. Aspects of the interaction have been reviewed elsewhere (e.g. Swanson, 1959; Cole, Hardy and Lewis, 1972; Kofrányi, 1972; Buttery and Boorman, 1976; Kielanowski, 1976; Lindsay, 1976; Lewis, 1978). Some points of relevance to nitrogen retention and responses to nitrogen intake will be made here.

The complexity of the interaction between protein and energy is such that a full quantitative description, integrating all aspects, has not been made. Miller and Payne (1963) extended the quantitative description of nitrogen retention represented in *Figure 8.2* to include values for nitrogen intake partitioned into that used for protein deposition and that used for metabolizable energy. This approach was found to have useful, and sometimes impressive, predictive value for rates of gain in rats and other mammals, including pigs and calves, and in chicks. Quantitative models of feeding systems will also include descriptions of responses to protein and energy in the diet (Dent, 1972; Whittemore, 1978).

Insufficient dietary energy will prevent full utilization of dietary nitrogen and impose a threshold on nitrogen retention which is lower than the maximum determined by body weight. This assumption is implicit in most treatments of the subject (Miller and Payne, 1963; Carr *et al.*, 1977). However, in addition to influencing the maximum nitrogen retention that can be achieved, there is a 'nitrogen-sparing' effect of dietary energy (Munro, 1951) which seems to operate throughout the range of nitrogen intakes (Fuller *et al.*, 1973). Thus energy supply can change the form of the nitrogen retention response curve (*Figure 8.2*) throughout its length. Fuller *et al.* (1973) fed rats in such a way as to vary protein intake, protein quality and energy (starch) intake. Their results showed continuous responses in nitrogen retention to increasing energy intake at all nitrogen intakes. These responses tended to a limit at lower starch intakes with proteins of poor quality than with proteins of good quality. The results were not expressed in a form to show the effect of energy intake on the relation between nitrogen retention and nitrogen intake. *Figure 8.8* shows the effect of energy intake on this relationship plotted from the equations given by the authors. These data show that the effect of energy is on the intercept of the response curve rather than its slope. The equations given by Fuller and Crofts (1977) for the pig are more complex and imply a small curvature in the response of nitrogen retention to nitrogen intake. Plots of these data for starch intakes of 20 and 60 g/pig per

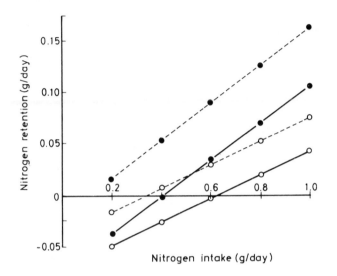

Figure 8.8   The effect of starch intake on the relation between nitrogen retention (N_R, g/day) and nitrogen intake (N_I, g/day) in the rat. Data calculated from the equations of Fuller *et al.* (1973) for good-quality protein ($N_R = -0.108 + 0.180N_I + 0.250S - 0.00070S^2 - 0.192W$) and poor-quality protein ($N_R = -0.048 + 0.113N_I + 0.236S - 0.00142S^2 - 0.439W$), where W = (body weight)$^{0.73}$ in kg and S = starch intake in g/day. Figure shows data for good (●) and poor quality (○) proteins at three (continuous lines) and six (broken lines) g starch/day for a 100 g rat.

day indicated essentially parallel curves. The implications of these relationships in biological terms are beyond the scope of this treatment, but the idea that energy intake can influence endogenous nitrogen losses is not consistent with the supposed absolute nature of these losses.

Figure 8.8 also shows that the effect of dietary energy on nitrogen retention is greater for a good-quality protein than for a poor-quality one, a fact commented on by the authors (Fuller *et al.*, 1973).

These data are the most complete description of this aspect of energy–protein interactions in the non-ruminant animal and extend the concepts of Allison, described earlier, and Miller and Payne (1963) in a logical manner. Unfortunately much of the work on this interaction has been empirical and has not been related to unifying concepts such as nitrogen retention. Another difficulty has been that many authors, in expressing responses in relation to concentrations of energy and protein (amino acids) in the diet, have failed to discern the effects of food intake. These can confuse the interpretation of responses, especially in poultry.

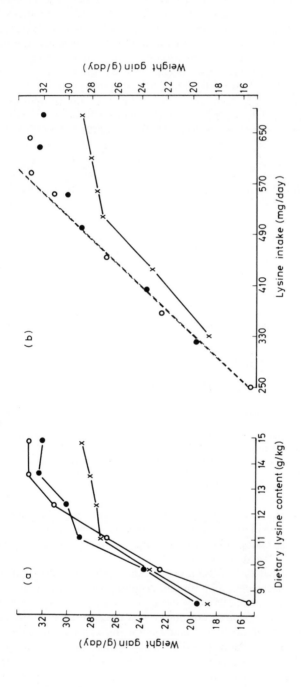

Figure 8.9 Responses in body-weight gain (g/day) to (a) dietary lysine content (g/kg) and (b) dietary lysine intake (mg/day) of chicks fed on diets of different energy contents: 12.6 (×), 13.9 (●) or 15.1 (○) MJ ME/kg. Broken line is fitted to means of data for linear ranges for the two diets of higher energy contents, pooled ($Y = 0.054X + 2.212$). (Data after Lewis, 1978)

D'Mello (1978) showed that the apparent differences due to dietary energy in requirements for sulphur amino acids for the chick, expressed as proportions of the diet, found by Boomgaardt and Baker (1973), were explicable as one response curve when these responses were plotted in relation to intake of sulphur amino acids.

Analysis on the basis of nutrient intake would seem obligatory to discern real effects of an energy–protein interaction. Lewis (1978) reported experiments in which the response of broiler chicks to dietary lysine was measured at three different dietary energy concentrations. *Figure 8.9(a)* shows the results plotted conventionally, and these data could be interpreted to give three values for the lysine requirement. *Figure 8.9(b)* shows that the responses with the two diets of higher energy content are essentially the same response, giving one value for lysine requirement in terms of intake. At the lowest dietary energy there is evidence of a marked limitation in the response to dietary lysine and this is evidence of a true effect of dietary energy on protein utilization for growth.

Table 8.4.   FOOD (g/CHICK PER DAY) AND ENERGY (kJ/ME/CHICK PER DAY) INTAKES OF CHICKS FED ON DIETS OF DIFFERENT LYSINE AND ENERGY CONTENTS FROM LEWIS, 1978)

Dietary lysine (g/kg)	Dietary energy (MJ/ME/kg)					
	12.6		13.9		15.1	
	food intake	energy intake	food intake	energy intake	food intake	energy intake
8.5	39.1	490	37.8	524	29.6	445
9.8	44.5	559	41.2	571	37.4	563
11.0	47.2	593	45.8	635	41.2	620
12.3	45.3	568	44.8	621	45.1	678
13.5	45.2	568	46.3	642	43.4	653
14.8	45.8	574	45.9	636	43.4	653

Table 8.4 shows the energy intakes of these chicks, indicating also that chicks fed on the lowest energy diet did not compensate for the lower dietary energy by increasing food intake sufficiently, and therefore the energy supply was inferior to that in chicks fed on the higher energy diets and evidently inadequate for maximum protein utilization.

Acknowledgements

My interest in this subject stems from my mentors in protein nutrition, the late Professor J.B. Allison, Professor H. Fisher and Professor D. Lewis. This interest was rekindled more recently by contact with Dr John Carr, who also did the hard work of collating the data on nitrogen retention in the pig. I am very grateful to my colleague Mr D.H.T. Yen and his supervisors Dr D.J.A. Cole and Professor D. Lewis for allowing me to quote their data on the lysine response of the pig before they appear elsewhere.

References

AGRICULTURAL RESEARCH COUNCIL (1975). *Nutrient Requirements of Farm Livestock* – No. 1, *Poultry,* Agricultural Research Council, London

AGUILAR, T.S., HARPER, A.E. and BENEVENGA, N.J. (1972). *J. Nutr.*, **102**, 1199–1208

ALLISON, J.B. (1957). *J. Am. Med. Ass.*, **164**, 283–289

ALLISON, J.B. (1959). In *Protein and Amino Acid Nutrition*, pp. 97–116 (Albanese, A.A., Ed.), Academic Press, New York

ALLISON, J.B. (1964). In *Mammalian Protein Metabolism*, Vol. 2, pp. 41–86 (Munro, H.N. and Allison, J.B., Eds), Academic Press, New York

ALLISON, J.B. and WANNEMACKER, R.W. (1965). *Am. J. Clin. Nutr.*, **16**, 445–452

ALLISON, J.B., ANDERSON, J.A. and SEELEY, R.D. (1946). *Ann. N.Y. Acad. Sci.*, **47**, 245–271

BATTERHAM, E.S. (1979). In *Recent Advances in Animal Nutrition – 1979*, pp. 11–22 (Haresign, W. and Lewis, D., Eds), Butterworths, London

BLOCK, R.J. and BOLLING, D. (1951). *The Amino Acid Composition of Proteins and Foods. Analytical Methods and Results*, C.C. Thomas, Springfield, Illinois

BOOMGAARDT, J. and BAKER, D.H. (1973). *J. Anim. Sci.*, **36**, 307–311

BOORMAN, K.N. (1979). In *Food Intake Regulation in Poultry*, pp. 87–126 (Boorman, K.N. and Freeman, B.M., Eds), British Poultry Science, Edinburgh

BUNCE, G.E. and KING, K.W. (1969a). *J. Nutr.*, **98**, 159–167

BUNCE, G.E. and KING, K.W. (1969b). *J. Nutr.*, **98**, 168–176

BUTTERY, P.J. and BOORMAN, K.N. (1976). In *Protein Metabolism and Nutrition*, pp. 197–206 (Cole, D.J.A., Boorman, K.N., Buttery, P.J., Lewis, D., Neale, R.J. and Swan, H., Eds.), Butterworths, London

CARR, J.R., BOORMAN, K.N. and COLE, D.J.A. (1977). *Br. J. Nutr.*, **37**, 143–155

CHUNG, E., GRIMINGER, P. and FISHER, H. (1973). *J. Nutr.*, **103**, 117–122

COLE, D.J.A. (1978). In *Recent Advances in Animal Nutrition – 1978*, pp. 59–72 (Haresign, W. and Lewis, D., Eds), Butterworths, London

COLE, D.J.A., HARDY, B. and LEWIS, D. (1972). In *Pig Production*, pp. 243–257 (Cole, D.J.A., Ed.), Butterworths, London

DENT, J.B. (1972). In *Pig Production*, pp. 259–277 (Cole, D.J.A., Ed.), Butterworths, London

D'MELLO, J.P.F. (1978). In *Recent Advances in Animal Nutrition – 1978*, pp. 1–15 (Haresign, W. and Lewis, D., Eds), Butterworths, London

FISHER, C. (1976). In *Protein Metabolism and Nutrition*, pp. 323–351 (Cole, D.J.A., Boorman, K.N., Buttery, P.J., Lewis, D., Neale, R.J. and Swan H., Eds), Butterworths, London

FISHER, C. (1980). In *Protein Deposition in Animals* (these Proceedings), Ch. 14 (Buttery, P.J. and Lindsay, D.B., Eds), Butterworths, London

FISHER, H. (1967). In *Newer Methods of Nutritional Biochemistry*, Vol. 3, pp. 101–124 (Albanese, A.A., Ed.), Academic Press, New York

FISHER, H. and SHAPIRO, R. (1961). *J. Nutr.*, **75**, 395–401

FISHER, H., GRIMINGER, P., LEVEILLE, G.A. and SHAPIRO, R. (1960). *J. Nutr.*, **71**, 213–220

FISHER, H., GRUN, J., SHAPIRO, R. and ASHLEY, J. (1964). *J. Nutr.*, **83**, 165–170

FREEMAN, C.P. (1979). *Br. Poult. Sci.*, **20**, 27–37

FULLER, M.F. and CROFTS, R.M.J. (1977). *Br. J. Nutr.*, **38**, 479–488

FULLER, M.F., BOYNE, A.W., ATKINSON, T. and SMART, R. (1973). *Nutr. Rep. Int.*, **7**, 175–180

GRIMINGER, P. and SCOTT, H.M. (1959). *J. Nutr.*, **68**, 429–442

HARPER, A.E., BENEVENGA, N.J. and WOHLHUETER, R.M. (1970). *Physiol. Rev.*, **50**, 428–558

HOLT, L.E., HALAC, E. and KAJDI, C.N. (1962). *J. Am. Med. Ass.,* **181**, 699—705

HURWITZ, S. and BORNSTEIN, S. (1973). *Poult. Sci.,* **52**, 1124—1134

KIELANOWSKI, J. (1976). In *Protein Metabolism and Nutrition*, pp. 207—215 (Cole, D.J.A., Boorman, K.N., Buttery, P.J., Lewis, D., Neale, R.J. and Swan, H., Eds), Butterworths, London

KLAIN, G.J., VAUGHAN, D.A. and VAUGHAN, L.N. (1962). *J. Nutr.,* **78**, 359—364

KOFRÁNYI, E. (1972). In *Protein and Amino Acid Functions*, pp. 45—56 (Bigwood, E.J., Ed.), Pergamon Press, Oxford

KUMTA, U.S. and HARPER, A.E. (1961). *J. Nutr.,* **74**, 139—147

KUMTA, U.S., HARPER, A.E. and ELVEHJEM, C.A. (1958). *J. Biol. Chem.,* **233**, 1505—1508

LAWRIE, R.A. (1974). *Meat Science*, 2nd edn, Pergamon Press, Oxford

LEWIS, D. (1978). In *Recent Advances in Animal Nutrition — 1978*, pp. 17—30 (Haresign, W. and Lewis, D., Eds), Butterworths, London

LINDSAY, D.B. (1976). In *Protein Metabolism and Nutrition,* pp. 183—195 (Cole, D.J.A., Boorman, K.N., Buttery, P.J., Lewis, D., Neale, R.J. and Swan, H., Eds), Butterworths, London

LINDSAY, D.B. and BUTTERY, P.J. (1980). In *Protein Deposition in Animals* (these Proceedings), Ch. 7 (Buttery, P.J. and Lindsay, D.B., Eds), Butterworths, London

McCANCE. R.A. (1977). In *Growth and Poultry Meat Production*, pp. 1—11 (Boorman, K.N. and Wilson, B.J., Eds), British Poultry Science, Edinburgh

MILLER, D.S. and PAYNE, P.R. (1963). *J. Theoret. Biol.,* **5**, 398—411

MILLWARD, D.J., GARLICK, P.J., JAMES, W.P.T., SENDER, P.M. and WATERLOW, J.C. (1976). In *Protein Metabolism and Nutrition*, pp. 49—69 (Cole, D.J.A., Boorman, K.N., Buttery, P.J., Lewis, D., Neale, R.J. and Swan, H., Eds), Butterworths, London

MITCHELL, H.H. (1959). In *Protein and Amino Acid Nutrition*, pp. 11—43 (Albanese, A.A., Ed.), Academic Press, New York

MITCHELL, H.H. (1962). *Comparative Nutrition of Man and Domestic Animals*, Vol. 1, pp. 129—191, Academic Press, New York

MORRIS, T.R. (1974). *Wld's Poult. Sci. J.,* **30**, 316—317

MUNRO, H.N. (1951). *Physiol. Rev.,* **31**, 455—488

PAQUAY, R., DEBAERE, R. and LOUSSE, A. (1972). *Br. J. Nutr.,* **27**, 27—37

ROGERS, Q.R. (1976). In *Protein Metabolism and Nutrition*, pp. 279—301 (Cole, D.J.A., Boorman, K.N., Buttery, P.J., Lewis, D., Neale, R.J. and Swan, H., Eds), Butterworths, London

RUFEGER, H. (1966). *Z. Tierphysiol. Tierernähr. Futtermittelk.,* **37**, 2—14

SCOTT, M.L., NESHEIM, M.C. and YOUNG, R.J. (1976). *Nutrition of the Chicken*, M.L. Scott and Associates, Ithaca, N.Y.

SIDRANSKY, H. and VERNEY, E. (1970). *Proc. Soc. Exp. Biol. Med.,* **135**, 618—622

SMITH, W.K. (1978). *Wld's Poult. Sci. J.,* **34**, 81—96

SMUTS, D.B. (1935). *J. Nutr.,* **9**, 403—433

SWANSON, P. (1959). In *Protein and Amino Acid Nutrition*, pp. 195—224 (Albanese, A.A., Ed.), Academic Press, New York

SWICK, R.W. and BENEVENGA, N.J. (1977). *J. Dairy Sci.,* **60**, 505—515

TOBIN, G. (1974). 'Amino acid interactions in the young cockerel', Ph.D. thesis, University of Nottingham

TOBIN, G., BOORMAN, K.N. and LEWIS, D. (1973). In *4th European Poultry Conference*, London, pp. 307—315

VAN ES, A.J.H. and BOEKHOLT, H.A.(1976). In *Protein Metabolism and Nutrition*, pp. 441–455 (Cole, D.J.A., Boorman, K.N., Buttery, P.J., Lewis, D., Neale, R.J. and Swan, H., Eds), Butterworths, London

WESSELS, J.P.H. and FISHER, H. (1965). *Br. J. Nutr.*, **19**, 57–69

WETHLI, E., MORRIS, T.R. and SHRESTA, T.P. (1975). *Br. J. Nutr.*, **34**, 363–373

WHITTEMORE, C.T. (1977). In *Recent Advances in Animal Nutrition – 1977*, pp. 158–166 (Haresign, W. and Lewis, D., Eds), Butterworths, London

9

HORMONAL CONTROL OF PROTEIN METABOLISM, WITH PARTICULAR REFERENCE TO BODY PROTEIN GAIN

VERNON R. YOUNG
Department of Nutrition and Food Science and Clinical Research Center, Massachusetts Institute of Technology

Summary

The hormonal control of protein metabolism is an extensive subject. Not only do numerous well-defined hormones affect the extent and rate of protein metabolism in animals, but there appears to be a variety of growth and differentiation factors which have hormone-like activity. The production of all these hormones and factors requires an elaborate regulatory mechanism. The endocrine system itself influences growth by regulating intake and by controlling the division of nutrients among various metabolic pathways. The cellular sites of action of hormones depend largely on the hormone and range from effects on cellular and subcellular membranes to alterations in the activity of enzymes and protein factors concerned with energy and nucleic acid metabolism and protein synthesis and breakdown. The mode of action of insulin, growth hormone, the thyroid hormones, the glucocorticoids, and steroids with anabolic activity, are discussed in detail. One of the factors which emerges from the considerations undertaken is that little fundamental knowledge exists in the area of hormonal regulation of protein metabolism, particularly in reference to growth and the use of hormones in animal protein production for improving the supply of human foods.

Introduction

The hormonal control of protein metabolism, with particular reference to growth, is too broad a topic to be reviewed comprehensively in this chapter. This is due to a number of reasons; first, growth itself is a complex and integrated, but still incompletely understood, process. It may be described, however, according to allometric relationships (Huxley, 1932; Brody, 1945) and involves formation and/or enlargement of metabolic units, in addition to differentiation and morphogenesis.

Secondly, it needs to be recognized that the endocrine system itself is a highly integrated system. Thus, an understanding of the factors that determine the total hormonal picture will be necessary in order to define precisely the role of hormones in the regulation of tissue and cell protein metabolism. This point is underscored by observations revealing interactions among hormones at common target gene sites. Examples of these are summarized in *Table 9.1*. These various findings indicate the probability that pre-translational control of cell function is achieved by a *specific balance among hormones* in the cell's environment (e.g. Oppenheimer, 1979). Furthermore, these interactions make it difficult to predict, in quantitative terms, the endocrine factors and other host and environmental conditions necessary for maximizing net organ and body protein gain in the management of farm livestock.

167

Table 9.1. EXAMPLES OF SOME INTERACTIONS AMONG HORMONES AT A COMMON SPECIFIC TARGET GENE

Interaction	Reference
1. α-2u globulin mRNA requires cortisol, T_3, testosterone, and growth hormone	Kurtz and Feigelson (1978)
2. GH mRNA requires T_3 and glucocorticoids	Shapiro *et al.* (1978)
3. T_3-induced malic enzyme blocked by glucagon in chick hepatocytes	Goodridge and Adelman (1976)
4. TSH synthesis by pituitary cells stimulated by oestrogen	Miller *et al.* (1977)

Thirdly, protein metabolism is intimately affected by, and related to, the status of energy, carbohydrate and lipid metabolism (Munro, 1964) in the whole organism, and these different phases of cellular metabolism and nutrient utilization are themselves influenced by hormones.

Finally, the general topic of hormones in relation to protein metabolism and animal production has been reviewed by various workers (e.g. Turner and Munday, 1976; Manchester, 1976; Young and Pluskal, 1977) and these are recommended for additional details.

Hence, this chapter will be restricted to a discussion of some selected topics that the reader might find stimulating and to help assess the specific manipulation in hormonal balance, within a given genotype, that might achieve a maximum efficiency in animal protein production, given limited land and food resources and a particular set of farm livestock management conditions.

The central importance of hormones in the regulation of protein metabolism and growth needs little emphasis here. However, it is now well recognized that a fundamental aspect of the regulation of metabolism, indeed the division of metabolic labour existing in pluricellular organisms, is the production by certain cells of special chemical signals that are utilized or manipulated by other cells (Rutter, 1978). Among these signals are the hormones produced in cells, usually at a site that is distant from their target cells.

Table 9.2. IMPACT OF DEFICIENCY OF VARIOUS HORMONES ON RAT GROWTH

Hormone	Growth response
Growth hormone	Depressed
Thyroxine	Depressed
Corticosteroid	Normal
Insulin	Depressed
Testosterone	Slowed
Oestrogen	Little change
Peptide tissue growth factors	?

The hormones usually included in a review of protein metabolism and body protein gain are listed in *Table 9.2*, perhaps with the exception of growth factors that have been added here in order to recognize their existence. Because the significance of these latter compounds in the farm animal is not yet understood, they will not be discussed in detail here. However, a listing of these hormone-like factors is given in *Table 9.3*. For the most part, they are important in the control of developmental processes, with many of them affecting proliferation of cells *in vitro* (Daugahady, 1977; Carpenter, 1978; Van Wyk and Underwood, 1978; Rudland and DeAsua, 1979). An intriguing possibility is that these factors may interact with hormones giving rise to synergistic growth and developmental responses.

Table 9.3. SOME GROWTH AND DIFFERENTIATION FACTORS WITH HORMONE-LIKE ACTIVITY*

Compound	Target cell
Nerve growth factor	Sympathetic nerves
Epithelial growth factor	Epidermal cells
Fibroblast growth factor	Mesodermal cells
Ovarian growth factor	Ovary
Myoblast growth factor	Myoblasts
Thrombopoietin	Thrombocytes
Thymosin	Thymocytes
Melanocyte stimulating	Melanocyte
Somatomedins	Cartilage
NSILA †	Various tissues

* Partial summary and slight modification of Rutter (1978).
† Non-suppressible insulin-like activity.

The following sections will briefly consider the regulation of the endocrine system itself, because this knowledge is essential to achieving a rational manipulation of the cell's hormonal environment for purposes of efficient animal production objectives. Some general aspects of hormone action will then be discussed, with emphasis on hormone receptors and tissue responsiveness, because of the potential importance that recent developments in this area may have for manipulation of body protein gain. Newer knowledge concerning the mode of action of 'growth-regulating hormones' will be reviewed. Finally, a summary of these topics will be made with the hope that this may help define systems aimed at a more effective production of human food from animal and farm livestock sources.

Regulation of the endocrine system

The transfer of chemical and metabolic information within the body is achieved by intra- and inter-cellular messengers which bridge short distances by local hormones or neurochemical transmission along nerve fibres. Hence, a large part of the endocrine system may, in fact, be regarded as an effector organ of the nervous system. Furthermore, as depicted in *Figure 9.1*, other environmental factors, as well as the metabolic state of tissues influence central nervous system (CNS) activity, and in turn, influence the endocrine system. This interactive loop must be considered in the *in vivo* regulation of tissue protein metabolism by hormones and in a comprehensive consideration of the use of hormones and hormone-like compounds in animal production.

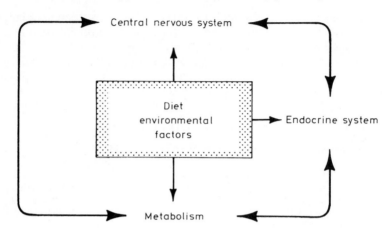

Figure 9.1   Schematic statement of the interrelationships between the central nervous system, the endocrine organs and the nutritional and metabolic status of the intact organism

Briefly, there are three control systems which enable the CNS to regulate endocrine functions (Young and Landsberg, 1977a). They are:

(1) Areas in the hypothalamus directly control the storage and release of posterior pituitary hormones.
(2) Hypothalamic centres regulate the secretion of anterior pituitary hormones either directly or by modifying feedback relationships of peripheral hormones under trophic control of the anterior pituitary.
(3) The sympatho-adrenal system, which is an extension of the hypothalamus, influences the secretion of other hormones.

With reference to the hypothalamus, there are a number of hormones, known to be synthesized in hypothalamic areas, that are oligopeptide in nature and transported to the anterior lobe of the pituitary by a specialized vascular system (Schally, 1978) (*Table 9.4*). Its regulatory influence extends to the secretion, release or inhibition of anterior pituitary and intermediate lobe hormones.

Catecholamine stimulation of other endocrine cells is another type of regulation of the endocrine system. According to Landsberg (1978), catecholamines appear to regulate peptide hormones that are not directly under the control of the pituitary and also thyroxine. The effects of catecholamines on metabolism and endocrine organs are transitory and result in rapid changes in effector cells.

Table 9.4. HYPOTHALAMIC REGULATORY HORMONES

Hormone	Affected hormone
Vasopressin	
Oxytocin	
Thyrotropin-releasing hormone (TRH)	TSH, prolactin
Gonadotropin-releasing hormone (GnRH)	FSH, LH release
Somatostatin (GH-TIH)	Inhibition of GH, insulin, glucagon
MIF = MSH – release inhibiting factor	MSH
MRF = MSH – releasing factor	MSH
Corticotrophin-releasing factor (CRF)	ACTH

As a consequence, the catecholamines can achieve a prompt mobilization of stored hormones and alteration in the rate of hormone secretion by various endocrine glands. Thus, profound changes in hormone balance are made possible by the actions of catecholamines and these occur, for example, with feeding after a fast, in vigorous exercise and under stress states. As also emphasized by Young and Landsberg (1977b) catecholamine-induced changes in hormone secretion can *anticipate* required changes in body hormonal balance. This is in contrast to the usual feedback loops that only respond to changes that have already taken place. Hence, the sympatho-adrenal system allows the CNS to initiate an integrated endocrine response in accord with the perceived metabolic needs of the organism as a whole.

Finally, it is worth pointing out that the effects of catecholamines on the various endocrine tissues tend to reinforce the direct effects of catecholamines on cellular metabolism and on the blood vascular system (*Figure 9.2*). The

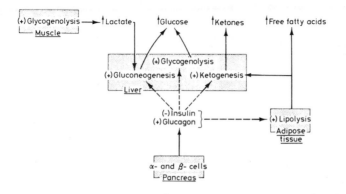

Figure 9.2   Metabolic processes affected by the catecholamines. (From Young and Landsberg, 1977b)

catecholamines have two principal effects on intermediary metabolism: (a) increased rate of energy utilization, and (b) enhanced mobilization of stored fuels for use in metabolizing tissues to accommodate increased metabolic demand. These effects are achieved by activating or inactivating enzymes, via cAMP- and cGMP-dependent systems, as well as by regulation of pancreatic endocrine function. Thus, catecholamines can influence, at least indirectly, the status of cellular protein metabolism by altering hormonal balance, energy substrate mobilization and substrate traffic patterns.

An additional point that should be made, in reference to the regulation of the endocrine system, is that the control of hormone secretion is complex. For example, there are many factors that influence insulin and glucagon secretion, as listed in *Table 9.5*. The importance of these factors is determined by the specific conditions of host. Thus, catecholamines only exert their effects during times of stress (Young and Landsberg, 1977a). Another example of the multiple factors influencing endocrine output balance can be given in reference to the vitamin D endocrine system (Christakos and Norman, 1978). As shown in *Table 9.6*, this system is influenced by many hormones, including growth hormone, insulin and thyroid hormone. Parenthetically, it might be pointed out that body protein

Table 9.5. MULTIPLE FACTORS GOVERNING SECRETION OF HORMONES (e.g. INSULIN AND GLUCAGON)

1. Glucose
2. Amino acids
3. Free fatty acids
4. Hormones (direct, indirect):
   gastrin, pancreozymin,
   secretin, growth hormone,
   glucagon, somatostatin,
   catecholamines

Table 9.6. INTERACTION OF THE VITAMIN D ENDOCRINE SYSTEM WITH OTHER HORMONES (Summarized from Christakos and Norman, 1978)

Hormone	Comment
Vitamin D $(1,25\text{-}(OH)_2D_3)$	Primary regulator of calcium (and phosphorus?) absorption
Calcitonin	Possibly modulates D-induced Ca absorption
Parathyroid hormone	Stimulates $1,25\text{-}(OH)_2D_3$ by kidney
Oestrogens	Possibly affects D metabolism
Growth hormone	Possibly affects D metabolism
Prolactin	Possibly affects D metabolism
Insulin	Diabetes $\rightarrow\downarrow$ production $1,25\text{-}(OH)_2D_3$
Thyroxine	Thyrotoxic patients; reduced $1,25\text{-}(OH)_2D_3$

gain depends not only on the exposure of protein synthesizing cells to an appropriate balance of hormones, but also upon the development of a skeletal system that serves as a structure to which tissues, especially skeletal muscles, can attach.

From the foregoing, it is apparent that the complexity of the regulation of body endocrine balance must be taken into account, if a rational approach to the hormonal manipulation of body protein gain in the whole organism is to be achieved.

General aspects of hormone action

The endocrine system influences growth by regulating nutrient intake (e.g. Baile and Forbes, 1974; Bray, 1978) and the division of nutrients among various metabolic pathways (*Figure 9.3*).

The possible cellular sites of action of hormones on protein metabolism are schematically depicted *Figure 9.4*. These range from effects on cellular and subcellular membranes, to alterations in the activity of enzymes and protein factors concerned with energy and nucleic acid metabolism and protein synthesis and breakdown. Thus, first we should examine briefly some of the factors underlying the qualitative and quantitative specificity of hormonal responses.

The major cellular effects of many, if not all, hormones are brought about initially by the recognition and interaction of the hormone with 'receptor' molecules in responsive cells (*Figure 9.5*). Depending upon the hormone, these receptors are located either on the cell membrane or within the cell. Following formation of the hormone–receptor (H–R) complex, there follows an alteration in the structural organization of the membrane or an interaction with effector

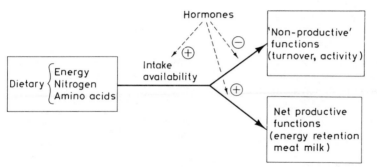

Figure 9.3   Schematic depiction of major influences of hormones on body processes

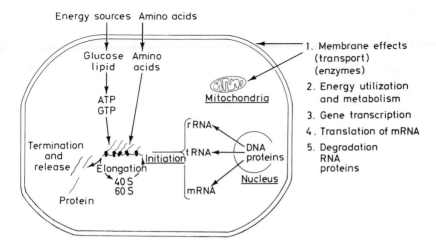

Figure 9.4   Schematic outline of the various sites at which hormones may affect the metabolism of their target cells

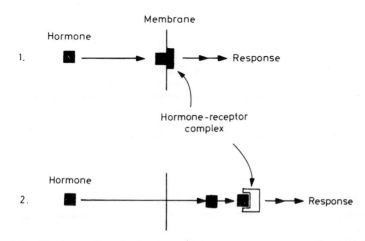

Figure 9.5   The interaction of a hormone with membrane or intracellular receptors is an initial basis for the responsiveness of cells to hormones

molecules, and thus leads to an orderly series of metabolic changes that result in cellular, organ and whole-body responses characteristic of the particular hormone in question.

There have been many excellent reviews concerned with hormone receptors and the mode of action of H–R complexes (e.g. Insel, 1978; Tell, Haour and Saez, 1978; Flier, Kahn and Roth, 1979; Catt *et al.*, 1979). These reviews serve as sources of more detailed information. However, of considerable importance, in reference to the overall topic of this book, is that the concentration of receptors and of effector sites are high in 'target' tissues and lower in hormone-independent tissues.

Hormones that exert their responses by interacting with membrane-bound receptors are listed in *Table 9.7.* Membrane receptors may be defined as the

Table 9.7. MEMBRANE-ACTIVE HORMONES (FROM INSEL, 1978)

Polypeptide hormones
    ACTH, angiotension, calcitonin, EGF, FSH, glucagon, LCG, LH, LH-RF, MSH, NGF, oxytocin, PTH, prolactin, GH, secretion, somatomedin, TSH, TSH-RF, ADH, insulin
Vasoactive amines
    Epinephrine, norepinephrine, dopamine, histidine, serotonin
Prostaglandins
Encephalins? (endorphins)
Acetylcholine?

specific recognition sites at which cell surfaces interact with hormones. A fundamental aspect of this recognition process is that a message is ultimately decoded and cellular function, in consequence, is perturbed by this interaction between hormone and its receptor. For many of these hormones the formation of the H–R complex brings about an activation of adenyl cyclase, via a coupling of transducing element, perhaps calcium. The cAMP generated is then used for the phosphorylation of protein substrates that bring about the observed biological response.

An important advance in this area of hormone action is the recognition that the interaction between the hormone and the receptor can be affected by various factors, the most important being the concentrations of hormone and receptors and the affinity of their interaction. Thus, just as hormone levels fluctuate under a variety of influences, the hormone receptor and its affinity for the hormone vary in response to influences within and outside the cell.

Table 9.8 summarizes the major factors that are known to affect the concentration and affinity of hormone receptors. Therefore, the H–R complex serves

Table 9.8. FACTORS THAT INFLUENCE THE *CONCENTRATION* OR AFFINITY OF HORMONE RECEPTORS (SLIGHTLY MODIFIED FROM FLIER, KAHN AND ROTH, 1979)

R concentration	R affinity	R concentration and affinity
Genetics	Ions	Homologous hormone
Growth rate (diet)	Temperature	Heterologous hormone
Cyclic AMP level		Antibodies to receptor
Cell cycle		
Cell differentiation		
Cell transformation		

as one point of control for the hormonal regulation of metabolism in target cells (see also *Table 9.9*). For example, there are changes in receptor affinity for its own hormone due to a cooperative interaction between the hormone and its receptor (Insel, 1978; Flier, Kahn and Roth, 1979). Cooperativity describes the change in affinity of receptors for the hormone with changes in the concentration of hormone. This process serves as a rapid control mechanism to prevent cells from overacting to high levels of hormones.

Table 9.9. HORMONAL REGULATION OF MEMBRANE RECEPTORS AND CELL RESPONSIVENESS

---

1. *Target cell responsiveness modulated by its target hormone*
   (a) Mechanisms
      (i) *Receptor affinity* changed due to cooperativity
      (ii) *Self-regulation of membrane receptor concentration*
         (A) ? by release of membranes
         (B) ? by internalization of H−R complex
      (iii) Modification of 'coupling system'
2. *Other hormones modulate*
   e.g. Oestrogens: ↑ TRH receptors

Secondly, the *concentration* of membrane receptors is affected by the level of hormone. This has been suggested as an especially important homeostatic regulatory mechanism in cell communication and function. Possible mechanisms include the release of receptors from the membrane, conformational changes in the receptor which renders it inactive, or by degradation either at the membrane level, or by internalization of the H−R complex (*Figure 9.6*). For some hormones, the loss of receptors is blocked by inhibitors of protein synthesis, indicating that

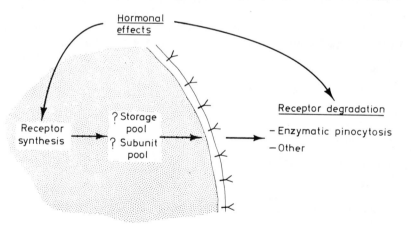

Figure 9.6   An outline of the regulation of receptor concentration by its hormone. (From Eastman, 1977)

receptor loss is a process requiring protein synthesis (Eastman, 1977). In summary, however, it seems probable that hormone receptors are lost in different ways and their recovery is also achieved by a number of possible mechanisms.

The coupling system may also be a site at which regulation of cell responsiveness to target hormones is achieved. Furthermore, there are many additional steps beyond the coupling system that may serve as sites for alterations in tissue responsiveness to hormones.

Finally, as also summarized in *Table 9.9*, in addition to the role played by the hormone in modulating the responsiveness of cells to their own stimulation, a significant function can also be attributed to other hormones which do not directly bind to the receptors being regulated. For example, oestrogens increase TRH receptors in both hepatocytes and adipocytes. However, little is known about the mechanism involved (Flier, Kahn and Roth, 1979).

These advances in our understanding of H–R physiology need to be borne in mind, because they are relevant to an assessment of the dose and rate of hormone administration that might be used for purposes of inducing protein metabolism and increased tissue and body protein gain in the whole organism.

Action of individual hormones

The mode of action of various hormones that are intimately linked with tissue protein metabolism, particularly skeletal muscle, should now be reviewed. Hence, a selected account of some advances will be presented because this topic has been reviewed recently (Young and Pluskal, 1977).

INSULIN

A difficulty that is faced in understanding the mode of action of insulin on tissue protein metabolism is that this hormone exerts a large number of diverse effects in cells (*Table 9.10*). In addition, these actions or cell responses to insulin have different temporal sequences, also as discussed by Goldfine (1978); there are rapid effects on membranes (glucose and amino acid transport), intermediate effects (changes in cytoplasmic enzyme activity and protein synthesis) and long-term effects (DNA synthesis and cell replication). Goldfine (1978) considers that

Table 9.10 ACTIONS OF INSULIN AT VARIOUS SUBCELLULAR LEVELS (FROM GOLDFINE, 1978)

Rapid
    Cell membrane
    Stimulation of transport
    Change of membrane potential
Intermediate
    Cytosol
        activation and inhibition of enzymes
    Endoplasmic reticulum
        activation and inhibition of enzymes
    Ribosome
        increased protein synthesis
    Mitochondria
        activation of enzymes
    Lysosome
        inhibition of protein degradation
Delayed
    Nucleus
        modulation of DNA and RNA synthesis

a comprehensive theory of insulin action must take into account not only the wide variety of cell functions that are related to insulin regulation, but also the time-dependency of these varied responses.

How insulin regulates a number of intracellular processes, including protein metabolism, is not known precisely, but one theory, depicted in *Figure 9.7*, is that the insulin—receptor complex generates a second message at the cell surface and this carries out all subsequent intracellular effects. This remains a hypothesis. Alternatively, it is possible that insulin itself enters target cells (Goldfine, 1978) and then the hormone or product of insulin interacts directly with intracellular structures.

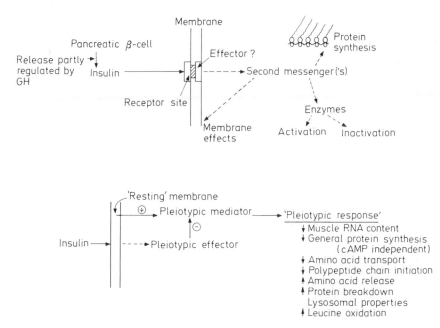

Figure 9.7   Scheme illustrating possible ways by which insulin influences protein metabolism in skeletal muscle cells. (Upper and lower panels are based on reviews by Fritz, 1972, and Tomkins and Geleherter, 1972, respectively, and taken from Young and Pluskal, 1977)

Steiner (1977) has proposed a receptor-induced internalization process that may help to explain the relationship between insulin degradation and insulin action, and also how insulin may enter the cell (*Figure 9.8*). Furthermore, Goldfine (1978) has suggested that there are nuclear-insulin binding sites that are biochemically and immunologically distinct from the receptors at the cell surface. Although the biological significance of these intracellular binding sites is unknown, the possibility exists that they are directly involved in mediating insulin action.

GROWTH HORMONE

As summarized in *Figure 9.9*, growth hormone affects amino acid transport, RNA metabolism and ribosomal aspects of protein synthesis, especially in

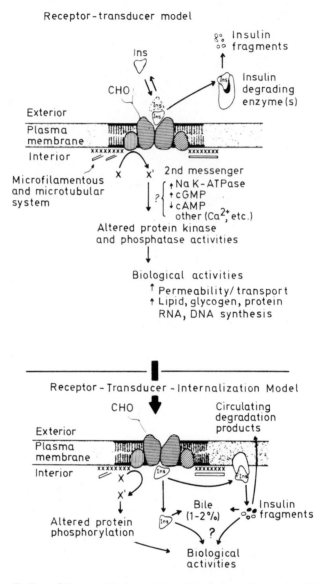

Figure 9.8   Outline of the possible interrelationships between the mode of insulin action and insulin degradation in target cells. (From Steiner, 1977)

muscle (Young, 1970; Young and Pluskal, 1977). In contrast to insulin, growth hormone does not appear to alter muscle protein breakdown (Goldberg, 1971; Flaim, Li and Jefferson, 1978a).

An early *in vitro* effect of growth hormone is to lower cAMP levels in the isolated rat diaphragm and from the work of Kostyo (1968) this change might mediate the rapid effects of growth hormone on protein synthesis and membrane transport in skeletal muscle. However, the *in vivo* significance of this

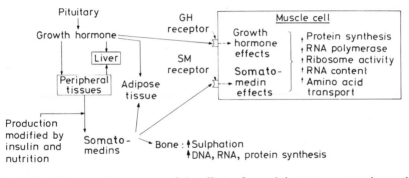

Figure 9.9   Diagrammatic summary of the effects of growth hormone on protein metabolism in muscle cells. (From Young and Pluskal, 1977)

observation, in relation to the net accretion of skeletal muscle protein, requires further exploration.

An understanding of the mode of action of growth hormone on muscle protein metabolism is complicated because of the existence of growth-promoting polypeptides under growth hormone control. These appear to be low molecular weight peptides – somatomedin A and C, non-suppressible insulin-like activity (NSILA-S) and multiplication stimulating activity (MSA). They are formed in various sites, such as the kidney and liver, and they exert an insulin-like action on their target tissues (Van Wyk and Underwood, 1978; Zapf *et al.*, 1978). Much remains to be learned about their physiological action *in vivo*, but a growth hormone-dependent factor from serum when added *in vitro* enhances protein synthesis in muscle (Uthne, 1975) and it is know that somatomedin C has its own primary binding sites in various tissues, including muscle (Van Wyk and Underwood, 1978). Furthermore, it is important to point out that insulin and nutritional state may modulate somatomedin generation (Chochinov and Daughaday, 1976; Yeh and Aloia, 1978; Phillips and Vassiloupoulou-Sellin, 1979).

These complex interrelationships among insulin, growth hormone, somatomedins and the nutritional states of the whole organism must be considered and investigated in more detail, in view of the interrelationships proposed between some anabolic agents and insulin and growth hormone (Buttery, Vernon and Pearson, 1978).

THYROID HORMONES

An endocrine factor that is attracting increased interest in reference to muscle protein metabolism is the iodothyronines. It is well known that thyroidectomy reduces growth in rats and a physiological replacement dose of about 2 µg/day returns growth toward normal values (e.g. see *Figure 9.10*). This response to thyroid hormone replacement reflects, as shown in *Figure 9.11*, an enhanced muscle protein synthesis (Flaim, Li and Jefferson, 1978b). Furthermore, based on our studies, utilizing N^τ-methylhistidine (3-methylhistidine) as an index of muscle protein breakdown (Young and Munro, 1978), we have observed a reduced rate of muscle protein breakdown in thyroidectomized rats, and with

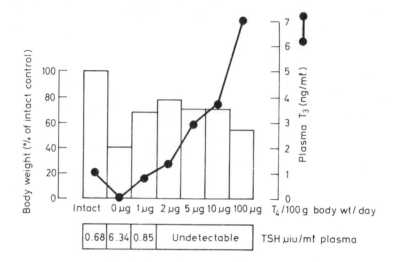

Figure 9.10   Effect of thyroidectomy (Tx) and various doses of thyroxine replacement on growth of rats, plasma thyroxine and TSH levels. Values based on 4–5 rats per group. (Unpublished results of M. Moreya *et al*.)

thyroxine replacement therapy there is an increased rate of muscle protein breakdown (*Figure 9.12*). Since a net gain of muscle and body protein is achieved with thyroid replacement, protein synthesis must be enhanced more than is breakdown. Thus, the turnover of muscle proteins is more rapid in the presence, than in the absence, of thyroid hormone.

It is not yet clear whether the effect of thyroxine on muscle protein metabolism is a direct consequence of thyroid action on the muscle or whether it is a

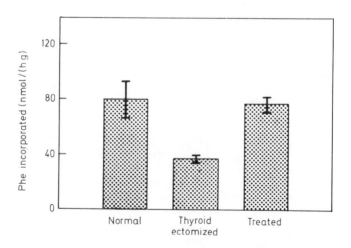

Figure 9.11   Effect of thyroxine on *in vivo* muscle protein synthesis in gastrocnemius muscle in rats. (Drawn from Flaim, Li and Jefferson, 1978b)

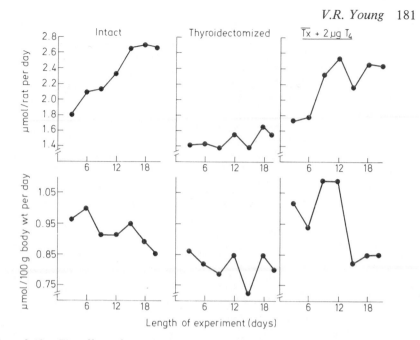

Figure 9.12   The effect of thyroidectomy (Tx) and a physiological replacement dose of thyroxine (2 μg T₄ per day) on muscle protein breakdown, as assessed by Nᵀ-methylhistidine excretion, in young rats. Each point mean for 4–5 rats. (Unpublished results of R. Burini *et al.*)

result of the responses of many other organs to variations in thyroxine level. As shown in *Figure 9.13* for most muscle groups that we have examined, the deficit in muscle mass in thyroidectomized rats, and increased size of muscle with a physiological replacement dose of the hormone, occur in proportion to changes in total body mass. On this basis, changes in thyroid hormone balance may not have a selective or specific effect of muscle protein metabolism and content but rather a more generalized effect on body energy and protein metabolism.

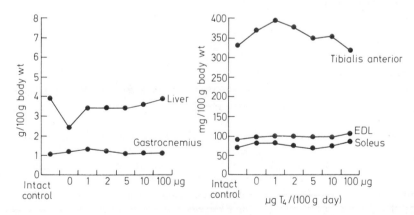

Figure 9.13   Relative weights for liver and various muscles in thyroidectomized (Tx) rats and various replacement doses of thyroxine replacement in Tx rats. Values based on 4–5 rats per group

Little is known about the mechanism of thyroid hormone action on protein metabolism; thyroid hormones bind to several subcellular constituents and, in particular, saturable nuclear binding sites have been demonstrated in several tissues (Sterling, 1979; Oppenheimer, 1979). Also a good correlation has been shown to exist between the relative affinities of thyroid hormone analogues for such sites and their relative biological properties (Sterling, 1979).

Binding sites have also been demonstrated in the mitochondrial fraction of certain tissues and these have been considered to be responsible for the thyroid hormone induction of calorigenesis. Thus, Sterling (1979) suggests that these mitochondrial receptors may be responsible for mediating the early effects of thyroid hormones on cellular energy metabolism, while the nuclear receptors mediate the later and more sustained effects of thyroid hormones on protein synthesis.

The existence of T_3-binding proteins in the cytosol is also firmly established but their function has not been defined (Sterling, 1979). However, they are not required for transport of thyroid hormones to nuclear binding sites. It has been suggested that they may help maintain a readily available intracellular pool of the hormone in a dissociable complex (Sterling, 1979).

Based on the recent review by Oppenheimer (1979), some major points concerning the mechanism of action of thyroid hormone are summarized in *Table 9.11*. First, a large body of evidence, largely circumstantial, suggests that

Table 9.11. SUMMARY OF STATEMENTS CONCERNING THE MECHANISM OF ACTION OF THYROID HORMONES

Nuclear mechanism
1. Nuclear site of initiation
2. Stimulation of transcription
Non-nuclear mechanism
    Mitochondria
    Which proteins?
    How do these proteins → physiological developmental and toxic effects
Local modulation of action
    Local cellular control
    Molecular basis unknown

the basic unit of thyroid hormone action is the T_3–nuclear receptor complex. This complex stimulates, either directly or indirectly, the formation of a diversity of mRNA sequences. Non-nuclear sites of initiation of thyroid hormone action cannot be ruled out, but Oppenheimer (1979) suggests that these extranuclear systems operate optimally at T_3 concentrations above the range characteristic of biological fluids.

However, as also indicated in *Table 9.11*, little is known about the proteins that are stimulated and how these proteins bring about the physiological, developmental and toxic responses to thyroid hormones (Bernal and Refetoff, 1977).

An intriguing aspect of thyroid action, discussed by Oppenheimer (1979), concerns the modulation of thyroid hormone action on the expression of selected genes. For example, under various conditions, as in starvation, certain T_3 characteristic responses are reduced, unaffected or actually augmented. An example of this is the response of α-glycerophosphate dehydrogenase (α-GPD) to T_3. This is unimpaired or actually increased in starved rats whereas, in contrast, the response to malic enzyme is impaired under these conditions in response to a dose of T_3 designed to saturate the nuclear sites. Hence, there is disparity between malic enzyme and α-GPD responses to T_3 balance when conditions of the host change, as in starvation. This and other examples (see Oppenheimer, 1979) of diversity in cell responses to T_3 serve to illustrate the principle of *local cellular* control in the expression of thyroid hormone action; the thyroid hormone signal appears to be capable of being locally modulated — either amplified or attenuated. In fact, it seems that in a given cell the expression of thyroid hormone action may be concomitantly amplified or attenuated, depending upon the gene product assayed and the metabolic and/or nutritional state of the host.

The molecular basis of such local modulation remains to be defined, but this phenomenon has also been stressed by Turner and Munday (1976) who have discussed the interaction between growth hormone and insulin in stimulating muscle protein synthesis. These investigators concluded that growth hormone dominates muscle protein deposition but this is achieved by the presence of insulin and thyroid hormones and in the absence of excessive glucocorticoids.

Local modulation is clearly of fundamental importance in determining the final integrated response of organs and the body as a whole to a given hormonal stimulus. However, it complicates prediction of growth and protein—metabolic responses to hormone manipulation.

Steroids

GLUCOCORTICOIDS

It is also necessary to consider briefly in this overview the steroid hormones, of which there are the catabolic or overall growth-inhibiting steroids and the anabolic androgenic steroids. With respect to corticosteroids, they can directly inhibit DNA and protein synthesis and cell replication (Loeb, 1976; Baxter, 1978). Although, as summarized in *Table 9.12*, glucocorticoids can inhibit

Table 9.12. A SUMMARY OF THE RELATIONSHIPS BETWEEN CORTICOSTEROIDS AND GROWTH — BASED ON REVIEWS BY LOEB (1976), BAXTER (1978) AND GRANNER (1979)

1 Low-doses inhibit cell replication (liver, heart, gastrocnemius, kidney)
2. Probably not related to inhibition of GH
3. Inhibit somatomedin?
4. By direct action? (glucose transport)
5. Major process for inhibition not known
6. Which cell types most important?

growth hormone production and perhaps somatomedin production it seems that this is not the only mechanism and that many possible mechanisms can be invoked to account for their adverse effects on body protein gain. It is possible that steroids may affect uptake of substrates, such as glucose, which, in turn, affect growth, or they may induce synthesis of inhibitory proteins or steroid—receptor complexes that block the synthesis of RNA. Furthermore, it is not known how the steroid inhibits linear growth in mass or which cell types are the most important targets of such effects.

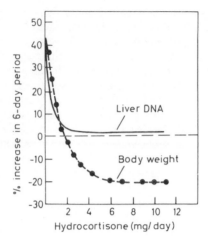

Figure 9.14   Effect of various doses of hydrocortisone in intact rats on body weight and liver DNA. (From Loeb, 1976)

Loeb (1976) has shown (*Figure 9.14*) that only low doses of hormone are necessary for inhibition of hepatocyte proliferation. Whether these studies provide an adequate model or set of circumstances to explain the role of gluco-corticoids in the regulation of normal somatic growth is unclear. It is important to emphasize that these studies, depicted in *Figure 9.14*, utilized intact rats and although so-called physiological dose was used to explore the effects of gluco-corticoids on growth, the effective glucocorticoid output in the treated rats would be approximately double the normal endogenous output. Furthermore, we (Tomas, Munro and Young, 1979) observed that growth in adrenalectomized rats was not affected by physiological replacement doses of corticosterone. Only when plasma levels of the steroid were raised to those characteristic of stress

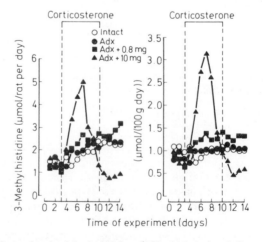

Figure 9.15   Effects of a physiological (0.8 mg/100 g body wt) and pharmacological (10 mg/100 g body wt) daily dose of corticosterone in adrenalectomized rats on N$^\tau$-methylhistidine excretion. Intact rats received the vehicle. Mean values based on 4 rats per point. (Unpublished results of S. Santidrian *et al.*)

states, was growth reduced. Based on measurements of N^T-methylhistidine excretion, this growth reduction was associated with an increased rate of muscle protein breakdown. The latter was unaffected by changes in plasma glucocorticoid levels within the normal range (*Figure 9.15*).

Thus, it is not yet clear whether glucocorticoids within normal ranges of output and plasma concentration inhibit somatic growth. This view contrasts with that of Loeb (1976) who suggests that low levels of endogenous steroid may exert a gentle, tonic suppressive effect on cell proliferation and, thereby, play a part in the regulation of normal growth.

Finally, it may be of interest to point out that the route of administration of the hormone influences the response, in terms of growth and the status of muscle protein breakdown. Thus, we have observed that a pharmacological dose of corticosterone (10 mg/100 g/day), given via intraperitoneal injection to adrenalectomized rats, does not inhibit growth to a marked extent (Santidrian *et al.*, unpublished results) or cause an increased rate of muscle protein breakdown. However, the same dose given subcutaneously has a profound, growth-inhibiting effect and it results in a markedly enhanced rate of muscle protein breakdown. It is apparent, therefore, that the route of hormone administration should be considered more critically in studies designed to explore the metabolic consequences of normal endocrine function in animal models.

ANABOLIC STEROIDS

There has been relatively little investigation into the mode of action of these hormones or their derivatives on the growth of skeletal muscle and tissues, other than sex organs and accessory sex tissues (Young and Pluskal, 1977). Therefore, any speculation concerning their mode of action must rely, at present, upon results of studies performed on tissues such as the prostate, oviduct and uterus. Although more is known about the mechanism of action of the oestrogens (e.g. Chan and O'Malley, 1978), the sex steroids appear to affect similar biochemical processes in their target organs, suggesting the possibility of similar modes of action. However, specifically whether the androgens and oestrogens bring about their myotrophic effects in similar or dissimilar ways is uncertain.

Based on studies of the major target tissues, testosterone, or dihydrotestosterone (DHT), interacts with cytoplasmic receptors prior to migration of the H–R complex to the nucleus where gene-transcription is regulated (Griffin and Wilson, 1978) (*Figure 9.16*). It has also been suggested that androgens may operate independently at both nuclear and extranuclear sites and that a dual role in the transcription and translation processes may serve to coordinate the regulation of gene expression in target cells (Liao *et al.*, 1975).

A summary of the effects of androgens on muscle is provided in *Figure 9.16* together with schemes depicting the possible mode of action of these steroids in muscle. However, the general applicability of the scheme shown here is not certain because little work has been devoted to an exploration of androgen action in skeletal muscle other than the levator ani muscle. Nevertheless, the following observations are relevant:

(1) Testosterone may directly act on skeletal muscle (Powers and Florini, 1975) and it may be the active form of the hormone (Baulieu, 1975).

Androgen on muscle

↑Muscle weight and protein mass, ↑body N retention
bind to cytosol protein, direct action of testosterone,
↑RNA synthesis (↑priming efficiency of DNA), ↑protein synthesis (↑ribosome activity),
oppose catabolic effect of glucocorticoids which ↓DNA and protein synthesis, and
increase myofibrillar protease and muscle protein breakdown (probably effect white fibres)

Direct transcriptional and translational actions

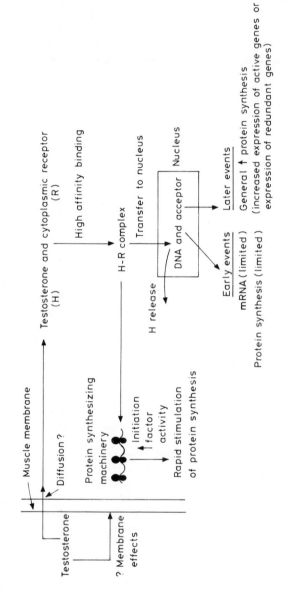

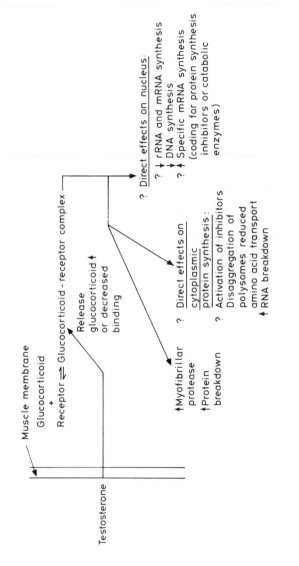

Figure 9.16 Schemes outlining the possible mode of action of testosterone on muscle protein metabolism. (Slightly modified from Young and Pluskal, 1977; the scheme in the lower portion of the figure is based on the suggestions of Mayer and Rosen, 1977)

(2) Testosterone receptors have been identified in skeletal muscle (Michel and Baulieu, 1976; Krieg, 1976) and there is a high concentration of binding sites in muscles that are more responsive to androgens (Dube, Lesage and Tremblay, 1976). It should be pointed out, however, that it is not known yet whether the binding of testosterone to the muscle androgen receptor is an essential process for its anabolic action.
(3) Testosterone stimulates RNA synthesis in skeletal muscle via an effect of nuclear chromatic (Breuer and Florini, 1966).

Based on these observations, the mechanism of action of androgen (testosterone) in relation to muscle protein gain may be broadly similar to that in *Figure 9.16*. Clearly, however, much further work will be required to define the details of its mode of action on muscle growth, particularly in the differentiated muscle.

An alternative mechanism, shown in the lower panel of *Figure 9.16*, has been suggested by Mayer and Rosen (1977) who concluded that androgens interfere with the binding of glucocorticoids to glucocorticoid receptors in muscle cytosol. Thus, these workers reasoned that glucocorticoids exert a catabolic response in muscle, associated with impaired synthesis of muscle DNA, RNA and protein, and that an androgen-induced reduction in the binding of glucocorticoids in the muscle cytosol would reduce the glucocorticoid-dependent catabolic effects in muscle. This alternative hypothesis is supported by the known antagonistic effects of anabolic steroids on the catabolic effects of glucocorticoids on muscle protein metabolism (see Young and Pluskal, 1977). This working hypothesis might be explored by conducting studies of glucocorticoid and testosterone replacement therapy in adrenalectomized-castrated rats. Furthermore, the suggestions by Buttery, Vernon and Pearson (1978) concerning the action of the synthetic steroid, trenbolone acetate, could also be examined in this way.

Conclusions

In this brief overview, selected aspects of the hormonal regulation of protein metabolism have been discussed, first in relation to the regulation of the endocrine system, and second in reference to the way hormones may bring about changes in tissue protein metabolism in the intact animal.

A number of points emerge from the foregoing:

(1) The regulation of the endocrine system is complex. In particular, hormones affect the levels of other hormones. Does this suggest that multiple hormonal treatments to achieve the gains in body protein, that would be physiologically possible and economically desirable, should be considered more thoroughly?
(2) Cell responsiveness is regulated by its target hormone and modulated by other hormones. Does this imply that there are more appropriate modes and rates of hormone administration that need exploring for purposes of achieving rapid, carcass (protein) gain and for maximizing feed utilization? The availability of sustained release systems, including those capable of delivery of large macromolecules (e.g. Langer and Folkman, 1977), offers an opportunity to explore this problem in greater depth.
(3) The sites and mechanism of action of hormones that promote growth and protein deposition differ among the various hormones. However, the precise

biochemical mechanisms for most, if not all, of the hormones that affect protein metabolism remain to be elucidated. A clearer understanding of the biochemical mechanisms responsible for the myotropic and somatic growth responses following changes in hormonal balance should help to identify more effective compounds or mixtures of compounds and define more precisely their rational application in farm animal industry.

(4) It must be recognized that the net protein metabolic response *in vivo* to an altered hormonal environment reflects the coordination of metabolic events at the subcellular and cellular levels as well as the cooperative, metabolic interactions among major body organs. Studies of the mode of action of hormones and of so-called anabolic agents must be related finally, to the complex, whole-body level.

It is sobering to learn that so little fundamental knowledge exists in the area of hormonal regulation of protein metabolism, particularly in reference to growth and to the use of hormones in animal protein production for improving the supply of human foods.

References

BAILE, C.A. and FORBES, J.M. (1974). *Physiol. Rev.,* **54**, 160–214

BAULIEU, E.E. (1975). *Mol. Cell. Biochem.,* 7, 157–174

BAXTER, J.D. (1978). *Kidney Int.,* **14**, 330–333

BERNAL, J. and REFETOFF, S. (1977). *Clin. Endocrinol.,* 6, 249–277

BRAY, G. (1978). *Proc. Nutr. Soc.,* 37, 301–309

BREUER, C.B. and FLORINI, J.R. (1966). *Biochem. Wash.,* 5, 3857–3865

BRODY, S. (1945). *Bioenergetics and Growth,* Van Nostrand–Reinhold, Princeton, N.J.

BUTTERY, P.J., VERNON, B.G. and PEARSON, J.T. (1978). *Proc. Nutr. Soc.,* 37, 311 315

CARPENTER, G. (1978). *J. Investig. Dermatol.,* **71**, 283–287

CATT, K.J., HARWOOD, J.P., AGUILERA, G. and DUFOU, M.L. (1979). *Nature, Lond.,* **280**, 109–116

CHAN, L. and O'MALLEY, B.W. (1978). *Ann. Int. Med.,* **89** (Pt 1), 694–701

CHOCHINOV, R.H. and DAUGHADAY, W.H. (1976). *Diabetes,* **25**, 997–1004

CHRISTAKOS, S. and NORMAN, A.W. (1978). *Mineral Electrol. Metab.,* 1, 231– 239

DAUGHADAY, W.H. (1977). *Clin. Endocrinol.,* 6, 117–135

DUBE, J.T., LESAGE, R. and TREMBLAY, R.R. (1976). *Can. J. Biochem.,* **54**, 50–55

EASTMAN, R.C. (1977). *Ann. Int. Med.,* **86**, 205–219

FLAIM, K.E., LI, J.B. and JEFFERSON, L.S. (1978a). *Am. J. Physiol.,* **234**, E38–E43

FLAIM, K.E., LI, J.B. and JEFFERSON, L.S. (1978b). *Am. J. Physiol.,* **235**, E231– E236

FLIER, J., KAHN, R. and ROTH, J. (1979). *New Engl. J. Med.,* **300**, 413–419

FRITZ, I.B. (1972). In *Biochemical Actions of Hormones,* Vol. II, Ch. 6, pp. 165– 214 (Litwack, G., Ed.), Academic Press, New York

GOLDBERG, A.L. (1971). In *Cardiac Hypertrophy,* pp. 301–314 (Alpert, N.R., Ed.), Academic Press, New York

GOLDFINE, I.D. (1978). *Life. Sci.,* **23**, 2639–2648

GOODRIDGE, A.G. and ADELMAN, T.G. (1976). *J. Biol. Chem.,* **251**, 3027–3032

GRANNER, D.K. (1979). In *Glucocorticoid Hormone Action*, Ch. 33, pp. 593–611 (Baxter, J.D. and Rousseau, G.G., Eds), Springer-Verlag, Berlin

GRIFFIN, J.E. and WILSON, J.D. (1978). *Clin. Obstet. Gynec.,* **5**, 457–479

HUXLEY, J. (1932). *Problems of Relative Growth*, Methuen, London

INSEL, P.A. (1978). In *International Review of Biochemistry*, Biochemistry and Mode of Action of Hormones II, Vol. 20, pp. 1–43 (Rickenberg, H.V., Ed.), University Park Press, Baltimore

KOSTYO, J.L. (1968). In *Growth Hormone*, pp. 175–182 (Pecile, A. and Miller, E.E., Eds), Int. Congr. Ser., Excerpta Medica Foundation, Amsterdam

KRIEG, M. (1976). *Steroids,* **28**, 261–274

KURTZ, D.T. and FEIGELSON, P. (1978). In *Biochemical Actions of Hormones*, Vol V, Ch. 11, pp. 433–455 (Litwack, G., Ed.), Academic Press, New York

LANDSBERG, L. (1978). In *The Year in Endocrinology, 1977*, pp. 291–343 (Ingbar, S.H., Ed.), Plenum Press, New York

LANGER, R.S. and FOLKMAN, J. (1977). *Polym. Preprints,* **18**, 379–384

LIAO, S., TYMOCZKO, J.L., CASTANEDA, E. and LIANG, T. (1975). In *Vitamins and Hormones*, Vol. 33, pp. 297–317 (Munsen, P., Diczfalusz, E., Glover, J. and Olson, R.E., Eds), Academic Press, New York

LOEB, J.N. (1976). *New Engl. J. Med.,* **295**, 547–552

MANCHESTER, K. (1976). In *Protein Metabolism and Nutrition*, pp. 35–47 (Cole, D.J.A., Boorman, K.N., Buttery, P.J., Lewis, D., Neale, R.J. and Swan, H., Eds), Butterworths, London

MAYER, M. and ROSEN, F. (1977). *Metabolism,* **26**, 937–962

MICHEL, G. and BAULIEU, E.E. (1976). In *Anabolic Agents in Animal Production*, Environmental Quality and Safety, Suppl. Vol V., pp. 54–59 (Lu, F.C. and Rendel, J., Eds), Thieme, Stuttgart

MILLER, W.L., KNIGHT, M.M. and GORSKI, J. (1977). *Endocrinol.,* **101**, 1455–1460

MUNRO, H.N. (1964). In *Mammalian Protein Metabolism*, Vol. I, Ch. 10 (Munro, H.N. and Allison, J.B., Eds), Academic Press, New York

OPPENHEIMER, J.H. (1979). *Science, N.Y.,* **203**, 971–979

PHILLIPS, L.S. and VASSILOUPOULOU-SELLIN, R. (1979). *Am. J. Clin. Nutr.,* **32**, 1082–1096

POWERS, M.I. and FLORINI, J.R. (1975). *Endocrinology,* **97**, 1043–1047

RUDLAND, P.S. and DEASUA, L.J. (1979). *Biochim. Biophys. Acta,* **560**, 91–133

RUTTER, W.J. (1978). In *Molecular Control of Proliferation and Differentiation*, pp. 5–10 (Papaconstantinou, J. and Rutter, W.J., Eds), Academic Press, New York

SCHALLY, A.V. (1978). *Science, N.Y.,* **202**, 18–28

SHAPIRO, L.E., SAMUELS, H.H. and YAFFE, B.M. (1978). *Proc. Natn. Acad. Sci. U.S.A.,* **75**, 45–49

STEINER, D.F. (1977). *Diabetes,* **26**, 322–340

STERLING, K. (1979). *New Engl. J. Med.,* **300**, 117–123 and 173–177

TELL, G.P., HAOUR, F. and SAEZ, J.M. (1978). *Metabolism,* **27**, 1566–1592

TOMAS, F., MUNRO, H.N. and YOUNG, V.R. (1979). *Biochem. J.,* **178**, 139–146

TOMKINS, G.M. and GELEHERTER, T.D. (1972). In *Biochemical Actions of Hormones*, Vol. II, Ch. 1, pp. 1–20 (Litwack, G., Ed.), Academic Press, New York

TURNER, M.R. and MUNDAY, K.A. (1976). In *Meat, Animals, Growth and Productivity*, pp. 197–219 (Lister, D., Rhodes, D.N., Fowler, V.R. and Fuller, M.F., Eds), Plenum Press, New York and London

UTHNE, K. (1975). In *Advances in Metabolic Disorders*, Vol. 8, pp. 115–126 (Luft, R. and Hall, K., Eds), Academic Press, New York
VAN WYK, J.J. and UNDERWOOD, L.E. (1978). In *Biochemical Actions of Hormones*, Vol. V, Ch. 3, pp. 101–148 (Litwack, G., Ed.), Academic Press, New York
YEH, J.K. and ALOIA, J.F. (1978). *Metabolism, 27*, 507–509
YOUNG, J.B. and LANDSBERG, L. (1977a). *Clin. Endocr. Metab., 6*, 657–695
YOUNG, J.B. and LANDSBERG, L. (1977b). *Clin. Endocr. Metab., 6*, 599–631
YOUNG, V.R. (1970). In *Mammalian Protein Metabolism*, Vol. 4, Ch. 40 (Munro, H.N., Ed.), Academic Press, New York
YOUNG, V.R. and MUNRO, H.N. (1978). *Fedn Proc., 37*, 2291–2300
YOUNG, V.R. and PLUSKAL, M.G. (1977). In *Protein Metabolism and Nutrition*, pp. 15–281 (Tamminga, S., Ed.), Europ. Assoc. Anim. Prod. Public, No. 2, Center for Agric. Public and Documentation, Wageningen, Holland
ZAPF, J., RINDERKNECKT, E., HUMBEL, R.E. and FROESCH, E.R. (1978). *Metabolism, 27*, 1803–1828

10

MANIPULATION OF PROTEIN METABOLISM, WITH SPECIAL REFERENCE TO ANABOLIC AGENTS

R.J. HEITZMAN
Agricultural Research Council, Institute for Research on
Animal Diseases, Newbury

Summary

Anabolic agents are used on a wide scale to improve the efficiency of meat production. They increased nitrogen retention and protein deposition in ruminants and pigs but their effects were not as consistent in rats, horses or poultry. The benefits in cattle and sheep were an increased daily live weight gain (DLG) and an improvement in feed conversion efficiency (FCE).

Anabolic agents have physiological functions similar to those of the sex steroids — the androgens and oestrogens. The correct choice of agents to obtain maximum responses was determined by the sex and species of the recipient animal. Usually the maximum response occurred in ruminants in which effective physiological concentrations of both androgens and oestrogens were present in the circulation.

In a few studies, carcass quantity and quality has been assessed following treatment with anabolic agents. The weight of carcass as a percentage of the whole animal was unaltered by treatment, although the weight of the carcasses was increased. There was a tendency for carcasses from treated ruminants, and pigs, to be leaner and less fatty. There was no evidence of significant changes in the quality of muscle and fat in treated animals.

The mechanism of action causing the increase of protein deposition by anabolic agents is not understood. It is possible that androgens and oestrogens act directly and/or indirectly on the muscle cell. The most likely actions of androgens are either a direct action involving specific androgen receptors within the muscle cell, or an indirect action by mediating the effects of corticosteroids or thyroid hormones. The role of oestrogens is also far from clear; a direct action has not been described and a proposed indirect action through growth hormone and insulin has been suggested. However, this was not observed in ruminants treated with both androgens and oestrogens, even though the growth responses to the two hormones appear to be separate and additive.

Introduction

The manipulation of protein synthesis in farm animals by anabolic agents affords a practical way of increasing the efficiency of meat production.

Anabolic agents are defined as substances which increase nitrogen-retention and protein deposition in animals. However, because they are usually substances with physiological properties similar to those of the sex steroids, a number of effects may be observed, including:

(1) Increased growth rate.
(2) Increased muscle mass.

(3) Improved feed conversion efficiency (FCE).
(4) Changes in fat distribution.
(5) Improvement of appetite.
(6) Improved muscular conditions to sustain heavy work or training.

Not all of these effects necessarily occur at one time, and the response depends upon the species, sex and age of the treated animal and the particular agents used. For example, androgens are used to increase growth rate and FCE in ruminants (Heitzman, 1979) to alter muscle : fat ratios in pigs (Fowler *et al.*, 1978), to improve performance in athletes (Johnson and O'Shea, 1969; Freed *et al.*, 1975) and to sustain appetite and training capacity in horses and dogs.

Chemical structure

Anabolic agents are either androgens, oestrogens or progestins. The structural formulae of some anabolic substances are shown in *Figure 10.1*. The androgens are related to the natural hormone testosterone or the synthetic 19-nortesto-sterone. Although testosterone is widely used as an anabolic agent it is a potent androgen and this may lead to undesirable side effects. Much research effort has been applied to changing the structure of the steroid molecule to produce andro-gens which possess greater anabolic : androgenic activity ratios, i.e. greater myotrophic index. Clearly this may be important in human medicine but seems less important in animals used for meat production where testosterone and trienbolone (with a myotrophic index 2–3 times greater than testosterone propionate) (Neuman, 1976) are both used extensively.

The oestrogenic substances used as anabolic agents are not all steroids, in fact only oestradiol-17β which is the natural female sex steroid possesses a steroid nucleus, while the other compounds are synthetic (see *Figure 10.1*) and at first glance structurally quite different from oestradiol. However, the spatial configur-ations of both types of compounds are similar. Consequently the cell may be unable to distinguish between oestradiol and the synthetic oestrogens.

Formulation

Anabolic agents may be administered either as feed additives, oil-based injections or as subcutaneous implants. In the UK, legislation restricts their use as feed additives, although a mixture of diethylstilboestrol (DES) and methyltestosterone is still administered as a feed supplement to pigs. Oil-based injections may be administered intramuscularly or subcutaneously. Some veal calves are injected with DES derivatives, although the main uses of these preparations are for the treatment of disease or in racehorses to maintain fitness and condition.

The subcutaneous implantation of anabolic agents as pellets at the base of the ear is the most widely used method of administration. The most commonly used preparations in Great Britain are listed in *Table 10.1*.

It is important, because of public health considerations, that implants are placed in the correct site at the base of the ear, so that after slaughter any resi-dues of the implant in the tissue immediately surrounding the implant may be disposed of. Donaldson (1977) showed that 15–20% of the original implant

Androgens

Testosterone

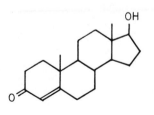

Nandrolin

R = phenylpropionate

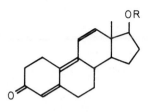

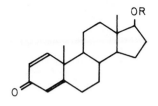

Boldane

R = undecanoate

Trienbolone acetate

R = acetate

Oestrogens

Oestradiol - 17β

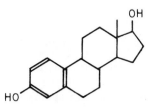

Zeranol

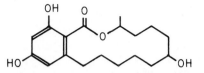

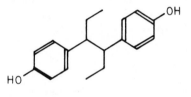

Hexoestrol

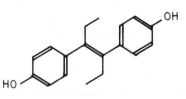

Diethylstilboestrol

Figure 10.1 Formulae of anabolic agents

Table 10.1 GROWTH PROMOTERS AVAILABLE FOR CATTLE AND SHEEP*

Chemical name	Trade name	Preparation	Animal
Trienbolone acetate	Finaplix	I	H,C
Diethylstilboestrol		I,F	S,V
Diethylstilboestrol dipropionate		O	V
Hexoestrol		I	S,L
Zeranol	Ralgro	I	S,H,V,L
Trienbolone acetate + oestradiol	Revalor Torelor	I I	S,V,L S,B
Trienbolone acetate + hexoestrol		I	S
Trienbolone acetate + zeranol		I	S,V
Testosterone + oestradiol	Implix BF	I	H,V
Progesterone + oestradiol	Implix BM	I	S
Testosterone propionate + oestradiol benzoate	Synovex-H	I	H
Progesterone + oestradiol benzoate	Synovex-S	I	S

* I, implants; O, oil-based injection; F, feed additive; H, heifer; C, culled cow; S, steer; V,
veal calf; B, bull; L, wether lamb.

dose of 300 mg trienbolone acetate remained unabsorbed in the ears of heifers which had been treated 84 days previously.

Use in farm animals for meat production

The purpose of using anabolic agents in farm animals is to improve meat production. Ideally the correct agent should improve growth rate, feed conversion efficiency and carcass quality and quantity. The introduction of anabolic agents occurred in the 1950s (Andrews, Beeson and Johnson, 1954; Clegg and Cole, 1954), when it was discovered that the synthetic oestrogen, DES, was an effective growth promoter in steers. This and other stilbene derivatives were very cheap. However, the potential health hazards associated with the use of some stilbene derivatives, especially DES, were recognized and during the 1970s several countries banned the use of these drugs. This has resulted in a series of new anabolic agents which include the natural sex steroids, the synthetic androgen, trienbolone, and a phyto-oestrogen, zeranol. They are, however, much more expensive to produce than the stilbene derivatives.

Use in cattle

Improvements in growth rate and FCE by the use of anabolic agents have been successfully demonstrated in cattle more frequently than in any other species. Anabolic agents are used mostly in intensive rearing systems producing meat from steers or veal calves; however, they may be used in most types of cattle and in a variety of management systems (for reviews see Heitzman, 1978; Scott, 1978).

It was suggested (Heitzman, 1976) that both androgens and oestrogens were necessary to achieve maximum growth rate potential. The concentrations of androgens in the circulation should be similar to those in bull plasma, and the plasma concentrations of oestrogens should not be less than those observed in the adult cow. In practice, therefore, the best responses are obtained when androgens are used in females (Best, 1972; Heitzman and Chan, 1974) and androgens combined with oestrogens are administered to steers. The latter case is demonstrated by the results for four separate studies shown in *Table 10.2*, which

Table 10.2. PERCENTAGE INCREASE OF AVERAGE LIVEWEIGHT GAIN IN STEERS ADMINISTERED ANABOLIC AGENTS

Anabolic agent *	Trial number †			
	(1)	(2)	(3)	(4)
TBA	10	9	11	24
Hexoestrol	11	25	22	–
Zeranol	–	15	17	18
TBA + hexoestrol	25	33	39	–
TBA + zeranol	–	–	–	51
Total no. of animals	60	1557	750	840

* TBA is trienbolone acetate, a synthetic androgen. Hexoestrol and zeranol are oestrogens.
† Data are from trials reported by (1) Heitzman, Harwood and Mallinson (1977), (2) Stollard *et al.* (1977), (3) Scott (1979), and (4) Roche and Davis (1977). The values are the percentage increase in average daily liveweight gain compared with untreated control steers.

show that the response of steers to combined implants of an androgen and oestrogen were greater than when either androgen or oestrogen were administered alone. The effect on FCE was parallel with the effect on growth rate, i.e. the largest improvements in FCE were observed in animals treated with those anabolic agents which produced the best responses in growth rate (Heitzman and Chan, 1974; Galbraith and Coelho, 1978).

The administration of synthetic oestrogens to veal calves by injection has been practised for a long time in Europe. More recently, however, there has been a large increase in the use of implants. The efficacy of different anabolic agents in the veal calf has been discussed by Van der Wal (1976). In both heifers and bull calves the maximum increases in growth rate and FCE occurred following administration of combined preparations of androgens plus oestrogens at 2–3 months of age.

Anabolic agents are not generally used in bulls, although there have been some reports of improved responses in bulls treated with oestrogens (Folman and Volcani, 1960) or with combined implants of androgens plus oestrogens (Grandadam *et al.*, 1973).

Several trials have been performed to investigate if repeated implantation on more than one occasion leads to even better results. In trials where oestrogens were used in steers, some authors reported some benefits (Perry *et al.*, 1970; An Forais Taluntais, 1973), while many others thought that no clear benefits were obtained (Jones, 1961; Macdearmid and Preston, 1969; Nicholson, Lespearance and McCormick, 1973; Nichols and Lespearance, 1973). In one small field trial with steers it was reported that reimplantation of steers three months after the first treatment with trienbolone acetate plus hexoestrol produced additional increases in live weight gain compared with control and those which had been implanted on only one occasion (Heitzman, Harwood and Mallinson, 1977).

CARCASS ANALYSIS

The evidence for an effect of anabolic agents on carcass size and quality is well documented for the effects of oestrogens, but information on combined implants is inconclusive. DES and hexoestrol reduce fat content in steers (Alder, Taylor and Rudman, 1964; Everitt and Duganzich, 1965; Macdearmid and Preston, 1969) but their effects in bulls are variable (Levy, Holzer and Folman, 1976). There is almost no information available on the effect on carcass characteristics of combined implants. A most recent study by Berende (1978) suggested that live weight gain, FCE and carcass weight were increased and fat content reduced in steers following treatment with trienbolone acetate plus oestradiol, while other studies reported that carcass weight was improved in steers treated with combined implants of treinbolone acetate plus zeranol (Roche and Davis, 1977), or trienbolone acetate plus hexoestrol (Galbraith and

Table 10.3. EFFECT OF AN ANABOLIC AGENT ON THE CARCASS OF VEAL CALVES (FROM VERBEKE *ET AL.*, 1976)*

	Control weight (kg)	Treated weight (kg)
Carcass	103.5	118.2†
3rd rib cut – total	3.48	4.09+
Total meat content	2.01	2.58†
Fat	0.83	0.84
Bone	0.64	0.67
Muscle, long dorsi	0.38	0.45†

* Eight bull calves were implanted with 140 mg trienbolone acetate plus 20 mg oestradiol at 8 and 4 weeks before slaughter at about 17 weeks of age.
† $P < 0.05$ compared with controls (8).

Watson, 1978). It is only possible therefore to conclude that there is a tendency for the use of anabolic agents to produce leaner and less fatty adult cattle. However, a study of Verbeke *et al.* (1976) has clearly demonstrated that when they gave high doses of anabolic agents to veal calves there was a significant increase in protein deposition and no change in fat and bone content in a selected rib cut compared with untreated controls. The data is summarized in *Table 10.3*.

Use in sheep

The use of growth promoters in sheep has been confined to a small-scale use of oestrogens in wether lambs and a few development trials using combined implants in wethers. Hexoestrol and zeranol have only yielded marginal benefits in trials (Meat and Livestock Commission – unpublished data) or no benefits (Vipond and Galbraith, 1978). However, in trials with wether lambs implanted with a combination of trienbolone acetate and oestradiol there were reports of increased live weight gain, carcass weights and FCE (Coelho, Galbraith and Topps, 1978; Szumowski and Grandadam, 1976).

Use in pigs

Boars tend to be leaner than females or castrate males and therefore the use of testicular hormones might be expected to improve their carcass quality. Van Weerden and Grandadam (1976) showed that oral administration of trienbolone acetate and ethynyl oestradiol or implantation of trienbolone acetate and oestradiol into castrate male pigs improved lean tissue deposition, reduced fat content and increased nitrogen-retention, growth rate and FCE. Similar studies by Fowler *et al.* (1978) showed that oral administration to pigs of trienbolone acetate or methyltestosterone combined with ethynyl oestradiol also reduced fat tissues in castrate males but not females. In farm practice, implants are not yet used in pigs; oral administration as a feed additive is the method of choice.

Use in poultry

The effects in poultry of anabolic agents are different from ruminants and pigs. According to Nesheim (1976) androgens, oestrogens and growth hormone are wihout effect as growth promoters. However, there has been a recent report that androgens may improve growth rate and FCE in turkeys (Ranaweera, 1977).

Mechanism of action of anabolic agents

The muscle mass of animals is determined by species, breed, age, sex, nutrition and hormonal status. The control of protein metabolism by hormones is not fully understood and many of the present concepts of control of protein deposition in muscle cells are based on the mechanism of action of hormones in other cell types, especially the prostate and the uterus. Also, most studies have been made using rats and there has been little information on protein synthesis in muscle cells of farm animals.

It is likely that androgens and oestrogens have separate mechanisms of action, and furthermore androgens have two separate effects in animals: first, a qualitative effect which causes differentiation of tissues, and secondly a quantitative effect altering the rate of biochemical processes within the tissues (Mainwaring, 1977).

The first mechanism affects differentiation during fetal or neonatal life when androgens secreted by the male cause qualitative changes in tissues which bring about the development of male characteristics. The changes are most obvious in the sexual and accessory sexual organs characteristic of males and females. However, it is also probable that some differentiation in non-sexual tissues such as liver, kidney and muscle may also occur during this period.

Although little is known about the exact differences between male and female type cells in peripheral tissues, Erikson (1974) reported differences in hepatic metabolism of steroids in male and female rats and Blyth *et al.* (1972) showed differences in hepatic binding sites for oestradiol and testosterone in male and female rats. Michel and Baulieu (1976) suggested that the number of androgen binding sites in rat muscle was different between males and females and that it would be interesting to study this further and especially in females treated with androgens or anabolic agents. Female rats treated with androgens during the neonatal period subsequently grew faster than untreated females (Brown-Grant, 1974; Perry and McCracken, 1978).

The second mechanism of action of androgens and other hormones involves quantitative changes where the hormone simply accelerates the events which proceed at slower rates in the absence of hormonal stimulus. This amplification effect may be described as

$$\text{Effect} = \rho \, [\text{H}] \, [\text{R}]$$

where [R] and [H] are the concentrations of receptors and hormones in/on the cell, respectively. What is not clear is, first, which receptors are involved, i.e. are they homologous and/or heterologous receptors and, secondly, whether the action involves the primary or a secondary hormone.

DIRECT ACTION OF ANABOLIC AGENTS ON MUSCLE CELL

It has been postulated that androgens and oestrogens have direct action within prostate and uterine cells, respectively (King and Mainwaring, 1974). The steroid enters the cell and forms a steroid–receptor complex with a homologous receptor which is translocated to the nucleus of the cell and, by mechanisms not yet known, influences protein metabolism (see Chapter 9). It is now known that muscle cells possess specific receptors for androgens (Jung and Baulieu, 1972; Michel and Baulieu, 1974), although the concentration of receptors in muscle is 1% in skeletal muscle and 10% in levator ani muscle compared with that in the prostate. The muscle receptors have high affinities for both testosterone and 17α-methyl-trienbolone which are potent anabolic agents, whereas the prostate receptor is specific for 5α-dihydrotestosterone. Thus it is possible that the differences between the effects of androgens in various tissues are determined by the specificity of the receptor for the androgen, and that the mechanisms of anabolic and androgenic action are not basically different.

It remains to be seen whether all anabolic androgens bind to the muscle receptor; certainly the oestrogen, DES, does not compete for binding to the androgen receptor. Thus while androgens may act through a receptor, they and the anabolic oestrogens may also act at other points in the complex processes of protein synthesis and degradation. Mayer and Rosen (1975) have suggested that androgens increase protein deposition by displacing the catabolic hormones, the corticosteroids, from their binding sites in the muscle cell, and thus decreasing protein degradation rates. More recently Mayer and Rosen (1978) have suggested that androgens can reduce or 'down regulate' the number of corticosteroid receptors in rat muscle cells. The reduction of these heterologous receptors would presumably also reduce protein catabolism.

The regulation of protein degradation as a control point of protein deposition has received considerable support from the studies of Vernon and Buttery (1976, 1978a, 1978b). They found that female rats treated with trienbolone had significantly reduced rates of protein degradation and synthesis. The reduction in the rate of degradation was greater than in the rate of synthesis; thus, the net effect was an increase in protein deposition at reduced fractional synthetic rates.

INDIRECT ACTION OF ANABOLIC AGENTS ON MUSCLE CELL

Anabolic agents may act indirectly at the muscle cell by changing the concentrations of other endogenous anabolic and catabolic hormones, e.g. GH, insulin, prolactin, thyroxine, triodothyroxine and corticosteroids. A summary of the overall effects in ruminants of androgens and oestrogens on plasma concentrations of hormones is given in *Table 10.4*. The implantation of oestrogens into

Table 10.4. EFFECT OF ANABOLIC AGENTS ON PLASMA HORMONES*

Exogenous hormone	*Endogenous hormone*				
	GH	insulin	thyroxine	FTI	prolactin
Androgen	=	=	−	−	=
Oestrogen	+	+	−	nm	=
Androgen + oestrogen	=	=	−	−	=

* Cattle or sheep were treated with an androgen (trienbolone acetate), an oestrogen (oestradiol-17β), or a combination of both steroids and plasma concentrations of endogenous hormones were measured for a period of two months after treatment. GH, growth hormone; FTI, free thyroxine index; =, +, −, no change, increase and decrease, respectively, of endogenous hormone concentration in the plasma of treated animals compared with untreated controls; nm, not measured.

sheep (Donaldson, 1977) and cattle (Trenkle, 1976) caused significant increases in the concentration of GH. The rise in GH was associated with a rise in the concentration of insulin. This combination of increased GH and insulin at the muscle cell is thought to increase protein accretion (Trenkle, 1976). The effects of exogenous GH and oestrogen are similar, and it was concluded that one of the main anabolic actions of oestrogens was through the increased production of GH and insulin. However, Donaldson (1977) has shown that when sheep are treated with combined implants of androgen and oestrogen, the androgen abolished the oestradiol-induced GH and insulin response even though these combined preparations produced the best growth responses.

The effects of androgens and oestrogens appeared to be independent and additive. The combined action of androgens and oestrogens resulted in a possible depression of thyroid function measured as a decrease of plasma thyroxine (T_4) and the free thyroxine index* (FTI) and no change in T_3 uptake in cattle and sheep (Donaldson, 1977; Heitzman, Donaldson and Hart, unpublished data). The FTI has been linked to the regulation of basal metabolism and consequently a depression of the FTI would imply a reduction of energy requirements. This could explain the observed improvement in feed conversion efficiency of animals treated with anabolic agents and with the reduction in protein fractional synthetic rates observed by Vernon and Buttery (1978b).

* FTI is the product of total T_4 concentration and T_3 uptake (% T_3 not bound to TBG) and reflects free T_3 and T_4 plasma concentrations.

Conclusions

The correct use of anabolic agents greatly increases the rate of protein deposition in some farm animals. The mechanism of action at the muscle cell is not understood but may be compared with that of androgens and oestrogens in other cells. The most recent studies have investigated the actions of anabolic agents in rat muscle cells and clearly this approach should be extended to farm animals.

References

ALDER, E.E., TAYLOR, J.C. and RUDMAN, J.E. (1964). *Anim. Prod.,* **6**, 47

ANDREWS, F.N., BEESON, W.M. and JOHNSON, F.D. (1954). *J. Anim. Sci.,* **13**, 99–107

AN FORAIS TALUNTAIS (1973). *Animal Production Research Report,* 45

BERENDE, P.L. (1978). In *3rd World Congress on Animal Feeding,* Madrid, Paper A-1-12

BEST, J.M.J. (1972). *Vet. Rec.,* **91**, 624–626

BLYTH, C.A., COOPER, M.B., ROOBOL, A. and RABIN, B.R. (1972). *Eur. J. Biochem.,* **29**, 293–300

BROWN-GRANT, K. (1974). *INSERM,* **32**, 357–376

CLEGG, M.J. and COLE, H.H. (1954). *J. Anim. Sci.,* **13**, 108–130

COELHO, J.F.S., GALBRAITH, H. and TOPPS, J.H. (1978). *Anim. Prod.,* **26**, 360

DONALDSON, I.A. (1977). 'The action of anabolic steroids on nitrogen metabolism and the endocrine system in ruminants', *Ph.D. thesis,* University of Reading

ERIKSON, H. (1974). *Eur. J. Biochem.,* **46**, 603–611

EVERITT, G.C. and DUGANZICH, D.M. (1965). *N.Z. J. Agric. Res.,* **8**, 370–376

FOLMAN, Y. and VOLCANI, R. (1960). *KTAVIM,* **10**, 179–182

FOWLER, V.R., STOCKDALE, C.L., SMART, R.I. and CROFTS, R.M.J. (1978). *Anim. Prod.,* **26**, 358–359

FREED, D.L.J., BANKS, A.J., LONGSTON, D. and BURLEY, D.M. (1975). *Br. Med. J.,* **1**, 472

GALBRAITH, H. and COELHO, J.F.S. (1978). *Anim. Prod.,* **26**, 360

GALBRAITH, H. and WATSON, H.B. (1978). *Vet. Rec.,* **103**, 28–30

GRANDADAM, J.A., SCHEID, J.P., JOBARD, A., DREUX, H. and BOLSSON, J.M. (1973). *J. Anim. Sci.,* **37**, 256

HEITZMAN, R.J. (1976). In *Anabolic Agents in Animal Production,* Suppl. V., pp. 89–98 (Coulston, F. and Corte, F., Eds), Thieme, Stuttgart

HEITZMAN, R.J. (1978). *Proc. Nutr. Soc.,* **37**, 289–293

HEITZMAN, R.J. (1979). In *Recent Advances in Animal Nutrition 1979,* pp. 133–143, Butterworths, London

HEITZMAN, R.J. and CHAN, K.H. (1974). *Br. Vet. J.,* **130**, 532–537

HEITZMAN, R.J., HARWOOD, D.J. and MALLINSON, C.B. (1977). *Abs. 69th Ann. Mtg Am. Soc. Anim. Sci.,* 44

JOHNSON, L. and O'SHEA, J.P. (1969). *Science, N.Y.,* **164**, 957

JONES, P.J. (1961). *Exp. Husb.,* **6**, 62

JUNG, I. and BAULIEU, E.E. (1972). *Nature, New Biol.,* **237**, 24

KING, R.J.B. and MAINWARING, W.I.P. (1974). In *Steroid-Cell Interactions,* Butterworths, London

LEVY, D., HOLZER, Z. and FOLMAN, Y. (1976). *Anim. Prod.,* **22**, 55–59

MACDEARMID, A. and PRESTON, T.R. (1969). *Anim. Prod.,* **11**, 419–422

MAINWARING, W.I.P. (1977). In *The Mechanism of Action of Androgens,* Springer-Verlag, New York

MAYER, M. and ROSEN, F. (1975). *Am. J. Physiol.,* **229**, 1381–1386

MAYER, M. and ROSEN, F. (1978). *Acta Endocrinol.,* **88**, 199–208

MICHEL, G. and BAULIEU, E.E. (1974). *C.R. Acad. Sci. Paris,* **279**, 421–427

MICHEL, G. and BAULIEU, E.E. (1976). In *Anabolic Agents in Animal Production,* Suppl. V, pp. 54–59 (Coulston, F. and Corte, F., Eds), Thieme, Stuttgart

NESHEIM, M.C. (1976). In *Anabolic Agents in Animal Production,* Suppl. V, pp. 110–114 (Coulston, F. and Corte, F., Eds), Thieme, Stuttgart

NEUMANN, F. (1976). In *Anabolic Agents in Animal Production,* Suppl. V, pp. 253–264 (Coulston, F. and Corte, F., Eds), Thieme, Stuttgart

NICHOLS, N.E. and LESPEARANCE, A.L. (1973). *Proc. Western Sec. Am. Soc. Anim. Sci.,* **24**, 304

NICHOLSON, L.E., LESPEARANCE, A.L. and McCORMICK, A.V. (1973). *Proc. Western Sec. Am. Soc. Anim. Sci.,* **24**, 304

PERRY, B.N. and McCRACKEN, A. (1978). *Br. J. Nutr.,* **37**, 109A

PERRY, T.W., STOB, M., HUBER, D.A. and PETERSON, R.C. (1970). *J. Anim. Sci.,* **31**, 789–793

RANAWEERA, P. (1977). 'The effects of trienbolone acetate in growing turkeys', *Ph.D. thesis,* University of Cambridge

ROCHE, J.F. and DAVIS, W.D. (1977). *Anim. Prod.,* **24**, 132–133

SCOTT, B.M. (1978). *ADAS Qly Rev.,* **31**, 185–216

STOLLARD, R.J., KILKENNY, J.B., MATTHIESON, A.A., STARK, J.S., TAYLOR, B.R., SUTHERLAND, J.E. and WILLIAMSON, J.T. (1977). *Anim. Prod.* **24**, 132

SZUMOWSKI, P. and GRANDADAM, J.A. (1976). *Rec. Vet. Med.,* **152**, 311–321

TRENKLE, A. (1976). In *Anabolic Agents in Animal Production,* Suppl. V, pp. 79–88 (Coulston, F. and Corte, F., Eds), Thieme, Stuttgart

VAN DER WAL, P. (1976). In *Anabolic Agents in Animal Production,* Suppl. V, pp. 60–78 (Coulston, F. and Corte, F., Eds), Thieme, Stuttgart

VAN WEERDEN, E.J. and GRANDADAM, J.A. (1976). In *Anabolic Agents in Animal Production,* Suppl. V, pp. 115–122 (Coulston, F. and Corte, F., Eds), Thieme, Stuttgart

VERBEKE, R., DEBACKERE, M., HICQUET, R., LAUWERS, H., POTTIE, G., STEVENS, J., VAN MOER, D., VAN HOOF, J. and VERMMERSCH, G. (1976). In *Anabolic Agents in Animal Production,* Suppl. V, pp. 123–130 (Coulston, F. and Corte, F., Eds), Thieme, Stuttgart

VERNON, B.G. and BUTTERY, P.J. (1976). *Br. J. Nutr.,* **36**, 575–579

VERNON, B.G. and BUTTERY, P.J. (1978a). *Anim. Prod.,* **26**, 1–9

VERNON, B.G. and BUTTERY, P.J. (1978b). *Br. J. Nutr.,* **40**, 563–572

VIPOND, J.E. and GALBRAITH, H. (1978). *Anim. Prod.,* **26**, 359

11

SOME IMPLICATIONS OF THE USE OF ANABOLIC AGENTS

B. HOFFMANN
*Institut für Veterinarmedizin des Bundesgesundheitsamtes
(Robert von Ostertag-Institut), Berlin, W. Germany*

Summary

Anabolic agents used in animal production can be classified according to their biological activity into compounds with oestrogenic, androgenic and gestagenic activity
and according to the structure into endogenous steroids, extraneous steroids and
non-steroidal compounds. Metabolism of endogenous steroids is well established, a
major principle being the formation of biologically less active metabolites, while the
metabolism of non-endogenous steroids is not fully understood. Differences between
compounds exist but it can be assumed that those with a high oral activity are only
to a small extent biodegraded. Tissue residues of anabolic steroids can be expected
in the ng/g to the pg/g range (confirmed by tracer studies and analytical approaches
using radioimmunoassay techniques). The legal frame regulating the practical application of anabolic agents varies between countries. Whereas in the USA more restrictive
measures have recently been taken, on a so-far fairly liberal market, other countries,
such as the Federal Republic of Germany, no longer prohibit the use of anabolic
agents in general (particularly oestrogens) but differ in their attitude towards compounds. It is concluded that endogenous hormones are natural constituents of edible
animals' tissues and that the amount of residues present in properly treated or untreated animals will not measurably contribute to the levels already seen in the
human. Conversely, residues of extraneous compounds have to be graded according
to normal toxicological parameters.

Introduction

Various groups of compounds which are basically different in their principal
biological activity are able to induce growth promotion in animals. By definition
these compounds lead to an increased protein synthesis which in turn depends
on the type, sex and age of the animal they are applied to (for review of the
literature see Lu and Rendel, 1976).

This chapter will restrict itself to those compounds which, apart from their
anabolic activity, also exert sex hormone-like activities. The clearance by the
American Food and Drug Administration for the use of the synthetic, non-
steroidal oestrogen, diethylstilboestrol (DES), for use in poultry in 1947 and in
cattle in 1954, marks the beginning of the widespread use of anabolic compounds
in animal production (Hoffmann *et al.*, 1975; Karg and Vogt, 1978).

Whereas in poultry the administration of oestrogens virtually has no effect on
growth promotion but rather on meat quality, characterized by an enhanced
deposition of fat in muscular tissue (see Hoffmann, 1978a), ample evidence has
been obtained to show that the induced protein deposition following the administration of these types of compounds to other species (like ruminants) is not

associated with changes in the carcass quality, apart from the possibility of the presence of hormonal residues (Verbeke *et al.*, 1976; Fischer and Schröder, 1976).

The use of any kind of drugs to food-producing animals requires consideration of the possible risks compared with the likely benefits. Today, the beneficial growth-promoting effects induced by anabolic agents, particularly in ruminants, are beyond any doubt (Heitzman, 1976, 1979). However, under certain conditions the risk to the consumer arising from the residues in edible animal tissues may outweigh the benefit. Owing to the initial lack of actual residue data, for many years the assessment of the risk was not so much based on scientific data but more or less on personal and public feelings. This situation has changed very little, in spite of recent advances in the characterization of the residue situation.

This chapter is an attempt to briefly review some of the basic aspects in the metabolism and residue formation of hormonally active anabolic agents and to outline to what extent practical consequences have been defined.

Classification of compounds used

As has been outlined elsewhere (Hoffmann and Karg, 1976), these materials can be classified according to their biological activity into compounds with oestrogenic, androgenic and gestagenic activity and according to their chemical structure into endogenous steroids, extraneous steroids and non-steroidal, extraneous compounds (*Table 11.1*). The table also shows that most of these preparations are used in ruminants and that with only two exceptions all of them contain an

Table 11.1. COMPOSITION OF VARIOUS ANABOLIC PREPARATIONS (FROM HOFFMANN, 1978b)

Determining component	Preparation oestrogenic	Preparation non-oestrogenic part	Type of animal production
Endogenous steroids	17β-oestradiol	Progesterone	Veal calf
	17β-oestradiol	Testosterone	Veal calf
Extraneous steroids	Oestradiol benzoate	Progesterone	Steer, heifer, lamb
	Oestradiol benzoate	Testosterone propionate	Steer, heifer, lamb
	Oestradiol monopalmitate		Poultry
	17β-oestradiol	Trienbolone acetate (TBA)	Veal calf/steer
		TBA	Heifer/cow
		Melengestrol acetate (MGA)	Heifer
Non-steroidal compounds	Diethylstilboestrol (DES)		Steer/heifer/lamb/veal calf
	DES	Testosterone	Steer/heifer
	DES	Methyltestosterone	Pig
	Hexoestrol		Steer/heifer/lamb
	Dienoestrol diacetate		Poultry
	Zeranol		Steer/heifer/lamb/veal calf

oestrogenic part. In contrast to the non-oestrogenic compounds such as trienbolone acetate, which is a potent androgen (Neumann, 1976), the single administration of oestrogens induces a rather high anabolic response, a point demonstrated by Van der Wal and Van Weerden (1976) in experiments with veal calves.

Table 11.2. CLASSIFICATION OF ANABOLIC AGENTS ACCORDING TO THEIR BASIC BIOLOGICAL ACTIVITY AND THEIR GROSS ORAL ACTIVITY *

Basic biological activity	Compound	Gross oral activity	
		weak	good
Oestrogenic	Oestradiol-17β †	x	
	Oestradiol benzoate †	x	
	Oestradiol monopalmitate†	x	
	Zeranol	x	
	Diethylstilboestrol		x
	Hexoestrol		x
	Dienoestrol		x
Androgenic	Testosterone †	x	
	Testosterone propionate †	x	
	Trienbolone acetate	x	
	Methyltestosterone		x
Gestagenic	Progesterone †	x	
	Melengestrol acetate (MGA)		x

* Derived from Hoffmann *et al.*, 1975, Rattenberger, 1976, *MGA Technical Manual*, Tuco, Division of the Upjohn Company, Kalamazoo, Michigan 49001.
† Endogenous steroid or equivalent to endogenous steroids.

In respect of their hormonal activity, anabolic agents not only show qualitative but also quantitative differences, which particularly concern their oral activity. As is evident from *Table 11.2*, all endogenous steroids have only little oral activity. With other compounds the response is varied, largely depending on their chemical structure. Thus generalizations like 'steroidal' or 'non-steroidal' should be avoided.

Metabolism

Metabolism of endogenous steroids has been well established in man and domestic animals (for review see Hoffmann *et al.*, 1975; Vermeulen, 1976; Velle, 1976). Perhaps one of the most important points is that biodegradation normally produces biologically less active metabolites. In cattle, 17α-epimerization seems to be the major metabolic pathway for both oestradiol-17β and testosterone (Martin, 1966; Velle, 1976). Apart from the possibility of peripheral transformation (Williams *et al.*, 1968) it has to be assumed that most of this 'catabolism of endogenous steroids' occurs in the liver, most likely during the first passage of the hormones through the organ. This would also explain their short biological half-life. Thus, in the case of endogenous steroids, enterohepatic circulation occurs with metabolites exerting little if any biological activity (Hoffmann, 1979).

In general in cattle most of the steroids are eliminated with the faeces, where 60—90% of the metabolites are found in the free form, while the steroids excreted from the urine were predominantly conjugated (Velle, 1976; Hoffmann and Karg, 1976). Recently, 17α-oestradiol glucopyranoside has been demonstrated to be a major oestrogen metabolite in cattle (Rao *et al.*, 1978).

Regardless of their origin (exogenous or endogenous), endogenous steroids enter the metabolic pathways outlined above. This is confirmed by the observation that conjugated oestradiol-17α is the major circulating oestrogen after implantation of oestradiol-17β, as is shown in *Table 11.3*.

Table 11.3. FREE AND CONJUGATED OESTRADIOL-17α, -17β AND OESTRONE IN PLASMA (MEAN VALUES; pg/ml OF 6 MALE CALVES TREATED WITH 20 mg OESTRADIOL-17β + 140 mg TRIENBOLONE ACETATE AND 2 CONTROL ANIMALS (FROM KARG *ET AL.*, 1976)

Oestrogen		Control	Treated animals (days after implantation)			
			34	42	19	71
Free	Oe 1*	4	17	16	19	71
	Oe 2β†	13	20	14	30	15
	Oe 2α‡	5	8	8	13	8
Conjugated	Oe 1*	8	80	60	54	91
	Oe 2β†	nd**	7	3	2	8
	Oe 2α‡	30	157	80	58	44

*   Oestrone
†   Oestradiol-17β
‡   Oestradiol-17α
** Not detectable.

Also, in the case of the extraneous androgenic steroid trienbolone acetate, the occurrence of a 'steroid-catabolism' to biologically less active metabolites could be demonstrated by Pottier *et al.*, 1978), who found 17α-OH-trienbolone to be the major metabolite.

Thus it seems justified to generalize that those anabolic agents exhibiting none or only a little oral activity in man or animals, are submitted to a 'catabolic' metabolism. In contrast, it seems reasonable to assume that the orally active anabolic agents are not, or are only partially, degraded and that the unchanged, biologically active molecule enters enterohepatic circulation. Indeed data derived by Aschbacher (1972), Aschbacher and Tacker (1974) and Aschbacher (1976) concerning the metabolism of DES in ruminants seem to confirm this concept, although not all questions in respect of some unidentified metabolites have yet been answered.

A final basic problem to be discussed is that apparently some anabolic compounds bind covalently to proteins. This was clearly demonstrated for oestrogens (Kuss, 1971; Blackburn, Tompson and King, 1976) and preliminary indications have been obtained that this might also happen in the case of trienbolone acetate (Ryan and Hoffmann, 1978). The biological significance of this phenomenon is not yet well understood. Owing to its low bioavailability the problem seems not to be so much the residue itself (Ryan and Hoffmann, 1978), but rather the fact that this phenomenon does occur, since binding with proteins, particularly with DNA as was shown for DES (Blackburn, Tompson and King, 1976), might be regarded as one of the first steps of carcinogenesis.

Formation of residues

The qualitative aspects of residue formation directly relate to the basic metabolic profile of each individual compound, as discussed above. The quantity of residue formation depends on various other pharmacokinetic parameters. In general, it is realistic to assume that elimation occurs rather rapidly, even after implantation or injection, as is shown for one example in *Figure 11.1*: 10 days after subcutaneous injection of 100 mg dienoestrol diacetate into the base of the ear

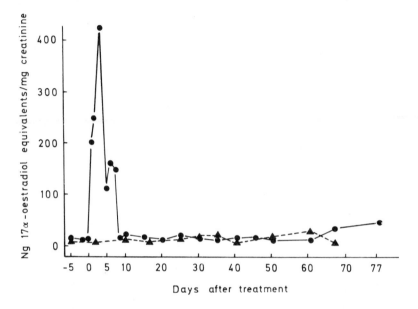

Figure 11.1   Urinary oestrogens (ng 17α-oestradiol equivalents/mg creatinine) in a female control calf (▲ – – – – ▲) and a female calf treated intramuscularly with 100 mg dienoestrol diacetate (● ——— ●). (From Rockel, Hoffmann and Kyrein, 1978)

of a veal calf, basal urinary oestrogen concentrations had been reached (determined by a radio-receptor assay) (Rockel, Hoffmann and Kyrein, 1978). Thus the residue concentrations in tissue after an adequate withdrawal period can be expected to be quite low, and studies with radio-labelled anabolic agents have clearly demonstrated that they will not exceed the p.p.t.- and low p.p.b.-range (Dunn *et al.*, 1977; Estergreen *et al.*, 1977). The only site of significant residue formation remains the site of injection or implantation (Vogt, Waldschmidt and Karg, 1970; Hoffmann *et al.*, 1975).

From the few methods available at present which are sensitive enough to measure these low residue concentrations, only radioimmunoassay (RIA) has been applied on a broader basis and only in cattle. The results obtained so far have been reviewed recently and are given, with some supplementary data concerning DES, in *Table 11.4*.

As is obvious the tissue concentrations of hormones vary not only between animals, depending on sex and reproductive status, but also within each animal. In the case of a high endogenous steroid production, such as testosterone in the bull, or after giving excessive dosages, as was done in the case of trienbolone,

Table 11.4. CONCENTRATIONS (pg/g) OF VARIOUS ANABOLIC AGENTS IN THEIR
FREE FORM IN TISSUES OF TREATED AND UNTREATED CATTLE

Compound determined	Animal	Tissue examined				Reference
		muscle	liver	kidney	fat	
Testosterone	Bull	535	749	2 783	10 950	Hoffmann and
	Heifer	92	193	595	250	Rattenberger (1977)
	Veal calf	16	39	256	685	
	Veal calf treated[1]	70	47	685	340	
Progesterone	Pregnant cow	–	–	–	336.2	Hoffmann (1978b)
	Heifer	–	–	–	16.7	
	Veal calf	–	–	–	5.8	
	Veal calf treated[2]	–	–	–	12.5	
Oestradiol-17β	Pregnant cow	370–860	–	–	–	Derived from
	Veal calf untreated and treated[3]	<100	<100	<100	<100	Hoffmann et al. (1975)
Oestradiol-17β	Steer	14.4	12.0	12.6	–	Henricks and
	Heifer	12.0	38.3	39.8	–	Torrence (1978)
Oestrone	Pregnant cow	0.12–2.09	–	–	–	Derived from
	Veal calf treated[3] and untreated	<100	<100	<100	<100	Hoffmann et al. (1975)
TBOH (trien-bolone)	Steer[4]	50	230	50	80	Hoffmann (1978b)
	Steer[5]	50	50	20	80	Hoffmann and Oettel
	Veal calf[3]	127	521	235	388	(1976)
	Veal calf[6]	797	3 467	2 563	2 580	
	Veal calf[7]	1 673	4 930	4 083	8 893	
DES	Veal calf[8]	100	–	–	–	Laschütza (1978)
	Veal calf[8a]	5 600	–	–	–	unpublished data
	Veal calf[9]	100	–	–	–	
	Veal calf[9a]	1 133 000	–	–	–	
	Cow[10]	4 400	–	–	–	
	Veal calf[11]	600	–	–	–	

[1] Slaughtered 77 days after implantation of 20 mg oestradiol-17β + 200 mg testosterone.

[2] Slaughtered 70 days after implantation of 20 mg oestradiol-17β + 200 mg progesterone.

[3] Slaughtered 70–77 days after implantation of 20 mg oestradiol-17β + 140 mg trienbolone acetate (TBA).

[4] Slaughtered 60 days after implantation of 40 mg oestradiol-17β + 200 mg trienbolone acetate.

[5] Implantation of 40 mg oestradiol-17β + 200 mg trienbolone acetate, implant removal after 60 days and slaughter after another 15 days.

[6] Slaughtered 70–77 days after impantation of 200 mg oestradiol-17β + 1400 mg trienbolone acetate (10-fold of normal dose).

[7] Slaughtered 70–77 days after implantation of 500 mg oestradiol-17β + 3500 mg trienbolone acetate (25-fold of normal dose).

[8] Slaughtered 4 days after intramuscular injection of 50 mg DES; general muscle.

[8a] Slaughtered 4 days after intramuscular injection of 50 mg DES; muscle injection site.

[9] Slaughtered 4 days after intramuscular injection of 100 mg DES; general muscle.

[9a] Slaughtered 4 days after intramuscular injection of 100 mg DES; muscle injection site.

[10] Slaughtered 7 days after implantation of 212 mg DES dipropionate; general muscle.

[11] Slaughtered 7 days after implantation of 200 mg DES dipropionate; general muscle.

highest residue concentrations are measured in fat. Otherwise, maximum hormone levels can be determined in liver and kidney. In particular, the kidney seems to bind testosterone rather specifically (Hoffmann and Rattenberger, 1977). The data obtained so far for endogenous anabolic steroids clearly indicate that under physiological conditions there are 100—1000-fold differences between animals. In the case of treatments with these endogenous anabolics, the residue levels seen in veal calves, steers and heifers are either not, or are only slightly, elevated above controls and are well at the bottom part of the physiological distribution curve.

It would appear (Hoffmann and Oettel, 1976; Hoffmann and Rattenberger, 1977) that the ratio of free to conjugated trienbolone and testosterone was lowest in liver (0.3—0.2), followed by kidney (1—2), muscle (5—15) and fat (36, to conjugated steroid not detectable). The percentage of conjugated testosterone was similar in treated and untreated animals.

Aspects on legislation

The aspects concerning the present situation in the USA and Europe have been discussed in detail recently (Perez, 1978; Karg and Vogt, 1978).

As was outlined for the USA: 'The Food, Drug and Cosmetic Act permits the approval of a carcinogenic animal drug provided that the use of the drug is efficacious, that it does not harm the animal, and that *no residue* will occur in food as determined by a method of analysis prescribed by the Secretary of Health, Education, and Welfare. The regulatory assay used to assure no residue will be the most sensitive reliable procedure available. Because FDA recognizes that the best available method may not be adequate to protect public health, minimum standards must be met for acceptability. These criteria are described in the Federal Register, February 22, 1977' (quoted from Perez, 1978).

These criteria, which are still at a state of discussion, seem to be tailored for the regulation of anabolic agents. A distinct difference is made between extraneous compounds and endogenous steroids. In the case of the latter compounds, tissue residue levels are only acceptable if they are not different from the norm of a certain animal population (i.e. steers, heifers or veal calves). Although it has become possible to determine these tissue levels (*Table 11.4*), a number of problems are still associated with the routine use of these radioimmunoassay techniques which require the utmost sensitivity and precision. In consequence, FDA has proposed a withdrawal of New Animal Drug Applications for products containing oestradiol benzoate, which because of its rapid hydrolysis can be classified as an endogenous oestrogen (*Food Chemical News*, 8 January, 1979).

Certainly some points questioning the reasonableness of this action could be raised. The levels seen in steers treated with 20 mg oestradiol benzoate are only slightly higher than those seen in untreated animals and still well below the 0.05 p.p.b. margin (Forchielli 1978, personal communication). Considerably higher levels, for example, were observed in untreated calves and pregnant heifers (see *Table 11.4*) which can be slaughtered without any restrictions. In addition, the biological significance of these residue levels has to be questioned, in particular when the daily production rate of oestrogens in the human is considered (Vermeulen, 1976).

For reasons of 'not being safe', FDA has also proposed a final withdrawal of DES in 1978 (Department of Health, Education, and Welfare, Doc. No. 76N-002). In this particular case, safety was not so much related to the methodological aspects of residue detection but rather to the still open questions concerning the characterization of the toxicological and pharmacological properties of DES as well as its pharmacokinetics and metabolism. Any extraneous anabolic agent intended for use in food animals will need a similar thorough characterization before it can be accepted for practical use. In this relation the expressions 'marker tissue' and 'marker residue' have been introduced (Perez, 1978) and also mathematical approaches for establishing a 'safe' residue of a carcinogenic drug have been proposed by the FDA (Kolbye and Perez, 1976; Perez, 1978).

In Europe the guidelines set up by the Commission of the European Communities prohibit the approval of compounds with hormonal and antihormonal activity as additives in animal feed in all member countries. In Belgium, the Netherlands, Luxembourg, Denmark and Italy, a strict ban exists on any application of hormonally active substances for fattening purposes.

While the official position in the USA has become more restrictive in recent years, in the Federal Republic of Germany the absolute ban on oestrogens from being used as anabolics has been loosened somewhat. The present regulation, according to paragraph 15 of the food law from 15 August, 1974, differs between oestrogens. This regulation (in force since 1 January, 1978) prohibits compounds with oestrogenic activity from being used in food animals during a certain age period, if their oral activity in the mouse uterus assay is higher than one-fifth of that exerted by DES. DES is banned totally. However, the other compounds not mentioned in that 'negative list' can theoretically be approved as veterinary drugs (which includes the indication 'growth promotion') if the application fulfills the requirement of the drug law.

It is hoped that due to this legal outlet the problem of the misuse of anabolic agents will be overcome, particularly that the use of orally highly active oestrogens, like the stilbene derivatives will be prevented. In fact, adequate methodology to test for the use of these compounds has been included in the meat inspection law (detection by TLC when urine is sampled and by radioimmunoassay when only muscular tissue is available; Bundesgesetzblatt, 3 February, 1978; see also Hoffmann, 1979).

In France on 27 November, 1976, a strict anti-oestrogenic law banned any use of oestrogens in food-producing animals. This ban includes all veterinary applications, with the exception of cycle regulation. However, for endogenous oestrogen concentrations in meat, upper physiological levels are given (0.2 p.p.b. in immature animals, 10 p.p.b. in animals of breeding age). A similar situation prevails in Ireland (Roche and Davis, 1976).

Within the European Community, only Great Britain allows a wider spectrum of anabolic preparations (including oestrogens), since they were licensed for veterinary purposes before September 1972. However, a committee has been appointed to reconsider the situation and to make new recommendations (Karg and Vogt, 1978).

Conclusions

Although the information available is still limited, certain conclusions in respect of residue concentrations and public health can be drawn. It has been clearly established that in man and animals endogenous steroids — regardless of their origin — enter metabolic pathways which lead to a rapid biodegradation and clearance from the organism. Endogenous steroids have to be regarded as natural environmental constituents. In respect to the daily production rate of sex steroids in the human, regardless of sex and age (Vermeulen, 1976), it cannot be expected that the endogenous hormones consumed with food of animal origin will measurably contribute to steroid levels in the human (Hoffmann, 1979). This applies to both treated and untreated animals. Only an implantation site, if consumed by accident, would represent a biologically effective dose. It is well known that even a long time after application rather high concentrations of hormone can be found at the implantation or injection site (Vogt, Wald-schmidt and Karg, 1970) and it is therefore strongly recommended that the site of application be automatically discarded at slaughter. These considerations are reflected in the new German food law.

When properly handled and if adequate withdrawal periods are observed, extraneous anabolic steroids or agents lead to only minute tissue residues. Also, some of these compounds seem to be rather effectively biodegraded. Neverthe-less, due to their 'non-physiological' status these residues have to be graded according to normal toxicological criteria.

In this regard the question of carcinogenicity, the metabolic fate and the type of tissue residues formed are of particular interest. The fact that various coun-tries have recently changed their legal frame for the use of anabolic agents in animal production — partly in the opposite direction — expresses the uncertainty in this area. Only if attempts to further elucidate the general aspects of safety are continued, using scientific principles, will it be possible to overcome these problems and to get public agreement and perhaps to establish a commonly accepted basis for the use of anabolic agents.

References

ASCHBACHER, P.W. (1972). *J. Anim. Sci.,* **35**, 1031—1035

ASCHBACHER, P.W. (1976). *J. Toxicol. Envmntl Hlth,* Suppl. 1, 45—59

ASCHBACHER, P.W. and TACKER, E.J. (1974). *J. Anim. Sci.,* **39**, 1185—1192

BLACKBURN, G.M., TOMPSON, M.H. and KING, H.W.S. (1976). *Biochem. J.,* **158**, 643- 646

DUNN, T.G., KALTENBACH, C.C., KORITNIK, D.R., TURNER, D.L. and NISWENDER, D.G. (1977). *J. Anim. Sci.,* **46**, 659—673

ESTERGREEN, V.L., LIN, M.T., MARTIN, E.L., MOSS, G.E., BRANEN, A.L., LEUDECKE, L.O. and SHIMODA, W. (1977). *J. Anim. Sci.,* **46**, 642—651

FISCHER, A. and SCHRÖDER, K. (1976). In *Anabolika in der Kälbermast,* Suppl. 6, pp. 59- 65 (Brüggemann, J. and Richter, O., Eds), Advances in Animal Nutrition *(Z. Tierphysiol. Tierernähr. Futtermittelk.)*

HEITZMAN, R.J. (1976). In *Anabolic Agents in Animal Production,* Suppl. V, pp. 89—98 (Coulston, F. and Corte, F., Eds), Thieme, Stuttgart

HEITZMAN, R.J. (1979). *J. Steroid. Biochem.,* **11**, 927—930

HENRICKS, D.M. and TORRENCE, A.K. (1978). *J. Ass. Off. Anal. Chem.,* **61**, 1280- 1283

HOFFMANN, B. (1978a). In *Rückstände in Geflügel und Eiern, DFG-Forschungs-bericht,* pp. 86—94, Harald Boldt, Boppard

214 Some implications of the use of anabolic agents

HOFFMANN, B. (1978b). *J. Ass. Off. Anal. Chem.*, **61**, 1263–1273

HOFFMANN, B. (1979). *J. Steroid. Biochem.*, **11**, 919–922

HOFFMANN, B. and KARG, H. (1976). In *Anabolic Agents in Animal Production*, Suppl. V, pp. 181–191 (Coulston, F. and Corte, F., Eds), Thieme, Stuttgart

HOFFMANN, B. and OETTEL, G. (1976). *Steroids*, **27**, 509–523

HOFFMANN, B. and RATTENBERGER, E. (1977). *J. Anim. Sci.*, **46**, 635–641

HOFFMANN, B., HEINRITZI, K.H., KYREIN, H.J., OEHRLE, K.L., OETTEL, G., RATTENBERGER, E., VOGT, K. and KARG, H. (1976). In *Anabolika in der Kälbermast*, Suppl. 6, pp. 80–90 (Brüggemann, J. and Richter, O., Eds), Advances in Animal Nutrition (*Z. Tierphysiol. Tierernähr. Futtermittelk.*)

HOFFMANN, B., KARG, H., VOGT, K. and KYREIN, H.J. (1975). In *Forschungsbereich der DFG; Rückstände in Fleisch und Fleischerzeugnissen*, pp. 32–59, Harald Boldt, Boppard

KARG, H. and VOGT, K. (1978). *J. Ass. Off. Anal. Chem.*, **61**, 1201–1208

KARG, H., CLAUS, R., HOFFMANN, B., SCHALLENBERGER, E. and SCHAMS, D. (1976). In *Nuclear Techniques in Animal Production and Health*, IAEA-SM 205/109, pp. 487–511, Vienna

KOLBYE, A.C. and PEREZ, M.K. (1976). *Anabolic Agents in Animal Production*, Suppl. V, pp. 212–218 (Coulston, F. and Corte, F., Eds), Thieme, Stuttgart

KUSS, E. (1971). *Hoppe-Seyler's Z. Physiol. Chem.*, **352**, 817–836

LU, F.C. and RENDEL, J. (1976). In *Anabolic Agents in Animal Production*, Suppl. V (Coulston, F. and Corte, F., Eds), Thieme, Stuttgart

MARTIN, R.P. (1966). *Endocrinology*, **78**, 907–913

NEUMANN, F. (1976). In *Anabolic Agents in Animal Production*, Suppl. V, pp. 253–264 (Coulston, F. and Corte, F., Eds), Thieme, Stuttgart

PEREZ, M.K. (1978). *J. Ass. Off. Anal. Chem.*, **61**, 1183–1191

POTTIER, J., COUSTY, C., HEIZMAN, R.J. and REYNOLDS, I.P. (1978). *J. Anim. Sci.*, in the press

RAO, P.N., CARR, W.M., MOORE, P.H.Jr. and GOLDZIEHER, J.W. (1978). *J. Toxicol. Envmntl Hlth*, **4**, 495

RATTENBERGER, E. (1976). 'Rückstandsanalytik von Testosteron im Gewebe vom Rind mit Hilfe des Radioimmonotests', *Diss. agr.*, Tech. Univ. München

ROCHE, F.J. and DAVIS, W.D. (1976). *Horm. Fd Res.*, 146–148

ROCKEL, P., HOFFMAN, B. and KYREIN, H.J. (1978). *Berl. Münch. Tierärztl. Wschr.*, **91**, 476–479

RYAN, J.J. and HOFFMANN, B. (1978). *J. Ass. Off. Anal. Chem.*, **61**, 1274–1279

VAN DER WAL, P. and VAN WEERDEN, E.I. (1976). In *Anabolic Agents in Animal Production*, Suppl. V, pp. 60–78 (Coulston, F. and Corte, F., Eds), Thieme, Stuttgart

VELLE, W. (1976). In *Anabolic Agents in Animal Production*, Suppl. V, pp. 159–170 (Coulston, F. and Corte, F., Eds), Thieme, Stuttgart

VERBEKE, R., DEBACKERE, M., HICQUET, R., LAUWERS, H., POTTIE, G., STEVENS, J., VAN MOER, D., VAN HOOF, J. and VERMEERSCH, G. (1976). In *Anabolic Agents in Animal Production*, Suppl. V, pp. 123–130 (Coulston, F. and Corte, F., Eds), Thieme, Stuttgart

VERMEULEN, A. (1976). In *Anabolic Agents in Animal Production*, Suppl. V, pp. 171–180 (Coulston, F. and Corte, F., Eds), Thieme, Stuttgart

VOGT, K., WALDSCHMIDT, M. and KARG, H. (1970). *Berl. Münch. Tierärztle. Wschr.*, **83**, 457–461

WILLIAMS, K.I.H., HENRY, D.H., COLLINS, D.C. and LAYNE, D.S. (1968). *Endocrinology*, **83**, 113–117

12

ENERGY COSTS OF PROTEIN DEPOSITION

A.J.H. VAN ES
*Research Institute for Livestock Feeding and Nutrition, 'Hoorn' Lelystad,
and Department of Animal Physiology, Agricultural University, Wageningen,
The Netherlands*

Summary

Energy costs of protein deposition have been estimated directly from biochemical considerations. Maximum efficiencies of the utilization of the metabolizable energy for protein deposition of about 85% and 80% for monogastrics and ruminants, respectively, have been derived. Lower efficiencies, however, may very well occur in the animal body in view of our lack of precise knowledge on the biochemistry of protein deposition and on rate of protein turnover.

An attempt has also been made to derive estimates of these energy costs of protein deposition indirectly. First, a survey of energy metabolism of homotherms is given. Next, energy costs for protein deposition by farm animals are derived by difference, i.e. by subtracting all other energy costs from total costs. All estimates obtained in this way lack precision, mainly because protein deposition forms such a small part of total energy metabolism, and estimates of energy costs of maintenance in growing animals and laying hens and of fat deposition in ruminants are not very accurate. Owing to lack of precise information no estimates can be derived of the energy costs of protein deposition for maintenance, for pregnancy purposes or for lactation of monogastrics. The estimate of the efficiency of the utilization of metabolizable energy for energy deposited as milk protein in the cow ranges from 65% to 55% and for energy deposited as egg protein in the hen from 30% to 87.5%. For energy deposited as tissue protein by growing homeotherms, the estimates range from 40% to 60%. Nearly all indirect estimates are considerably lower than the maximum values estimated directly. They tend to be slightly lower for growing animals.

Introduction

The energy costs of protein deposition can be estimated directly and indirectly. The direct method makes use of knowledge of the biochemistry of protein synthesis and of the rate of protein turnover. As information on this rate is still far from precise, this method often fails (Waterlow, Garlick and Millward, 1978). The indirect method, in fact, is a difference method. Total energy needs are estimated and from these all energy needs other than for protein deposition are subtracted. The precision of the results obtained with this method clearly depends heavily on the precision of the separate estimates, the more so because the difference is usually small compared to the total energy needs.

Both methods of estimation will be discussed, but most attention will be paid to the second.

Direct method of estimating energy costs of protein deposition

Linking the various amino acids in the correct order for the synthesis of a protein requires a fair amount of free energy stored in ATP (Campbell, 1977). Energy in the form of ATP is needed not only for the amino acids to be activated and coded, and for the actual synthesis at the ribosome, but also for the synthesis of the mRNA and the tRNA. In higher animals mRNA and tRNA are assumed to be used many times, so probably their synthesis does not add much to the other ATP costs. For some proteins, more amino acids have to be linked together than the final protein contains (Campbell, 1977). In this case some amino acids at one end of the peptide chain appear to be needed only for passage through a membrane, after which they are split without giving rise to ATP. The total number of moles of ATP needed to synthesize one peptide bond is estimated at about five. On average, 1 mol amino acids as present in proteins, i.e. without the mol H_2O lost by peptide formation, weighs about 115 g and contains about 2.7 MJ. The complete oxidation of 1 mol glucose (heat of combustion 2814 kJ/mol) by the animal is considered to yield 38 mol ATP, so that the synthesis of 1 mol ATP requires 74 kJ. The same holds true for di- and polysaccharides which can be hydrolysed to glucose or fructose.

Owing to the biochemical pathways followed (including in the case of protein the costs of urea synthesis), ATP yields are some 5% lower for energy in fat and 10–20% lower for energy in volatile fatty acids (VFA) and in protein (corrected for loss of urea energy with the urine). These figures apply to the energy of these substances at the tissue level after subtraction of energy losses in faeces, urine and methane, from gross energy of the feed. For animals with little or no fermentation in the gastro-intestinal tract this energy is equal to the metabolizable energy (ME). For ruminants it is close to 0.9 ME because the rumen microorganisms convert some energy of the feed into heat which cannot yield ATP and which is included in ME (see page 218).

The theoretical ME costs for protein synthesis now can be calculated in nonruminants, if there is little or no gastro-intestinal fermentation. To synthesize 100 g of average protein containing 2400 kJ, we first need 2400 kJ amino acid–ME for building blocks and next about $5 \times (100/115) \times (74$ to $90) = 320$ to 390 kJ ME for peptide formation. Thus the theoretical efficiency is $2400 \times 100/(2400 + 320$ to $390) = 88\%$ to 86%. This calculation applies for ME as measured in the producing animal. Often ME values of feeds are given after correction to zero nitrogen retention, i.e. they apply then to the non-productive state. In that case all nitrogen is supposed to be finally excreted as urea or uric acid. Such protein–ME, if used as the protein building block, has a 20% higher energy value, as no nitrogen is excreted as urea or uric acid. Thus the calculated efficiency becomes, in that case, $2400 \times 100/(1900 + 320$ to $390) = 108\%$ to 105%.

In the case of ruminants, the theoretical efficiency is only about $2400 \times 100/(2400 \times 1.1 + 370 \times 1.1) = 79\%$, due to the fact that ruminant–ME contains some 10% fermentation heat and, furthermore, mainly VFA and amino acids, but hardly any glucose. Higher theoretical efficiencies also hold true here for ME corrected to zero nitrogen balance.

All these theoretical efficiency figures are based on the estimated requirement of 5 mol ATP for the synthesis of 1 mol peptides from amino acids. The possibility exists that because of so far unknown reasons more ATP is needed; thus, the efficiency estimates may be too high.

Protein deposition is not equal to protein synthesis but the net result of the processes of synthesis and degradation. Degradation of protein to amino acids does not yield ATP. Thus to arrive at the actual energy costs of protein deposition it may be necessary to account for the rate of renewal or turnover of protein. However, this renewal of protein can be considered to be part of the animal's maintenance because it is not immediately connected with protein deposition itself and continues as long as the protein exists. However, there is some evidence that in the growing animal the rate of turnover of all or part of the protein of the whole body increases with the rate of protein deposition (Arnal, 1977; Waterlow, Garlick and Millward, 1978). The energy costs of the increase of this rate, say from 3% to 8% per day, certainly should be written on the account of protein deposition. For the 12 kg protein of a pig of 75 kg this would mean an additional daily synthesis of $(8-3) 120 = 600$ g protein requiring some $6 \times (320$ to $390) = 1.9$ to 2.3 MJ ME. Thus, this would reduce the above calculated efficiency of 88% to 86% for a deposition of 200 g protein to $2 \times 2400 \times 100/(2 \times 2400 + 2 \times 320$ to $390 + 1900$ to $2300) = 65\%$ to 61%. On such increases of turnover rates we are poorly informed with regard to the growing animal and even less with regard to the lactating or egg-laying animal.

It will be clear that at present the direct method cannot give a precise quantitative estimate of the size of the energy costs of protein deposition because there is insufficient information on ATP costs of peptide formation and rate of protein turnover.

Indirect method of estimating energy costs of protein deposition

GENERAL

With this difference method all energy needs other than for protein deposition are estimated and subtracted from total energy needs. Therefore, the method requires a good understanding of the animal's energy metabolism, which will be discussed first.

SURVEY OF ENERGY METABOLISM OF HOMOTHERM ANIMALS

By means of a bomb calorimeter, the calorific value or (gross) energy content of food, faeces, urine, milk, eggs, tissues, etc., can be determined. The precision of the method is high. In the case of wet samples some special precautions of sample preparation are required, but these hardly decrease the precision. Thus it is not difficult to make a balance of energy input and output as far as it concerns matter for an animal, provided that the average daily quantities of ingested food and of faeces, urine, milk, eggs, etc., are precisely known.

However, there are two other sources of energy loss: energy lost as heat and energy lost as methane. Data on these losses can be obtained from direct and indirect calorimetry. The former method measures all heat loss from the animal directly and requires complicated equipment. The measurement often interferes somewhat with the animal's normal behaviour. The second method calculates the animal's heat production from its average daily O_2 consumption, CO_2 and CH_4 production and urinary N loss. Methane energy losses can also be derived from it. The method allows the animal to behave nearly normally, provided some precautions are taken. Unfortunately, this is not always done. Over periods of a day, on average, there is no net accumulation or loss of heat in the animal body, so heat loss is equal to heat production, which links the two methods.

The scheme of *Figure 12.1* gives information on terms used in energy balance work. As losses with faeces vary from food to food and those with urine and methane, although being small, also decrease the energy value of the feed for the animal, the ME is often used as the starting point of discussions on energy utilization. ME, of course, should be seen as the energy of the chemical substances absorbed from the gastro-intestinal tract into the blood – monosaccharides, fats and fatty acids, amino acids – in which case losses of energy of detoxication products excreted with faeces and urine are already accounted for.

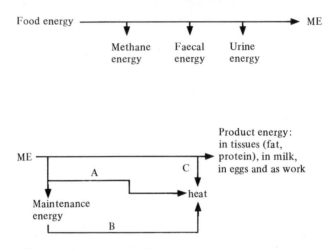

Figure 12.1   Scheme of energy metabolism

In ruminants, most of the carbohydrates are fermented in the fore-stomachs, a process giving volatile fatty acids and methane as energy-containing compounds. These acids are absorbed in the blood, and the methane is lost by eructation (the reason its energy is subtracted in calculating ME). The fermentation process also produces some heat. This is not subtracted in calculating ME, because it is very difficult to measure. Ruminant ME, therefore, does not only consist of chemical substances: some 10% is heat energy. This heat is seldom of any use for the animal as ruminants have a high heat production and only at temperatures below 10 °C may they gain any advantage from it.

All items of the energy balance of *Figure 12.1* may be measured separately, except for maintenance energy and each of the three main sources of heat A, B and C. In fact, C is the sum of the separate heat losses due to tissue, milk and egg synthesis and work. This means that it is not possible simply to derive from this balance the amount of ME which was needed for the deposition of protein in the tissues, in the milk, in the eggs or for protein synthesized for maintenance purposes. Fortunately, energy balance experiments, mainly with farm animals, have produced a lot of information on the amount of ME needed for each of these purposes. The usual procedure in such experiments has been to study the change in ME needs when the size of one kind of production is changed, while it is assumed that maintenance and all the others stay as they are. While discussing ME costs of protein deposition of various kinds, we shall see to what extent it is indeed possible to keep these other items unchanged.

These considerations lead to the following model to estimate the ME costs of protein deposition in an animal which does not perform external work:

$$\text{Total ME} = \text{Maintenance--ME} + a \cdot E_{\text{fat}} + b \cdot E_{\text{carbohydrate}} + x \cdot E_{\text{protein}}$$

where E_{fat}, $E_{\text{carbohydrate}}$ and E_{protein} are the energy depositions as fat, carbohydrate and protein, respectively, a and b are constants, and x is the constant looked for. If x were to be found equal to one, this would mean a 100% conversion of the ME of protein; a higher figure would mean that more ME was needed to deposit the energy content of protein than the ME value of the protein itself. In the first case it was obviously enough to supply the amino acids, whereas in the second case, besides this kind of ME more was needed to link these amino acids.

It will be clear that reliable values of x may only be obtained provided that E_{protein} can be measured rather precisely and has a size which is not too small compared to the other variables, the sizes of which should be known rather precisely. In particular, it is necessary that the estimate of maintenance--ME be accurate, because this kind of ME requirement is so large compared to the other requirements. If maintenance--ME is estimated with the model by means of regression, it usually is related to metabolic weight ($W^{0.75}$). Obtaining a correct estimate of maintenance--ME by means of regression depends heavily on the suitability of the material used for that purpose (van Es, 1972).

ENERGY COSTS OF PROTEIN DEPOSITION IN MATURE ANIMALS

In mature animals, protein is synthesized for maintenance, as well as in lactation, egg production and pregnancy. Enzyme and hormone production, tissue renewal, mucus skin and hair production are the main *maintenance* processes which require a steady supply of new protein. Unfortunately the size of the total number of peptide linkages needed − for these ATP is required and thus ME − for this protein synthesis is only very poorly known. Metabolic faecal nitrogen and endogenous urinary nitrogen do not tell us much about it. They only show the quantity of protein which, due to the maintenance metabolism, is completely degraded and excreted. A considerable part of the proteins, however, is degraded only to amino acids which are used again for protein synthesis. Even if we were to have precise information on the total amount of peptide linkages required for maintenance, e.g. from nitrogen turnover studies, it would not help us much to derive indirectly their ME costs because we know only very roughly the size of the energy required for the various other maintenance processes.

Also, the protein deposition during *pregnancy* does not help us much to estimate ME needs for protein deposition. Even in the last weeks of pregnancy daily protein deposition in fetus, uterus and other reproductive tissues is small: less than 50 kJ/$W^{0.75}$ in cattle and also in sows, compared to close to 200 kJ/$W^{0.75}$ in 25 kg milk of a cow (W = body weight in kg). The size of this protein deposition changes daily, which makes measuring the energy required for it even more complicated. Moreover, pregnancy often changes the behaviour of the mother; in addition, there are changes in her endocrine system. Both aspects may alter the maintenance requirements, but for neither is there precise quantitative information available. The estimation of the ME needs for maintenance of the fetus is another unsolved problem.

In *lactation*, too, the information is often far from sufficient for our purpose. On milk production and composition in women there is sufficient information, but the average daily quantity is rather low. Estimates of reasonable precision of ME requirement for milk production are lacking; the available estimates come from intake studies which for our purpose are far too imprecise. Using estimates of maintenance from measurements before the onset of pregnancy and lactation as information for maintenance during lactation may also lead to errors because many women reduce their physical activity when pregnant or lactating.

Daily milk protein secretion by the sow can be high, of the order of 150 $kJ/W^{0.75}$, but here both on the quantity of milk produced daily and on the total ME needs for milk synthesis no precise data are known so far.

The many results of energy balance studies with lactating cows (see van Es, 1975) do make it possible to study the costs of protein deposition in a mature animal. They showed that the cow's ME needs increase rectilinearly with increasing milk energy production. The majority of more than 1000 separate energy balances show an efficiency of utilization of ME for the production of the energy of milk of 60%. A minority is said to indicate values of 65%. It has to be added that rather high maintenance requirements were derived for the latter, so the reliability of the 65% is not great. We shall use only the lower value in our calculations.

Milk with 4% fat, containing 3000 kJ per kg and requiring above maintenance 5000 kJ ME per kg, consists of 50% fat energy, 25% lactose energy and 25% protein energy. In most balance studies milk of such a composition was produced. This means that the separate costs of ME for the synthesis of each of the three components cannot very well be derived by means of multiple regression, the more so because fat and protein are positively correlated, although not very strongly. Thus, to arrive at ME costs for milk protein deposition by difference the ME costs of fat and lactose synthesis have to be derived in a different way, i.e. from data on biochemical pathways and on efficiencies of body fat production by the mature non-producing ruminant. Unfortunately such data are not very precise (van Es, 1976; van Es and van der Honing, 1979).

In ruminating animals most of the fat is synthesized from acetate absorbed from the fore-stomachs rather than from glucose absorbed from the small intestine, as is the main case in the non-ruminating animal. The synthesis of higher fatty acids requires NADPH as hydrogen donor. The ruminant lacks some ways of NADPH synthesis which non-ruminating animals have, so a greater part of it has to be supplied from glucose via the pentosephosphate pathway. Unfortunately, little glucose is usually absorbed from the gastro-intestinal tract in the ruminating animal. Most glucose has to be synthesized by gluconeogenesis from propionate, absorbed from the fore-stomachs, or from glucogenic amino acids, probably energetically a less efficient way. Glucose is also needed for the synthesis of milk lactose. It is so far not clear whether gluconeogenesis from propionate supplies enough glucose for lactose and NADPH synthesis or not, nor if NADPH synthesis from NADH at the isocitrate level of the TCA cycle plays an important role.

Most balance trials with mature ruminants which are fattened gave efficiencies of utilization of ME for body fat synthesis which were below the values that were biochemically expected. The discrepancy between measured and expected efficiency seemed to increase with lower quality of the ration. This might be due to the fact that from low-quality rations hardly any glucose will be absorbed, whereas the propionate content of the absorbed VFA will also be low. As a result, NADPH supply might be difficult. In the lactating ruminant one would expect even greater problems with NADPH supply because of the glucose needed for lactose synthesis. It is true that productive lactating cows usually receive rations of high quality. Maybe their high feeding levels allow some of the starch of the rations, usually present in fair quantities, to pass on to the small intestine to be absorbed as glucose. Good dairy rations would probably have a 50–55% efficiency of ME utilization for fat synthesis in mature, non-lactating ruminants.

Biochemically, lactose synthesis from ME in the ruminant means a conversion of propionate or glucogenic amino acids via glucose to lactose. The energetic efficiency for the former process, taking into account that ME contains some 10% fermentation heat, is about 75%. For the other process it will be of the same size; after deamination glucose synthesis will probably be more efficient than from propionate but the detoxication of the ammonia and urea excretion requires some energy.

Using the efficiency figures given above for production of milk energy, body fat energy and lactose energy we arrive at the following calculation based on milk energy containing 50% fat energy and 25% protein energy, and 25% lactose energy:

$$60 = 0.5(50 \text{ to } 55) + 0.25 \cdot x + 0.25 \cdot 75$$

so that $x = 65\%$ to 55%. This estimate is also low compared to the theoretical value of about 80% given on page 216. It should be stressed that the assumptions made in our calculation of the lactating cow are subject to doubt.

Finally, *egg production* by the hen gives us another possibility to estimate the energy costs of protein deposition. With this animal several energy balance studies have been performed [a (Waring and Brown, 1965, 1976; Tasaki and Sasa, 1970; van Es *et al.*, 1970; Burlacu *et al.*, 1974), b (Grimbergen, 1970; Hoffmann and Schiemann, 1973)]. Energetically, egg production does not increase total metabolism very much. A hen of 2 kg producing 8 eggs in 10 days only requires about 1.6 times the amount of ME needed for maintenance. Thus for our purpose it is extremely important to have a precise estimate of the maintenance needs. However, in some of the studies the maintenance estimate came from non-laying hens; whether such data also apply to the laying state is questionable. In other experiments regression techniques were used on data with high and low or no egg productions. Such data lie far apart and the maintenance requirement may have changed during the interval. In view of the rather low total metabolism, it will be clear that slightly higher maintenance estimates, e.g. around 480 kJ/$W^{0.75}$ in the first five studies (a) mentioned above, as opposed to close to 420 kJ/$W^{0.75}$ in the last two (b), resulted in considerably higher estimates of the efficiency of the utilization of ME above maintenance – around 80% versus about 60% for the two groups of studies, respectively.

We assume that egg lipid deposition occurs with the same energetic efficiency of 75–80% as body fat deposition in the mature monogastric animal. Such efficiencies of body fat deposition can be expected biochemically (Nehring and

Schiemann, 1966; Armstrong, 1969), and have been found repeatedly in balance experiments (Schiemann *et al.*, 1971). The energy of an egg consists of 40% of protein energy; nearly all the other energy is due to lipid. Using the two estimates of 60% and 80% for egg energy production above maintenance, the efficiency (x) of deposition of protein energy may be derived as follows:

$$80 = 0.6 \ (75 \text{ to } 80) + 0.4x; \ x = 80 \text{ to } 87.5$$
$$60 = 0.6 \ (75 \text{ to } 80) + 0.4x; \ x = 30 \text{ to } 37.5$$

These results show clearly that the available information on the laying hen's energy metabolism is not accurate enough for deriving a precise estimate of the ME costs of egg protein deposition.

ENERGY COSTS OF PROTEIN DEPOSITION IN GROWING ANIMALS

Total energy metabolism in young rapidly growing homeotherms liberally fed highly digestible and palatable rations can be as high as three times their maintenance energy metabolism. The rate decreases with increasing age and, understandably, with lower net energy intake. With advancing age also, the kind of energy deposited changes. In early growth, 60% of the energy content of deposited tissue is as protein and the remainder as fat. Soon afterwards this percentage decreases under *ad lib.* feeding, e.g. to 35% in a chicken of 1.5 kg, to 25% in a veal calf of 150 kg, to 20–15% in a pig of 100 kg and to 15–10% in an early maturing steer of 500 kg.

It will be clear that, for our purpose, results obtained with animals in their first stage of growth are most suited. Again we need a good estimate of the ME costs for maintenance and for body fat deposition. There does not seem to be much objection against using for fat deposition the efficiency figures from mature animals. To obtain a good estimate of the maintenance requirement is much more difficult, especially for animals in their first stage of growth (van Es, 1972; Thorbek and Henckel, 1976; Pullar and Webster, 1977). Fasting such animals or feeding them only maintenance rations may change their behaviour and thus their maintenance requirements. Using maintenance estimates derived from mature animals seems also incorrect because young animals are more active and more susceptible to stress.

The latter observation makes it preferable to perform measurements with a few young animals together, instead of separate ones. Best estimates might theoretically be obtained with young growing animals kept in groups of two or more and fed in one experiment *ad lib.* and in another 10–20% less. From the results, the maintenance requirement could then be calculated by extrapolation. Again, this method is not foolproof. First, owing to the extrapolation the accuracy is not high. Secondly, at the lower feeding level the energy deposition usually differs from that at the *ad lib.* level — it contains less fat. Therefore, it might be incorrect to assume the same efficiencies of the utilization of ME for these two kinds of energy deposition. Thirdly, a food restriction of 10–20% might change the animal's behaviour and thus maintenance requirement. In groups of chickens we found some indications for the first phenomenon but it did not influence maintenance very much.

In most earlier attempts to derive the energy costs of protein deposition during growth, not too much attention was paid as to whether the material which was used was suited for this purpose. Results of balance or slaughter trials of animals fed *ad lib.* were used and average daily ME intake was regressed on metabolic body weight (as an estimate of maintenance), on average daily protein and on average daily fat deposition (Kielanowski, 1976; Pullar and Webster, 1977). Most of these estimates gave for fat deposition energetic efficiencies which were close to those found in mature animals and to those expected biochemically. However, the estimates for protein deposition, 40–60%, were far below the expected 80–85%; they showed considerable variation. These protein efficiency figures might be biased because in the studies used, maintenance requirements per metabolic weight might have decreased and the ratio of protein to fat deposition decreased with advancing age.

Pullar and Webster (1977) tried to avoid these biases. They worked with two strains of rats – a normal and a fatty strain – to obtain a fair difference in protein deposition at the same feeding level. Energy balances were measured for small groups of these rats at two body weights, each time at a high and a lower feeding level. While interpreting the results it was only assumed that within strains at the two feeding levels at the same body weight, maintenance requirement was equal and that ME utilization for protein and fat deposition did not differ between animals or treatments. Thus the following model could be used:

$$ME = A_i + a \cdot E_{\text{protein}} + b \cdot E_{\text{fat}}$$

where A_i stands for the maintenance requirement of the two strains of animals at two body weights ($i = 1$ to 4). Also this study, somewhat surprisingly, gave similar low efficiencies for protein-energy deposition (44%) as most of the earlier less suited experiments. The plan of this study did not exclude all biases. The assumption of equal maintenance requirement at the same body weight for the two levels of feeding may not be correct. Energy costs of protein deposition in the fatty strain may have differed from that in the other strain. Moreover, it was somewhat unexpected that protein to fat ratios did not always decrease at a higher feeding level. Even so, it is very improbable that these possible biases were the reason for the large difference between the biochemically expected maximum efficiency of 80–85% and the value of 44% derived in this study.

Conclusion

The direct methods of estimating the costs of protein deposition in producing animals fail because there is insufficient information on the ATP requirement for peptide formation and on the rate of protein turnover. Maximum estimates of the efficiencies of protein energy formation from ME are near 85% and 80% for monogastrics and ruminating animals, respectively. Evidence on these efficiencies from energy balance experiments, although far from precise due to lack of information on other energy costs and/or due to nearly unavoidable biases, indicates much lower efficiencies.

There is a slight indication that protein deposition in mature animals is less costly than in growing animals, estimates of energetic efficiencies varying from 55% to 65% for dairy cows, from 30% to 87.5% for laying hens, versus from

40% to 60% in growing animals. This might be due to protein turnover being less important with regard to energy costs in the former animals. Another possibility might be that mRNA and tRNA are utilized more often before they have to be renewed, i.e. are more stable, in the case of secretory proteins (milk) than in the case of tissue proteins (growth, eggs), as was suggested by Hoffmann and Schiemann (1973). These authors derived from their results with laying hens a very low energetic efficiency of protein deposition, as was mentioned earlier.

References

ARMSTRONG, D.G. (1969). In *Handbuck der Tierernährung I,* pp. 385–414 (Lenkeit, W., Breirem, K. and Crasemann, E., Eds), Paul Parey, Hamburg

ARNAL, M. (1977). In *Proc. 2nd Int. Symp. Protein Metabl. and Nutr. EAAP,* pp. 35–37 (Tamminga, S., Ed.), Pudoc, Wageningen

BURLACU, G., MOISA, D., IONILA, D., TASCENO, V. and BALTAC, M. (1974). In *Proc. 6th Symp. Energy Metabl. EAAP,* pp. 265–268 (Menke, K., Lantzsch, M. and Reichl, J., Eds), University of Hohenheim, Hohenheim

CAMPBELL, P.N. (1977). In *Proc. 2nd Int. Symp. Protein Metabl. Nutr. EAAP,* pp. 12–14 (Tamminga, S., Ed.), Pudoc, Wageningen

GRIMBERGEN, A.H.M. (1970). *Neth. J. Agric. Sci.,* 18, 195–206

HOFFMANN, L. and SCHIEMANN, R. (1973). *Arch. Tierernähr.,* 23, 105–132

KIELANOWSKI, J. (1976). In *Protein Metabolism and Nutrition,* pp. 207–216 (Cole, D.J.A. *et al.,* Eds), Butterworths, London

NEHRING, K. and SCHIEMANN, R. (1966). In *Handbuch der vergleichenden Ernährungslehre,* pp. 581–683 (Hock, A., Ed.), VEB Fischer Verlag, Jena

PULLAR, J.D. and WEBSTER, A.J.F. (1977). *Br. J. Nutr.,* 37, 355–363

SCHIEMANN, R., NEHRING, K., HOFFMANN, L., YENTSCH and CHUDY, A. (1971). *Energetische Futterbewertung u. Energienormen,* p. 344, VEB Deutscher Landwirtschaftsverlag, Berlin

TASAKA, I. and SASA, Y. (1970). In *Proc. 5th Symp. Energy Metabl. EAAP,* pp. 197–200 (Shurch, A. and Wenk, C., Eds), Juris Druck & Verlag, Zurich

THORBEK, G. and HENCKEL, S. (1976). In *Proc. 7th Symp. Energy Metabl. EAAP,* pp. 117–120 (Vermorel, M., Ed.), G. de Bussac, Clermont-Ferrand

VAN ES, A.J.H. (1972). In *Handbuch der Tierernährung II,* pp. 1–54 (Lenkeit, W. and Breirem, K., Eds), Paul Parey, Hamburg

VAN ES, A.J.H. (1975). *Livest. Prod. Sci.,* 2, 95–107

VAN ES, A.J.H. (1976). In *Principles of Cattle Production,* pp. 237–253 (Swan, H. and Broster, W.H., Eds), Butterworths, London

VAN ES, A.J.H. and VAN DER HONING, Y. (1979). In *Feeding Strategy for the High Yielding Dairy Cow,* pp. 68–89 (Broster, W.H. and Swan H., Eds), Granada, St Albans

VAN ES, A.J.H., VIK-MO, L., JANSSEN, H., BOSCH, A., SPREEUWENBERG, W., VOGT, Y.E. and NIJKAMP, H.J. (1970). In *Proc. 5th Symp. Energy Metabl. EAAP,* pp. 201–204 (Shurch, A. and Wenk, C., Eds), Juris Druck & Verlag, Zurich

WARING, J.J. and BROWN, W.O. (1965). *J. Agric. Sci.,* 65, 139–146

WARING, J.J. and BROWN, W.O. (1967). *J. Agric. Sci.,* 68, 149–155

WATERLOW, J.C., GARLICK, P.J. and MILLWARD, D.J. (1978). *Protein Turnover in Mammalian Tissues and in the Whole Body,* pp. 591 and 753, North-Holland, Amsterdam

13

PREDICTION OF PROTEIN DEPOSITION IN RUMINANTS

J.C. MACRAE
P.J. REEDS
Rowett Research Institute, Aberdeen

Summary

Any discussion concerning the possibility of predicting protein deposition (i.e. net accretion of body tissue protein plus those proteins exported by the body, such as milk and/or wool) presumes that these parameters can be accurately measured. This is not necessarily the case in larger animals such as ruminants, and initial consideration will be given to the problems associated with and the limitations of the various techniques available for actually measuring protein deposition.

In practical terms the prediction of protein deposition *per se* might appear to be little more than an interesting academic exercise, as in most situations the need is to determine the level of dietary nitrogen required to support a given level of production; in most practical situations this is set by the energy intake of the animal. However, if we are ever to achieve a universally applicable model which can furnish this information, a detailed understanding of the mechanisms which govern the digestive and metabolic processing involved in the conversion of dietary nitrogen into animal protein must be achieved. Accordingly as assessment will be made of the current state of our knowledge regarding, first, the relationship between dietary nitrogen intake and the amount of amino nitrogen arriving at the small intestine and, secondly, the mechanisms which are involved in the utilization of this amino nitrogen by the animal.

Some of the amino nitrogen arriving at the small intestine of ruminants will be of dietary origin, but on most types of diet a large proportion will have undergone a complicated fermentative conversion to microbial protein. Factors which influence this process include the solubility (degradability) of the dietary protein, the efficiency of microbial capture of NPN and the contribution of endogenous nitrogen to the process. These will be discussed in an attempt to assess whether we know sufficient information to start to predict the amount of amino nitrogen arriving at the small intestine.

Assessment of our present knowledge about the metabolism of amino acids absorbed from the small intestine will be considered in two ways, i.e. how amino acids are utilized first by individual tissues and secondly by the various pathways in the whole animal. In the former, attention will be paid to the movement of amino acid carbon between various major tissues and in the latter, the emphasis will be upon the interactions between dietary energy and protein deposition.

Finally some attempt will be made to assess the strengths and weaknesses of the methods currently available for predicting the dietary nitrogen requirements of ruminants.

Introduction

Any discussion of the possibility of predicting protein deposition in ruminants in the early months of 1979 might appear to be a daunting exercise, because as yet there is by no means sufficient information available to fulfil such a remit. However, the major advances in our knowledge of ruminant digestion and metabolism which have been achieved over the past decade now make it possible to suggest at least the elements of a model which should ultimately allow such a prediction. In this chapter, therefore, we see our task as perhaps threefold. First, to consider the metabolic factors which we can use as the building bricks of a universally applicable prediction model; second, to assess the knowledge which is available to service this model; and third, to consider why we will lack information and how we can best achieve this in the future.

The Agricultural Research Council will be publishing shortly its second edition of *The Nutrient Requirements of Farm Livestock* (ARC, 1980) which will bring before the practising agricultural community the major conceptual advances made by research workers in this field over the past 10–15 years. It is probably fair to say that the new scheme has been designed mainly with the dairy industry in mind. Undoubtedly the two fundamental innovations it will include are (1) an appreciation that an index of useful protein must consider amounts of amino acids absorbed from the small intestine, and (2) a recognition that under practical circumstances, such as those pertaining to the lactating dairy cow, the performance of the animal probably will have been set by its intake of energy; thus the nitrogen (N) requirements of both host animals and rumen microbes must be assessed relative to the available energy. This should improve the precision of calculations of N requirements for the animal beyond that available from previous schemes based originally upon digestible crude protein (DCP) and later upon the concept of available protein (ARC, 1965).

However, most researchers, including those members of the ARC Working Party who formulated the new proposals, realize that much more information is required before it is safe to assume that the factors included in the new scheme are static and before any universally applicable relationships can be achieved. Indeed the proposals will include a strong recommendation that 'the new approach should be regarded as a framework for future research efforts and a means of focusing attention on those factors for which data are required', rather than an unalterable prescription for the ruminant nutrition industry.

In this chapter we intend to discuss the longer-term development of a more general prediction system which can be used in the dynamic situation of ruminant nutrition. To accomplish this task it will, of course, be necessary to consider in detail the newer knowledge of digestive and metabolic mechanisms which pertain to the utilization of dietary N by the ruminant, and so unfortunately it is inevitable that some parallels will be drawn between this discussion and the new recommendations. Suffice it to say that the ARC (1980) proposals were formulated at least two years ago on data available up to and including 1976; in fact, they were released in a preliminary form by Roy *et al.* (1977). Since then more data have become available which perhaps suggest that the emphasis on certain measurements might need to be changed if we are to produce a generally applicable equation for a prediction system which can change relative to, say, the nutrition of the animal.

Overall prediction of protein deposition

The substrates for protein accretion and secretion in the mammal are the amino acids which are absorbed from the small intestine. In simple-stomached animals, ingested protein passes directly into the stomach where it immediately comes under the influence of the gastric and intestinal proteolytic enzymes of the animal. Thus from a knowledge of the quantity and amino acid composition of ingested protein and the availability of the individual amino acids (i.e. digestibility anterior to the caecum and colon; which is not very different from true digestibility measurements – Armstrong, 1976) it is possible to make a fairly accurate assessment of the amounts of amino acid available for protein deposition.

This is not so in ruminant species, where ingested protein is subject to microbial attack, altering the amount and composition of the protein which arrives at the small intestine. This is perhaps the main reason why an empirical approach of the type encompassed in the original DCP or ARC (1965) systems, and which serves very successfully to predict energy retention in ruminants (see Blaxter, 1969; Blaxter and Boyne, 1978; Webster, 1978), cannot be used in predictions of protein deposition. Any discussion of the factors which control protein deposition in ruminants must, in fact, be divided into two separate parts; firstly, a consideration of the factors which govern the supply of amino acids to the small intestine and then, secondly, a consideration of the factors which govern how these amino acids are utilized by the animal. These two parts, which together constitute a 'causal approach' (Baldwin, Koong and Ulyatt, 1977) will now be considered separately.

Predicting the relationship between dietary nitrogen and amino acids arriving at the small intestine

Originally, the ruminant's pre-gastric microbial fermentation allowed it to forage on poorer quality indigenous herbages containing large amounts of complex polysaccharides upon which mammalian digestive enzyme systems have little effect. However, on the better quality feeds currently given to ruminants this fermentation can be less of an advantage. For example, most diets used in intensive animal production systems contain little poor-quality roughage and therefore less structural material, and so the microbes utilize instead non-structural entities of the diet (e.g. soluble carbohydrates and protein) and so influence considerably the nature of these compounds before they can be used by the host animal. As a result the quantity and quality of the amino acids arriving at their site of digestion in the host animal can be very different from that consumed.

These complications to protein digestion were first realized some 25–30 years ago, when researchers observed that the proteolytic properties of rumen contents led to ammonia production in the rumen and that this ammonia either supported an increase in microbial protein or was absorbed directly from the rumen, converted to urea in the liver and either excreted by the kidneys into the urine or recycled back into the rumen via the saliva (see Annison and Lewis, 1959; Barnett and Reid, 1961). It was not until later, following the development of techniques for the collection of abomasal and duodenal contents (see Faichney, 1975; MacRae, 1974, 1975), which allowed workers to measure the amounts of digesta entering the small intestines of ruminants, that the quantitative significance to the host animal of this pre-gastric fermentation became apparent. Indeed it was the late 1960s/early 1970s before data of the type presented in *Table 13.1* started to suggest that the flow of amino acid to the small intestine

Table 13.1. DIETARY FACTORS AFFECTING THE AMOUNTS OF NITROGENOUS CONSTITUENTS REACHING THE SMALL INTESTINE (TAKEN FROM MACRAE, 1978)

(a) N INTAKE

Diet (g/24 h)	Hay	Hay + maize	Hay + maize + soya bean	Hay + soya bean	Forage oats maturity		
					III	II	I
N intake	5.1	7.3	16.4	24.8	6.2	28.0	38.0
NAN* reaching SI†	8.5	11.6	16.4	17.4	9.9	23.9	31.1
NAN app. absorbed from SI	4.6	7.0	11.3	12.1	7.1	17.8	22.6
		Clarke, Ellinger and Phillipson (1966)			Hogan and Weston (1969)		

(b) SOLUBLE CARBOHYDRATE (CHO) CONTENT OF RATION

Diet (g/24 h)	Forage oats Maturity I		Chopped lucerne	Chopped dried grass
N intake	37.3	38.0	30.9	20.6
Sol. CHO intake	142	35	74	173
NAN reaching SI	38.3	31.1	23.8	21.4
NAN app. absorbed from SI	30.3	22.6	16.2	13.7
	Hogan and Weston (1969)		Hogan and Weston (1967a)	MacRae et al. (1972)

(c) SOLUBILITY OF N IN DIET

Diet (g/24 h)	Perennial ryegrass			Clover		
	fresh	frozen	dried	fresh (white)	frozen (red)	dried and wafered (red)
N intake	26	26	27			
Amino acid intake				127	127	123
N reaching duodenum	20	25	32.5			
Amino acid reaching duodenum				80	133	148
N app. absorbed from SI	13	15.5	20			
Amino acid absorbed from SI				50	95	79
	MacRae and Ulyatt (1974)	Beever et al. (1969)		Macrae and Ulyatt (1974)	Beever, Thomson and Harrison (1971)	

* NAN, non-ammonia N
† SI, small intestine

was dependent on factors such as the N intake of the animal, the soluble carbohydrate content of the diet and the solubility of the dietary protein. Once such information became available workers started to question the validity of DCP and available protein measurements as indices of amino acid availability for ruminants. Later still, following the development of the use of DAPA (Hogan and Weston, 1967b; Hutton, Bailey and Annison, 1971), RNA (Smith and McAllan, 1970, 1971) and ^{35}S (Roberts and Miller, 1969; Beever *et al.*, 1974) as markers for microbial protein, it was possible to attempt to quantify the contribution of microbial protein to the N fraction of digesta entering the small intestine.

There are now many measurements in the literature (see ARC, 1980) which relate (a) the flow of microbial N to the digestion of OM anterior to the small intestine, or (b) the flow of microbial and hence non-microbial N into the small intestine, thus giving an indication of the amount of dietary N which is used by the rumen microbes. Unfortunately in order to predict the flow of amino acids at the duodenum we need to know not only what proportion of the total N in duodenal digesta is of microbial, undigested dietary N, or endogenous N origin (this last fraction has been little studied, but is usually assumed to be only a small proportion of total duodenal N, e.g. 1–2 g N/day – Phillipson, 1964; Hogan and Weston, 1967b), but also how these different fractions change in association with alterations in the nature of the diet, with different intakes of the same diet, or with the physiological status of the animal. We do not yet possess this information and so it is perhaps pertinent to consider what exactly we need to know and how we can best achieve this knowledge.

There are two fundamental processes which occur in the rumen. Firstly dietary N is 'degraded' to non-protein nitrogen (NPN) and then this NPN is utilized by the microbes to produce their own body protein, which may subsequently leave the rumen along with any undegraded dietary N and be digested by the enzymes secreted into the lower GI tract. Although these two processes are obviously not independent of each other, it is convenient to consider them separately.

'DEGRADABILITY'

The degradability of dietary protein, i.e. the fraction of the protein which is broken down to NPN by the microbes, is taken to be a characteristic of the diet itself. In particular, it is thought to be in some way related to the solubility of the dietary protein and also, especially in the case of the less well degraded materials, to the time which the protein spends in the rumen.

Different authors have calculated 'degradability' in different ways, for example some have based their calculations on the flow of duodenal total N (N_D), while others have used the flow of non-ammonia N ($NAN_{(D)}$). Similarly, some have allowed for a small endogenous N component (E_D) in the non-microbial fraction, which others have ignored. This situation led ARC (1980) to adopt the over-simplified formula

$$\text{Degradability} = 1 - \frac{N_D - M_D}{I_N}$$

(where M_D is the microbial N flow at duodenum and I_N the intake of N), thus accepting the inaccuracies which ensue when neither endogenous N nor ammonia N components of duodenal digesta N are taken into account.

Even when the more apposite expression

$$1 - \frac{NAN_{(D)} - M_D + E_D}{I_N}$$

is used to calculate degradability there still could appear to be certain difficulties associated with obtaining a precise, reliable measurement based on the currently available methodology. These difficulties arise, first, from the fact that the different markers used for measuring microbial N flow appear to give widely differing values and, secondly, from the fact that even when an endogenous N component is considered there is no easy way of actually measuring it and so it is usually assumed to be only $1-2$ g N/day in sheep (Phillipson, 1964; Hogan and Weston, 1967a), whereas in fact data are now accruing to suggest that it is probably much higher than this value.

Comparison of microbial markers

Table 13.2 summarizes data from two studies which compared the measurement of duodenal flow of microbial N using different markers (DAPA, RNA, ^{35}S and ^{15}N). Agreement between the markers was poor in both studies. On the hay and silage diets (Beever *et al.*, unpublished), DAPA and RNA gave flow rates almost double those obtained with ^{35}S and ^{15}N. Similar discrepancies have been reported by other workers using just two different markers (see Hume, 1975;

Table 13.2. COMPARISON OF DIFFERENT MICROBIAL MARKERS

Diet	*Percent duodenal digesta N of microbial origin*				*Reference*
	DAPA	RNA	^{35}S	^{15}N	
Barley + urea	0.80	0.98	0.92	ND	
Barley + soyabean	0.47	0.70	0.64	ND	Ling and Buttery (1978)
Barley + fish meal	0.42	0.56	0.54	ND	
Hay	0.80	0.80	0.43	0.46	
Silage	0.76	0.93	0.46	0.45	Siddons *et al.* (1979)

Tamminga, 1978). If such discrepancies are consistently occurring in measurements of microbial N flow made on the same samples, then the values reported in the literature might well be as much a characteristic of the method used to obtain them as they are of the situation to which they relate. This methodological situation is clearly not satisfactory and it warrants a very critical evaluation if rates of microbial N flow are to be used (as they must be) in prediction equations.

Recent findings which place reservations on the current values used for endogenous protein secretion anterior to the duodenum will be discussed in the next section.

EFFICIENCY OF CAPTURE OF NPN BY MICROBES

Until recently it was assumed that peptides and amino acids derived from degraded dietary protein were deaminated to ammonia by the microbes before they were utilized for microbial protein production. However, recent studies using ^{15}N have suggested that bacteria can derive between 30% and 50% of their protein from sources other than ammonia; presumably amino acids or peptide N (Nolan, 1975; Nolan, Norton and Leng, 1976; Nolan and MacRae, 1976).

Amounts of NPN which the microbes can utilize are thought to be related directly to the amount of available energy which they can generate during the digestion of other dietary components. *Figure 13.1* gives a stylized representation of different types of fermentation which could alter the efficiency of this process.

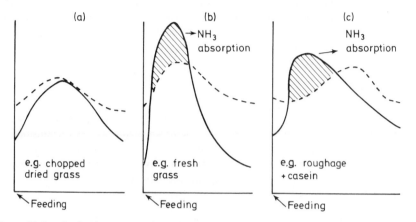

Figure 13.1   Stylized representation of ways in which the release of NPN and energy during fermentation can alter the efficiency of microbial utilization of 'degraded protein': (a) N-limiting diet; (b) energy-limiting diet; (c) imbalanced diet (—— = ammonia, - - - = energy; abscissa, time; ordinate, rate of release of substrate)

Figure 13.1(a) would represent a normal roughage-type diet, where N is limiting (i.e. with crude protein content of less than 8–9%). Here the release of NPN by the microbes is well matched to the release of energy and so the microbes are able to capture most of the NPN available.

In *Figure 13.1(b)* the NPN released from fresh herbage, which has a high protein content, much of which is soluble (Beever, Thomson and Cammell, 1976), is very rapid and so considerable amounts of ammonia are absorbed directly from the rumen. Data in *Table 13.1* illustrate the magnitude of this ammonia loss (25% of apparently digested N of fresh ryegrass). However, when Beever *et al.* (1969) fed ryegrass which had been freeze-stored, the protein solubility of which was very much lower than the fresh grass (MacRae, 1970; MacRae, Campbell and Eadie, 1975), they reported only 5% of the apparently digested N disappearing anterior to the duodenum. Indeed, in a later experiment (Beever, Cammell and Wallace, 1974), when fresh or freeze-stored herbage was given to sheep, a 26% increase in microbial protein production and a 15% increase in amino acids entering the small intestine was observed in the sheep given the frozen herbage. It has been argued that the failure of the rumen microbes to capture the released NPN could place lambs grazing fresh pasture in a 'protein limiting state' (MacRae, 1976).

Figure 13.1(c) probably represents the situation which occurs when readily available N supplements such as casein or urea are given to animals fed basal roughage diets. For example, MacRae *et al.* (1972) found that when acid-precipitated casein (9 g N/day) was given as a supplement to sheep fed dried grass (supplying 21 g N/day), the equivalent of only 54% of the supplemental N arrived at the duodenum. When the casein was protected from rumen degradation by treatment with formaldehyde, the equivalent of 98% of the extra N reached the duodenum. Thus on diets limited by N it is thought that the capture of NPN is very efficient and indeed ARC (1980) have assumed an efficiency of 100% for such diets. However, there are many situations where the capture of the degraded NPN is clearly much lower than 100%.

Interesting as these observations are, they are only balances between N intake and the amounts of N components arriving at the duodenum. While they have formed the basis of the hypotheses above, it is arguable whether they give any sound quantitative data on either the degradation or the capture of NPN by the microbes, because they provide no information on the inputs of N anterior to the duodenum which are not derived from the diet. In order to be able to make any systematic study of degradability and capture of NPN, especially if the long-term aim is a study of the *dynamics* of rumen fermentation, there is little alternative but to try to develop methods which can estimate the endogenous N contribution anterior to the duodenum. One way of attempting to resolve this problem is using tracer techniques. Data from investigations which have used these tracers are already proving to be very rewarding and are beginning to provide an insight into the intermediary processes involved in rumen fermentation.

Use of tracer techniques to study rumen fermentation

Observations on the kinetics of ^{15}N in the rumen (Nolan and Leng, 1974) have allowed the workers in Armidale, Australia, to develop a very interesting model for the fermentation processes in sheep given lucerne cubes (Nolan and Leng, 1972; Nolan, 1975; Nolan, Norton and Leng, 1973, 1976; Mazanov and Nolan, 1976). With more general use such techniques should be able to provide valuable quantitative data, not only on the kinetics of ammonia metabolism in the rumen but also on the amounts of urea and non-urea N which are recycled into the rumen over a wide range of dietary situations.

One type of simple experiment involving ^{15}N which could provide useful data for future models is that which was described briefly by Nolan and MacRae (1976). Two sheep, each given lucerne cubes (16 g N/day) and prepared with a rumen cannula plus duodenal and ileal re-entrant cannulae, were given 36 h intra-ruminal infusions of ^{15}N-labelled ammonium sulphate. The ^{15}N-enriched duodenal digesta were removed from the animals throughout the period and replaced with equal amounts of previously collected non-labelled digesta. This procedure was adopted to simplify the interpretation of the kinetics of the labelling of rumen ammonia, by preventing the digestion of ^{15}N microbial protein in the small intestine and any subsequent return of ^{15}N label to the rumen. *Figure 13.2* gives a stylized representation of the relationships with time of ^{15}N enrichment in rumen ammonia, rumen bacteria and duodenal NAN. From this type of data three sets of basic information can be calculated:

(1) Knowing the infusion rate of ^{15}N, it can be calculated that the ammonia production rate on this diet was 12 g N/day.
(2) The relative enrichment of bacterial N and ammonia N suggest that only

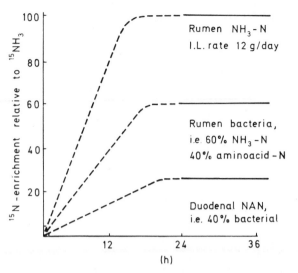

Figure 13.2   Schematic diagram of the [15] N-enrichment of rumen ammonia, rumen bacteria and duodenal non-ammonia N during a 36 h continuous infusion of [[15] N] -ammonia sulphate into the rumen of sheep given lucerne

60% of the bacterial N was derived from rumen ammonia, the rest presumably coming from amino acid or peptide N.

(3) From the relative enrichments of total duodenal NAN and the rumen bacterial N it can be calculated that 40% of duodenal NAN was bacterial N. With this information and a knowledge of the NAN flow at the duodenum (17 g NAN/day) and taking into account the amount of urea N being recycled to the rumen (1.5–2 g N/day; obtained on a similar diet in an earlier study (Nolan, Norton and Leng, 1973)), it is possible to make several other interesting calculations:

(4) 40% of the duodenal NAN (7–8 g N/day) was bacterial, but only 60% of this came from rumen ammonia (i.e. approximately 5 g N/day). Thus if ammonia production was 12 g N/day, then 7 g of ammonia N/day must have been absorbed directly into the blood stream.

(5) Knowing this, it is possible to calculate the endogenous protein N secreted anterior to the duodenum, on the basis that over a 24 h period the inputs to the rumen (viz. feed + recycled urea + endogenous protein secretion) must equal the outputs from the rumen (viz. ammonia absorption + duodenal flow), i.e.

16 + 2 + Endogenous protein secretion = 17 + 7
∴ Endogenous protein N secretion anterior to the duodenum = 6 g N/day.

This value is much higher than the 1–2 g N/day attributed to gastric secretion (Phillipson, 1964). However, it is not unusually high when compared with the other limited data now available.

Table 13.3 gives, as far as is known, the only five values on endogenous protein secretion anterior to the duodenum yet available. The value for lucerne is that

Table 13.3. SECRETION OF ENDOGENOUS PROTEIN N ANTERIOR TO THE DUODENUM

Diet	Intake (g N/day)	NH_3 absorption (g N/day)	Urea recycling (g N/day)	Endogenous protein secretion (g N/day)	Reference
Lucerne cubes	16	7.0	1.5	6.0	Nolan and MacRae (1976)
Agrostis festuca	7	ND	0.9	3.0	MacRae *et al.*
Heather	5	ND	1.1	2.3	(1977, 1979)
Hay	11	3.5	1.5	6.6	Nolan *et al.* (un-
Silage	19.5	10.0	1.5	4.6	published)

derived above. The *Agrostis festuca* and heather data were obtained in Blackface sheep which, because they were consuming only very low levels of N (7 and 5 g N/day, respectively), had duodenal flows of NAN considerably higher than intake (duodenal flow = 11 and 8.5 g NAN/day, respectively). ^{14}C-labelled urea and ^{14}C-labelled bicarbonate were used to calculate the recycling of urea to the rumen (MacRae *et al.*, 1977, 1979) and endogenous protein was calculated as the non-urea addition. These values would be the minimum amount of endo-genous N, because if there was any direct absorption of ammonia from the rumen then more endogenous N would be needed to balance the model. The values for the hay and silage diets were obtained in a recently conducted and more elegant repeat of the lucerne experiment, employing automated machines for the transfer of duodenal digesta (Canaway and Thomson, 1978) from six sheep on each diet (J.V. Nolan, D.E. Beever, J.C. MacRae and R.C. Siddons, unpublished data). Clearly, data in *Table 13.3* would suggest that more endo-genous non-urea N is secreted anterior to the duodenum than was previously thought.

It is not possible to determine into which organ the endogenous N is secreted. However, recent observations on the turnover of rumen epithelial cells (mitotic indices as high as 1%; Tamate and Fell, 1977; Sakata and Tamate, 1978) and on the amounts of non-microbial N in the rumen contents of sheep maintained solely on intra-ruminal infusions of volatile fatty acids and abomasal infusions of casein (approximately 3 mg N/ml; Wallace *et al.*, 1979) would suggest that the N contained in sloughed rumen epithelial cells could readily account for the values reported in *Table 13.3* (5–10 g N/day on normal diets).

This amount of endogenous protein secretion will alter considerably some of the conventionally measured parameters. For example, in the experiment discussed above conventional 'degradability' measurements of the lucerne (ARC, 1980) would be 45%. If, however, 6 g of the duodenal NAN was endogenous protein rather than undegraded dietary N, then the 'degradability' could in fact be 81%. It is likely that some of the endogenous protein will be degraded to ammonia and so the 'degradability' will lie somewhere between 45% and 80%; however, it does illustrate yet another uncertainty of the *in vivo* 'degradability' measurement. A second point worthy of mention is that the lucerne feed used above had a crude protein content of 12%. This is higher than the 8–9% gener-ally accepted as representing the boundary of 100% capture of NPN, but not so very much considering that the capture was only 40%. Undoubtedly, studies on the relationships between NPN capture and the diet need far more emphasis.

There is one further very important aspect of the higher endogenous protein flow at the duodenum. Conventionally, the duodenal flow data for the above experiment would have indicated that in the sheep given lucerne (16 g N/day) there was little wastage of protein N anterior to the duodenum (17 g NAN/day at the duodenum). However, if 6 g of this duodenal NAN represented a 'futile cycle' of endogenous protein N, then the net amount of amino acid N derived from the diet (either microbial N or undegraded dietary protein) was 25–30% lower than would have been assumed. This finding may well provide a possible explanation as to why workers who have tried to measure uptake of amino acids in the portal vein have failed to account for all the amino acids which disappear from the small intestine; this point will be discussed in more detail when the question of availability of amino acids in the small intestine is considered in the next section.

The foregoing discussion at least suggests why we cannot as yet predict the amounts of amino acid arriving at the duodenum. Clearly we need to do much more work before we can start to put forth prediction equations, and we need a considerable change in emphasis in some of our current studies. First, we need to sort out the discrepancies between microbial markers with which we presently live and, secondly, we need to develop a much more general expertise in tracer techniques. The techniques involved in the lucerne experiment described above were relatively simple and, given that a research group has access to equipment for the measurement of ^{15}N- and ^{14}C-labelled tracers there now seems little excuse for continuing to spend large sums of money on experiments which, for example, are designed to collect more observational data solely on duodenal flow rates. These data give little insight into the complicated processes involved in determining the amounts of microbial, undigested dietary and endogenous protein N arriving at the small intestine, and until we have some idea of the dynamic relationships between these processes we will not be able to predict the flow of amino acids at the duodenum with any precision.

Utilization of amino acids which enter the small intestine

Half of the amino acids which make up the mixture of microbial, undigested dietary and endogenous proteins which enter the small intestine can be synthesized by mammalian enzymes. The magnitude of the metabolic changes which these amino acids subsequently undergo in the tissues can be very considerable (e.g. see Heitmann, Hoover and Sniffen, 1973). The nitrogen of the non-essential amino acids can be transferred to other amino acids, both within a single tissue and by movement to other tissues, and the carbon skeleton of the newly synthesized non-essential amino acids may be derived from the carbon of glucose or of the lipid of the body. Consequently the interpretation of the dynamics of the metabolism of the non-essential amino acids solely in terms of protein metabolism is probably difficult, and it is unlikely that the deposition of body protein is limited by their availability.

In contrast, the processes which influence the metabolism of the essential amino acids will have a considerable influence upon the ability of the animal to deposit protein. Therefore, in this section the main consideration will be directed towards the metabolism of the essential amino acids.

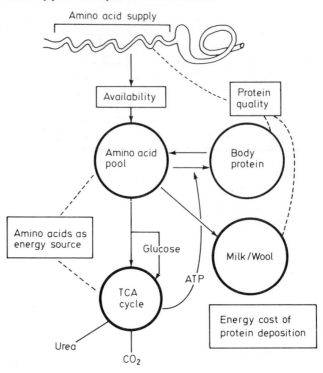

Figure 13.3   Schematic representation of the factors which can influence the utilization of absorbed amino acids

There are probably four major factors which influence the processes involved in the utilization of the amino acids which enter the small intestine. These are illustrated schematically in *Figure 13.3*. The first two of these factors, namely the efficiency of absorption of amino acids from the small intestine and the quality of the amino acid mixture which is absorbed, determine the amount and composition of the amino acid mixture which is available for deposition and will be considered first.

ABSORPTION

There are basically two ways of measuring amino acid absorption in ruminants. The first and most extensively used method involves the measurement of the disappearance of the amino acids between the duodenum and the ileum (Clarke, Ellinger and Phillipson, 1966). Only two research groups, at Cornell and Kentucky, have reported extensive observations of amino acid absorption measured by the alternative method — amino acid uptake into the hepatic portal venous blood (HPVB). Comparisons of the two methods are discussed below.

At the outset it is important to establish the degree of variability which exists in the relative efficiencies of absorption of different amino acids and how this is affected by the diet which the animal consumes. *Table 13.4* summarizes 21 separate measurements of amino acid absorption, obtained by the first method, in animals which were fed a wide variety of diets providing a three-fold range of intake of N. The diets also contained a variety of sources of protein including some which had been protected in such a way that they escaped degradation in the rumen. Two important points arise from this data.

First, the availability of all the amino acids ranged only from 70% to 80% and although it might be possible to distinguish arginine (78%), leucine (77%) and methionine (76%) as being somewhat better absorbed than the others, the

Table 13.4. AVAILABILITY OF INDIVIDUAL ESSENTIAL AMINO ACIDS FROM SMALL INTESTINE (data from five separate studies on 21 different diets incorporating a three-fold range of N intakes and including protected protein supplements)*

Amino acid	Mean availability (%)	Coefficient of variation (%)
Arginine	78	8
Histidine	72	7
Isoleucine	70	10
Leucine	77	6
Lysine	73	6
Methionine	76	12
Phenylalanine	69	14
Threonine	72	7
Valine	74	7
Arginine, leucine methionine	78	8
Rest	72	12

* Data from Clarke, Ellinger and Phillipson (1966); MacRae *et al.* (1972); Sharma, Ingalls and Parker (1972); Coehlo da Silva *et al.* (1972b); Armstrong and Hutton (1975).

differences are small. Cysteine is not included in the table because few workers have reported values for it. This is possibly because there is some controversy as to the most appropriate method of analysis. One recent report (Armstrong, 1979) suggests that the availability of this amino acid may be *considerably* lower than 70%.

The second point relates to the magnitude of the coefficients of variation of the means for individual amino acids. These are generally less than 10%, a value which is probably within the limit of the accuracy of this method. Two amino acids, however, showed greater variability than this; methionine for which the difficulty of assay may provide a partial explanation and phenylalanine whose mean value includes data for five diets (Armstrong and Hutton, 1975) which all gave low values.

The values presented in *Table 13.4* are for the whole diet. Protein supplements, particularly those which escape fermentation in the rumen (Ørskov, Fraser and McDonald, 1971) may give higher apparent availabilities. Nevertheless, from these data it is probably reasonable to assume as far as any prediction equations are concerned that a constant value (75%) can be attributed to the efficiency of absorption of amino acids from the small intestine.

The alternative measurement of the addition of amino acids to the HPVB is a particularly difficult procedure, being heavily dependent upon the measurement of the blood flow in the hepatic portal vein. Although the few measurements of the uptake of amino acids in the HPVB do not allow any consideration of the effects of the source of protein or the level of feeding on the portal flow of amino acids, it is clear from the data which are available that much less amino acid appears in the HPVB than disappears from the small intestine. The differences are particularly marked for glutamate and asparate (Wolff, Bergman and Williams, 1972; Hume, Jacobson and Mitchell, 1972) but they are also apparent for the essential amino acids.

Until Professor Bergman reported his recent results (Chapter 4) there were no direct comparisons of amino acid availability determined by the two methods. *Table 13.5* compares the only two previously reported extensive studies of the

Table 13.5. DIFFERENCES BETWEEN APPARENT ABSORPTION AND NET UPTAKE OF AMINO ACIDS*

Diet	*N intake (g/day)*	*Absorption of amino acid N (g/day)*		
		small intestine	portal vein	difference (g/day)
Chopped dried grass[1]	21	13.7		
Chopped alfalfa[2]	21		7.8	5.9
Pelleted lucerne[3]	26	14.2		
Pelleted alfalfa[4]	26		6.6	7.6

[1] MacRae *et al.* (1972).
[2] Hume, Jacobson and Mitchell (1972).
[3] Coehlo da Silva *et al.* (1972a).
[4] Wolff and Bergman (1972).

appearance of amino acids in the HPVB with two studies of the disappearance from the small intestine selected because the diet and level of N which were fed were similar. In both comparisons a difference of 6–7 g N/day was apparent.

The interpretation of these differences must be tentative and several authors have attributed them to possible 'metabolism (catabolism implied) by the gut mucosa'. However, it is possible that a proportion of the disappearance from the small intestine represents the digestion and reabsorption of endogenous protein secreted into the gastro-intestinal tract anterior to the duodenum. If this is so then the amino acids required for the net synthesis of this endogenous protein, be it digestive secretions or sloughed mucosal cells, must be supplied from the blood supplying the gastro-intestinal tract. In effect amino acids are removed from the arterial input to the portal drained viscera, to support the synthesis of these proteins, and are later reabsorbed from the small intestine. The reabsorption of endogenous protein secreted anterior to the duodenum appears as a loss from the small intestine but not as a gain in the HPVB, because the latter measurement will by definition contain the loss from the arterial input in the arteriovenous difference.

Compared with the total flow of amino acids through the mesenteric circulation the removal of amino acids from the arterial input need not result in a large fall in the concentration of amino acids across the portal drained viscera. Thus if we consider the movement of leucine in the sheep, endogenous losses of *protein* anterior to the duodenum could easily amount to 30 g (i.e. 5 g N) (see *Table 13.3* and above) and these proteins probably contain on average 8% leucine by weight. The amount of leucine required to supply the endogenous loss is therefore about 18 mmol/day. Daily hepatic portal blood flow is approximately

1800 l/day (Wolff, Bergman and Williams, 1972) with an average arterial concentration of leucine of 148 mmol/l (Bergman and Heitman, 1978). Thus about 266 mmol leucine arrive at the intestine in the arterial circulation and the loss of leucine therefore represents only $18/266 = 7\%$ of the arterial flow of leucine – a very small difference.

At present the measurements of the disappearance of amino acids from the small intestine is the only reasonable alternative for extensive studies in ruminants. Interpretation of these results will depend upon a knowledge of the amount of endogenous protein which is included in the duodenal digesta. It is possible, of course, that this may not be a critical factor but it would seem appropriate to attempt to identify its magnitude and how it changes with the diet as an alternative to extensive measurements of amino acid uptake into the HPVB.

PROTEIN QUALITY

In non-ruminant animals the amino acid mixture arriving at the duodenum is in large part a reflection of the composition of the dietary protein, and hence the rate of protein deposition (other factors being equal) will be limited by the composition of the dietary protein. Manipulation of this composition by supplementation of the diet with specific amino acids has, in these species, been the subject of much research (for example, see Chapters 8 and 14 of this publication).

In the earlier part of this chapter it was pointed out that dietary proteins, unless suitably protected, are extensively degraded in the rumen and support a gain of microbial protein which ultimately leaves the rumen along with undegraded dietary and endogenous protein. There are therefore several questions which are pertinent to the present discussion:

(1) How variable is the amino acid composition of the duodenal digesta of ruminants?
(2) Can the composition be materially altered and if so in what respect?
(3) Does the composition of the amino acid mixture change during absorption?
(4) How does the composition of the mixture of absorbed amino acids relate to the amino acid requirement of the animal?

Figure 13.4 presents data which suggest answers to these questions.

In the histograms the composition of the mixture of amino acids (omitting cysteine) has been normalized to lysine = 1, although this is not meant to imply that lysine is the first limiting amino acid in the ruminant. The histogram in *Figure 13.4(a)* compares nine measurements of the composition of duodenal digesta in animals which were receiving basal diets (roughage + urea) with or without unprotected casein, soya bean or rapeseed supplements (dark) with three measurements in which the animals were receiving the basal diet supplemented with protein treated to reduce its availability for microbial fermentation (light) and one in which the supplement was fish meal. The important points are that the coefficient of variation of ratios are very small, in all cases less than 8%, and that with the 'protected' supplements, all of which led to an improvement in weight gain and N balance which were ascribed to an improvement in protein quality, the only significant change is in the level of methionine. It

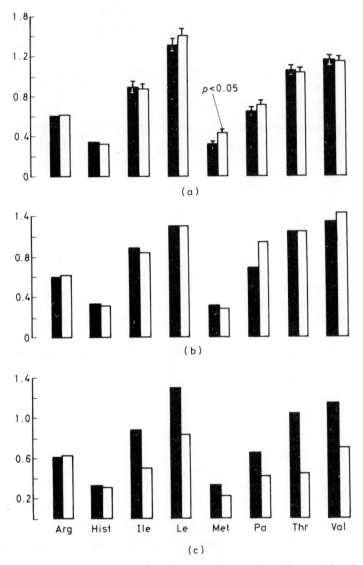

Figure 13.4   Comparisons of the composition of the mixture of essential amino acid, normalized to lysine = 1 in: (a) the duodenum of animals fed unprotected (■) and protected (□) supplements of protein; (b) the disappearance from the small intestine (■) and gain in the portal blood (□); (c) the gain in the portal blood (■) and that of tissue protein (□)

would appear, therefore, that with the exception of this amino acid (which has been proposed as being the first limiting – Chalupa, 1976) little manipulation of 'protein quality' is possible.

The data already given in *Table 13.4* suggest that the pattern of available amino acids is not changed during absorption as the relative efficiencies of absorption of the different amino acids are very similar. This conclusion is largely confirmed by the histogram in *Figure 13.4(b)*, in which the composition of the loss of amino acids from the duodenum is compared with that of the uptake into the HPVB.

In order to relate the pattern of the available amino acids to their requirement, it is obviously necessary to have some measure of the relative amino acid requirements of the animal. One way of estimating these (as opposed to the allowance, which depends upon a number of factors) of an animal is by the so-called 'factorial method'. In this the amino acid composition of the protein gain is measured and the minimum requirement for a given rate of gain can be calculated (Williams *et al.*, 1954; Armstrong and Annison, 1973; Chalupa, 1976). The histogram in *Figure 13.4(c)* compares the composition of the mixture of amino acids which are absorbed into the HPVB (from Wolff *et al.*, 1972; Hume, Jacobson and Mitchell, 1972) with the average composition of the protein of the sheep, including wool. It can be seen that the supply of arginine, histidine, methionine and of course lysine are very close to that required for the minimum requirement, while the other amino acids are supplied in abundance; it is, however, dangerous to use these two sets of data to draw any firm conclusions.

Taken in general, the comparisons in *Figure 13.4* indicate that unless care is taken to bypass the rumen by a variety of methods (see, for example, Ørskov, 1976) there is little room for manipulation of the composition of the mixture of amino acids arising from digestion. It follows from this that the primary limiting factor to protein deposition in the ruminant will be the total amount of protein available for net absorption by the animal.

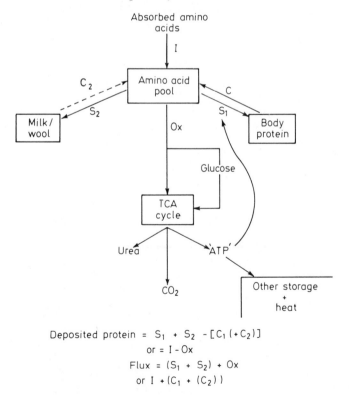

Deposited protein $= S_1 + S_2 - [C_1 (+ C_2)]$
or $= I - Ox$
Flux $= (S_1 + S_2) + Ox$
or $I + (C_1 + (C_2))$

Figure 13.5 A simplified view of the relationship between protein and energy metabolism

INTERACTION BETWEEN PROTEIN AND ENERGY

Apart from 'protein quality', the other major factor which will affect the efficiency with which duodenal amino acids can be utilized for protein deposition will be the energy supply available to the animal. Indeed it would be incorrect as well as over-simplistic to study protein metabolism as a process which is independent of energy metabolism. *Figure 13.5* shows a representation of protein metabolism in which an attempt is made to show the areas in which protein and energy could interact.

Role of amino acids as substrates for gluconeogenesis and the production of energy

Figure 13.5 suggests that in a steady state where the deposition of protein is zero the catabolism of the amino acids represents the only pathway of irreversible loss. This diagram, however, represents an over-simplification of ruminant metabolism particularly where it relates to low levels of feeding. In the ruminant animal, a high proportion of N loss, under conditions of zero or negative N balance, occurs by the faecal route. This is, presumably, due to a combination of the microbial capture of endogenous intestinal losses of protein and the inefficiency of reabsorption. In any investigation of the control of the deposition of apparently absorbed protein the definition of the division of 'basal' N losses between the intestine and amino acid catabolism is critical to the interpretation of the relationship between protein synthesis (as with the exception of the excretion of N^t-methyl histidine there is no satisfactory method of measuring the breakdown of protein), amino acid catabolism and the deposition of protein in the growing animal. There is a need to measure all three factors, particularly amino acid catabolism.

In the non-ruminant animal, total amino acid catabolism can be estimated reliably from the excretion in the urine of urea and ammonia. Such an approach to the measurement of 'tissue' amino acid catabolism is not possible in the ruminant because of the complexities which are introduced by the absorption of ammonia from the rumen and the hind-gut and the recycling of urea N. However, the catabolism of at least some of the essential amino acids can be quantified by measurement of the rate of catabolism in their carbon chains, i.e. by measuring the labelling of CO_2 following the administration of a ^{14}C-amino acid. To some extent the study of the metabolism of the carbon of the essential amino acids is more satisfactory than measurements of N metabolism, as it is the inability of the animal to synthesize the carbon skeletons of the essential amino acids which confers upon them their particular role in the nutrition of the animal. This particular approach may be advantageous in other ways.

It has been recognized, but not investigated extensively (see Armstrong and Annison, 1973; Chalupa, 1976; Miller, 1978; Mathers and Miller, 1979) that the measurement of changes in amino acid catabolism as the proportion of a dose of ^{14}C-amino acid excreted as $^{14}CO_2$, provides a method for estimating the allowance of the amino acid. According to this view, when the allowance is exceeded the proportion of the dose excreted as CO_2 will rise (see Armstrong and Annison, 1973). Alternatively, if the measurement of the labelling of respiratory CO_2 is combined with a measurement of the flux of the amino acid (see Chapter 3), then the total amount of that amino acid which has been catabolized can be estimated. Then with a knowledge of the amount of amino acid which was absorbed (and this is a prerequisite for any attempt to measure the efficiency with which dietary protein is utilized), the gain of body protein can be estimated as intake minus (irreversible) catabolism. There are limitations to this appraoch. For example, it is imperative that the amino acid is labelled in such a position that the rate of production of $^{14}CO_2$ truly reflects the rate of the pathway of catabolism of the amino acid. This limits the amino acids which may be used and effectively rules out this method for measurements of non-essential amino acid metabolism. In addition, the method relies upon the absolute accuracy of the measurement of the flux of the amino acid and specifically upon a close similarity between the specific activity of the amino acid in the blood and that of the precursor for the catabolism of the amino acid.

It is surprising how few systematic studies of the carbon catabolism of amino acids have been made in ruminants. Even among the estimates which are available there are few studies in which any attempt has been made to measure changes which occur when conditions are altered, the majority of the reports being confined to measurements made under one particular nutritional or physiological condition. Unfortunately also, the majority of the studies have been confined to measurements made with non-essential amino acids; these are summarized in *Table 13.6*. It is, of course, the essential amino acids which will

Table 13.6. GLUCONEOGENESIS FROM AND CATABOLISM OF NON-ESSENTIAL AMINO ACIDS

Amino acid	Percent flux to glucose	Percent flux to CO_2
Glutamate[1]	15	–
Glutamate[3]	22	44
Glutamate[4]	11	58
Aspartate[1,2]	30	–
Alanine[1,2]	25	–
Serine[2]	4	–
Serine[4]	11	24
Glycine[2]	7	–

[1] Bergmann and Heitman (1978).
[2] Wolff and Bergman (1972).
[3] Egan, Moller and Black (1970).
[4] Heitman, Hoover and Sniffen (1973).

be important in a consideration of the possibility of the limitation of protein deposition via the catabolism of amino acids. There are four published reports of the catabolism and gluconeogenic contribution of essential amino acids (Egan, Moller and Black, 1970; Egan and MacRae, 1978; Morton, Lindsay and Buttery, 1978; Lindsay, 1979). Typical data are given in *Tables 13.6* and *13.7*. It can be seen from the limited data which are available, that with the possible exceptions of alanine, aspartate and glutamate the contribution of any given amino acid to glucose synthesis is minimal. More importantly for the present discussion, the catabolism of amino acid carbon either to CO_2 or into gluconeogenesis is only a small proportion of the flux of the amino acid. This is particularly true for the essential amino acids.

Table 13.7. APPARENT ABSORPTION, IRREVERSIBLE LOSS RATE AND PERCENTAGE TRANSFER OF C TO GLUCOSE AND CO_2 FOR FIVE ESSENTIAL AMINO ACIDS (from Egan and MacRae, 1979)

	Number of sheep	Uptake from small intestine (g C/day)	Amino acid irreversible loss (g C/day)	Percent amino acid irreversible loss of C contributed to:	
				glucose	CO_2
Methionine	4	0.8	2.6	<1.0	5–8
Threonine	3	1.1	3.1	3	10–12
Isoleucine	4	1.8	5.0	<1	14–18
Leucine	3	2.8	7.0	–	12–20
Lysine	3	1.7	4.0	–	9–12

Table 13.8. CONTRIBUTION OF THREONINE TO GLUCOSE

Condition	Glucose production (g C/day)	Percent threonine flux to glucose	Threonine to glucose (g C/day)	Percent glucose from threonine
Basal[1]	77	2.9	0.22	0.2
Phloridzin[1]	100	4.5	0.35	0.3
Pregnant[2]				
105 days	150	3.7	0.45	0.3
120 days	210	4.8	0.47	0.2
140 days	260	3.8	0.54	0.2

[1] Egan and MacRae (1978).
[2] MacRae and Egan (unpublished).

It should be re-emphasized that the majority of the data in *Table 13.6* and *13.7* were obtained under single conditions and the possibility remains that amino acid catabolism might vary when the demand for energy or glucose is altered and that amino acid catabolism could *become* limiting under certain specific circumstances. The only extensive study of such changes is summarized in *Table 13.8*, where the catabolism of threonine to glucose or CO_2 was measured under two changing circumstances: following the administration of phloridzin (Egan and MacRae, 1978) and during pregnancy (MacRae and Egan, unpublished). In the wether sheep given phloridzin, where glucose production increased rapidly, but probably protein deposition remained unaltered, the small contribution of threonine to glucose and CO_2 production did appear to increase (+50% and +56%, respectively). However, in the pregnant ewes, where the demand for glucose was much higher than even with phloridzin, but where there was also presumably an increase in protein deposition, there was very little change in the contribution of the threonine flux to glucose plus CO_2 (approx. 12–15% in all cases). These data would appear to suggest that in the physiological state of pregnancy protein synthesis has a prior claim on essential amino acids, which might in turn suggest that in a situation where an animal is depositing protein, amino acid catabolism, in itself, is unlikely to limit this deposition. Such a conclusion is, of course, speculative and needs substantiation.

Energy requirement for protein deposition

It is now well established that the protein pool of the body is never static but is continually turning over, i.e. protein synthesis and breakdown continue even when N balance is zero or negative. The deposition of protein is therefore the resultant of these two processes, and it is possible for a change in protein deposition to result from a change in protein synthesis, breakdown or in both processes. Because protein synthesis continues even when N balance is zero, it follows that for any given level of protein deposition a considerable excess of protein synthesis is occurring, and although this will not necessarily directly affect the efficiency of utilization of dietary protein, it can influence the amount of energy required to deposit the protein.

The energy requirement for protein turnover has two implications for the nutritional physiology of the animal. First, protein turnover at energy maintenance or even under a situation of negative energy balance will be a contributor to heat production. Because of the uncertainty of the so-called theoretical cost of protein synthesis the contribution of total protein synthesis to heat production is in itself unknown. However, if we take a minimum estimate of 4.6 kJ/g of protein synthesized (which is based upon the assumption that each mole of protein requires five ATP equivalents for its synthesis), the data in *Table 13.9* suggest that in adult animals fed at such a level that they were in zero energy balance, protein synthesis, and by implication its contribution to heat production, is remarkably constant over a wide variety of species, amounting to some 15% of total heat production at maintenance.

Table 13.9. BODY PROTEIN SYNTHESIS MEASURED FROM THE FLUX OF LEUCINE OR TYROSINE (all animals receiving a maintenance intake of energy)

Species	Weight (kg)	Grammes protein synthesized per $kg^{0.75}$	Minimum contribution to heat production* (%)
Rat	0.2	13.7	14
	0.35	13.6	14
Pig	32	19.8	20
Sheep	50	13.3	14
Man†	70	13.6	14
Cow	600	14.8	15

* Assuming the energy cost of protein synthesis is a minimum of 4.6 kJ/g (i.e. 5 ATP/mol of peptide) and maintenance heat production is 450 $kJ/kg^{0.75}$.
† Data of James *et al.* (1976); tyrosine flux. All other data Lobley and Reeds (unpublished).

The second implication of protein turnover for the efficiency of growth is that the relationship between the total amount of protein synthesized to that deposited will presumably influence the energy which must be provided in the diet to support a given level of protein deposition. This amount of energy is a matter of conjecture, as the two major approaches to this problem (see Chapter 12) have given quite different estimates for the energy requirement for protein deposition. However, a further question arises. In most treatments of this subject it is generally assumed that the energy requirement for protein deposition is constant between species, or between diets (Pullar and Webster, 1977). The measurement of protein synthesis combined with the simultaneous measurement of protein deposition should allow us to answer the question whether the relationship between protein synthesis and protein deposition varies between diets or under different physiological circumstances. Comparisons of studies of normally growing pigs (P.J. Reeds, M.F. Fuller and G.E. Lobley, unpublished observations) with those in children recovering from severe protein/energy malnutrition (Golden, Waterlow and Picou, 1977), indicates that the increment in protein synthesis per unit of increment in deposition is different between these two situations of growth, the ratio of the increment in synthesis to the increment of deposition being approximately 2 in the pig and 1.4 in the human infant. Clearly these are preliminary studies and we require much more data from a variety of species in order to answer the question whether there is a fixed relationship between protein synthesis and protein deposition, and hence the provision of energy for the deposition of protein.

Conclusion

It should be clear from the foregoing discussion that if we are to achieve a dynamic model for protein deposition we will need to first identify the areas of uncertainty in our current concepts and secondly direct our research resources to designing experiments which will provide data which can be analysed in the context of the dynamic nature of the problem. Obviously, as already pointed out, there are areas of investigation into the digestion of protein by ruminants that require a radical change in emphasis. Equally we must also concentrate the limited resources and expertise which are available for studying the metabolic aspects into areas which are relevant to nutritional and physiological circumstances which are normally encountered and to the metabolism of those amino acids which might have some bearing on a limitation to protein deposition (i.e. the essential amino acids). Above all, in both the digestive and the metabolic areas we must gain a greater awareness of the need to plan 'experiments' rather than collect observational data made in particular static situations. Only, then, armed with this type of information, will it be easier to carry out response-type analysis and so achieve some precision in a universally applicable dynamic model.

References

ANNISON, E.F. and LEWIS, D. (1959). *Metabolism in the Rumen*, p. 92, Methuen, London

ARC (1965). *The Nutrient Requirements of Farm Livestock*, No. 2 Ruminants, ARC, London

ARC (1980). In *The Nutrient Requirements of Farm Livestock*, ARC, London, in the press

ARMSTRONG, D.G. (1976). *Ubers Tierernahrg*, 4, 1–24

ARMSTRONG, D.G., WALKER, A.C.U. and WEEKEN, T.E.C. (1979). In *Protein Metabolism in the Ruminant*, pp. 21–28 (Buttery, P.J., Ed.), ARC, London

ARMSTRONG, D.G. and ANNISON, E.F. (1973). *Proc. Nutr. Soc.*, 32, 107–114

ARMSTRONG, D.G. and HUTTON, K. (1975). In *Digestion and Metabolism in the Ruminant*, pp. 432–447 (McDonald, I.W. and Warner, A.C.I., Eds), University of New England Publishing Unit, Armidale, Australia

BALDWIN, R.L., KOONG, L.J. and ULYATT, M.J. (1977). In *Microbial Ecology of the Gut*, pp. 347–391 (Clarke, R.T.J. and Bauchop, T., Eds), Academic Press, London

BARNETT, A.J.G. and REID, D.L. (1961). *Reactions in the Rumen*, p. 107, Arnold, London

BEEVER, D.E., CAMMELL, S.B. and WALLACE, A. (1974). *Proc. Nutr. Soc.*, **33**, 73A

BEEVER, D.E., THOMSON, D.J. and CAMMELL, S.B. (1976). *J. Agric. Sci., Camb.*, **86**, 443–452

BEEVER, D.E., THOMSON, D.J. and HARRISON, D.G. (1971). *Proc. Nutr. Soc.*, **30**, 86A

BEEVER, D.E., HARRISON, D.G., THOMSON, D.J., CAMMELL, S.B. and OSBOURN, D.F. (1974). *Br. J. Nutr.*, **32**, 99–112

BEEVER, D.E., NOLAN, J.V., SIDDONS, R.C., McALLAN, A.B. and MacRAE, J.C. (1979). *Annls Rech. Veterin.*, **10**, 286–287

BEEVER, D.E., THOMSON, D.J., PFEFFER, E. and ARMSTRONG, D.G. (1969). *Proc. Nutr. Soc.*, **28**, 26A

BERGMAN, E.N. and HEITMAN, R.N. (1978). *Fedn Proc.*, **37**, 1228–1232

BLAXTER, K.L. (1969). In *4th Symposium of Energy Metabolism of Farm Animals*, pp. 21–30 (Blaxter, K.L., Kielanowski, J. and Thorbek, G., Eds), EAAP Publication No. 12

BLAXTER, K.L. and BOYNE, A.W. (1978). *J. Agric. Sci., Camb.*, **90**, 47–68

BUTTERY, P.J. and ANNISON, E.F. (1976). In *From Plant to Animal Protein*, pp. 111–122, Reviews in Rural Science II (Sutherland, T.M., McWilliams, J.R. and Leng, R.A., Eds), University of New England Publishing Unit, Armidale, Australia

CANAWAY, R.J. and THOMSON, D.J. (1978). 'Automatic sampling equipment for digestion studies with sheep', *Technical Report No. 23*, Grassland Research Institute, Hurley

CHALUPA, W. (1976). In *From Plant to Animal Protein*, Reviews in Rural Science II (Sutherland, T.M., McWilliams, J.R. and Leng, R.A., Eds), University of New England Publishing Unit, Armidale, Australia

CLARKE, E.M.W., ELLINGER, G.M. and PHILLIPSON, A.T. (1966). *Proc. Roy. Soc., Ser. B,* **166**, 63–79

COEHLO DA SILVA, J.F., SEELEY, R.C., THOMSON, D.J., BEEVER, D.E. and ARMSTRONG, D.G. (1972a). *Br. J. Nutr.* **28**, 43–61

COELHO DA SILVA, J.F., SEELEY, R.C., BEEVER, D.E., PRESCOTT, J.H.D. and ARMSTRONG, D.G. (1972b). *Br. J. Nutr.*, **28**, 357–371

EGAN, A.R. and MacRAE, J.C. (1978). *Proc. Nutr. Soc.*, **37**, 15A

EGAN, A.R. and MacRAE, J.C. (1979). *Proc. Vth International Symposium on Ruminant Physiology*, Clermont Ferrand, in the press

EGAN, A.R., MOLLER, F. and BLACK, A.L. (1970). *J. Nutr.*, **100**, 419–428

EL-SHAZLY, K. (1952). *Biochem. J.*, **51**, 647–653

FAICHNEY, G.F. (1975). In *Digestion and Metabolism in the Ruminant*, p. 277 (McDonald, I.W. and Warner, A.C., Eds), University of New England Publishing Unit, Armidale, Australia

GOLDEN, M., WATERLOW, J.C. and PICOU, D. (1977). *Am. J. Clin. Nutr.*, **30**, 1345–1349

HEITMAN, R.N., HOOVER, W.H. and SNIFFEN, C.J. (1973). *J. Nutr.*, **103**, 1587–1593

HOGAN, J.P. and WESTON, R.H. (1967a). *Aust. J. Agric. Res.*, **18**, 803

HOGAN, J.P. and WESTON, R.H. (1967b). In *Physiology of Digestion and Metabolism in the Ruminant*, pp. 474–485 (Philipson, A.T., Ed.), Oriel Press, Newcastle-upon-Tyne

HOGAN, J.P. and WESTON, R.H. (1969). *Aust. J. Agric. Res.*, **20**, 347

HUME, I.D. (1979). In *From Plant to Animal Protein*, pp. 79–84 (Sutherland, T.M., McWilliam, J.R. and Long, R.A., Eds), Reviews in Rural Science II, University of New England Publishers Unit, Armdale, Australia

HUME, I.D., JACOBSON, D.R. and MITCHELL, G.E.Jr. (1972). *J. Nutr.*, **102**, 495–506

HUTTON, K., BAILEY, F.J. and ANNISON, E.F. (1971). *Br. J. Nutr.*, **25**, 165–173

JAMES, W.P.T., GARLICK, P.J., SENDER, P.M. and WATERLOW, J.C. (1976). *Clin. Sci. Mol. Med.*, **50**, 525–532

LINDSAY, D.B. (1979). *Biochem. Soc. Trans.*, **6**, 1152–1156

LING, J.R. and BUTTERY, P.J. (1978). *Brit. J. Nutr.*, **39**, 165–179

MacRAE, J.C. (1970). *N.Z. Jl Agric. Res.*, **13**, 45–50

MacRAE, J.C. (1974). *Proc. Nutr. Soc.*, **33**, 147–154

MacRAE, J.C. (1975). In *Digestion and Metabolism in the Ruminant*, pp. 261–276 (McDonald, I.W. and Warner, A.C., Eds), University of New England Publishing Unit, Armidale, Australia

MacRAE, J.C. (1976). In *From Plant to Animal Protein*, pp. 93–98, Reviews in Rural Science II (Sutherland, T.M., McWilliam, J.R. and Leng, R.A., Eds), University of New England Publishing Unit, Armidale, Australia

MacRAE, J.C. (1978). In *Ruminant Digestion and Feed Evaluation*, Ch. 6 (Osbourn, D.F., Beever, D.E. and Thomson, D.J., Eds), A.R.C., London

MacRAE, J.C. and ULYATT, M.J. (1974). *J. Agric. Sci., Camb.*, **82**, 309–319

MacRAE, J.C., CAMPBELL, D.R. and EADIE, J. (1975). *J. Agric. Sci., Camb.*, **84**, 125–131

MacRAE, J.C., MILNE, J.A., WILSON, S. and SPENCE, A.M. (1979). *Br. J. Nutr.*, **42**, 525–534

MacRAE, J.C., ULYATT, M.J., PEARCE, P.D. and HENDTLASS, J. (1972). *Br. J. Nutr.*, **27**, 39–50

MATHERS, J.C. and MILLER, E.L. (1979). In *Protein Metabolism in the Ruminant*, pp. 3.1–3.11 (Buttery, P.J., Ed.), ARC, London

MAZANOV, A. and NOLAN, J.V. (1976). *Br. J. Nutr.*, **35**, 149–174

MILLER, E.L. (1978). In *Ruminant Digestion and Feed Evaluation*, pp. 15.1–15.9 (Osbourn, D.F., Beever, D.E. and Thomson, D.J., Eds), A.R.C., London

MORTON, J.L., LINDSAY, D.B. and BUTTERY, P.J. (1978). *Proc. Nutr., Soc.*, **37**, 7A

NOLAN, J.V. (1975). In *Digestion and Metabolism in the Ruminant*, pp. 416–431 (McDonald, I.W. and Warner, A.C., Eds), University of New England Publishing Unit, Armidale, Australia

NOLAN, J.V. and LENG, R.A. (1972). *Br. J. Nutr.*, **27**, 177–194

NOLAN, J.V. and LENG, R.A. (1974). *Proc. Nutr. Soc.*, **33**, 1–8

NOLAN, J.V. and MacRAE, J.C. (1976). *Proc. Nutr. Soc.*, **35**, 110A

NOLAN, J.V., NORTON, B.W. and LENG, R.A. (1973). *Proc. Nutr. Soc.*, **32**, 93–98

NOLAN, J.V., NORTON, B.W. and LENG, R.A. (1976). *Br. J. Nutr.*, **35**, 127–147

ØRSKOV, E.R. (1976). *Wld Rev. Nutr. Diet.*, **26**, 225–257

ØRSKOV, E.R., FRASER, C. and McDONALD, I. (1971). *Br. J. Nutr.*, **25**, 243–252

PHILLIPSON, A.T. (1964). In *Mammalian Protein Metabolism*, Vol. 1, pp. 71–103 (Munro, H.N. and Allison, J.B., Eds), Academic Press, New York

PULLAR, J.D. and WEBSTER, A.J.F. (1977). *Br. J. Nutr.*, **37**, 355–363

PURSER, D.B. (1976). In *From Plant to Animal Protein*, pp. 39–46, Reviews in Rural Science II (Sutherland, T.M., McWilliams, J.R. and Leng, R.A., Eds), University of New England Publishing Unit, Armidale, Australia

ROBERTS, S.A. and MILLER, E.L. (1969). *Proc. Nutr. Soc.*, **28**, 32A

ROY, J.H.B., BALCH, C.C., MILLER, E.L., ØRSKOV, E.R. and SMITH, R.H. (1977). In *Protein Metabolism and Nutrition*, E.A.A.P. Publ. No. 22, pp. 126–129

SAKATA, T. and TAMATE, H. (1978). *Res. Vet. Sci.*, **24**, 1–3

SHARMA, H.R., INGALLS, J.R. and PARKER, R.J. (1972). *Can. J. Anim. Sci.*, **54**, 305–313

SIDDONS, R.C., BEEVER, D.E., NOLAN, J.V., McALLAN, A.B. and MACRAE, J.C. (1979). *Ann. Rech. Vet.*, **10**, 286–289

SMITH, R.H. and McALLAN, A.B. (1970). *Br. J. Nutr.*, **24**, 545

SMITH, R.H. and McALLAN, A.B. (1971). *Br. J. Nutr.*, **25**, 181

TAMATE, H. and FELL, B.F. (1977). *Vet. Sci. Communs*, **1**, 359–364

TAMMINGA, S. (1978). In *Ruminant Digestion and Feed Evaluation*, Ch. 5 (Osbourne, D.F., Beever, D.E. and Thomson, D.J., Eds), ARC, London

WALLACE, R.J., CHENG, K-J., DINSDALE, D. and ØRSKOV, E.R. (1979). *Nature, Lond.*, in the press

WEBSTER, A.J.F. (1978). *Wld Rev. Nutr. Diet.*, **30**, 189–226

WILLIAMS, H.H., CURTIN, L.V., ABRAHAM, J., LOOSLI, J.K. and MAYNARD, L.A. (1954). *J. Biol. Chem.*, **208**, 277–286

WOLFF, J.E. and BERGMAN, E.N. (1972). *Am. J. Physiol.*, **223**, 445–460

WOLFF, J.E., BERGMAN, E N. and WILLIAMS, H.H. (1972). *Am. J. Physiol.*, **223**, 438–446

14

PROTEIN DEPOSITION IN POULTRY

C. FISHER

Agricultural Research Council, Poultry Research Centre, Edinburgh

Summary

Although protein deposition is a key element of productivity in poultry it has not been directly investigated to the extent found in some other species. This reflects the relative constancy of the composition of the output (meat and eggs) and the fact that the value of the output is not generally related to its gross composition. When this is not the case, indirect measures of composition, e.g. carcass grade, have been used. However, the topic merits attention because (a) the potential for protein deposition may be an important limiting and controlling factor in poultry productivity, and (b) the development of general theories of productivity and of the effects of environmental factors, especially nutrition, requires that the individual chemical components of production are considered separately.

In this chapter, protein deposition in poultry will be discussed under four headings: (1) *Rate of protein deposition in poultry production.* During normal commercial growing periods average protein deposition varies among species such that turkeys (0–84 days) > ducks (0–49 days) > broilers (0–56 days). Since most of the variation is in growth rate, not composition, expressing protein growth on a metabolic body weight (W) basis gives a relatively constant figure of about 6 g protein/day per $W^{0.75}$ kg for all species. Rate of protein deposition in laying hens is lower, about 3.5 g/day per $W^{0.75}$ kg.
(2) *Factors affecting protein deposition in growing chickens.* The effects of breed and age will be considered.
(3) *Factors affecting protein deposition in laying hens.* Changes in the rate of protein deposition during a laying cycle will be described and factors which have a major influence discussed. The nature of the response made by the hen to differences in dietary protein supply will be outlined.
(4) *The relationship between dietary protein supply and deposition.* In birds receiving adequate energy the major determinant of protein deposition is the dietary supply of protein (amino acids). Some models which describe the relationships involved will be compared and their predictive value assessed. An analysis of the efficiency with which dietary protein is utilized shows that, although gross utilization is low, the prospects for major improvements are not great. The possible ways of achieving such improvements will be outlined.

Introduction

Although protein deposition is a key element of productivity in poultry, it has not been directly investigated to the extent found in some other species. This probably reflects the relative constancy of the protein content of the output, both meat and eggs, and the fact that the monetary value of the product is not related in any general way to its gross composition. When this is not the case, indirect measures of composition, such as carcass grade, are used. As a consequence total productivity alone has been reported in most experimental work and most studies of carcass composition are in technological terms, e.g. yields of edible meat, etc., and not chemical. However, the topic merits attention for two main reasons. First, the potential for protein deposition may be an important, and limiting, controlling factor in poultry productivity. Secondly, the development of general theories of productivity and of the effects of environmental factors, especially nutrition, requires that the components of production can be described individually. In their construction of a theoretical model of growth in the pig, Whittemore and Fawcett (1976) have defined 'potential protein deposition' as a key controlling element and it seems likely that similar assumptions will be required to provide a general description of growth in poultry.

This chapter is in five sections and contains a description of protein deposition in poultry of different types, together with a more detailed discussion of the growing broiler chicken and of the laying hen. In the final sections the relationship between dietary protein supply and protein deposition and the efficiency with which dietary protein is utilized are discussed.

The following symbols are used:

$BW, \Delta BW$ = body weight (kg) and body weight gain (g/day).
$BW^{0.75}$ = metabolic body weight (kg).
$P, \Delta P$ = protein mass (g) and gain in protein mass (g/day).
$P\%, \Delta P\%$ = protein content of body weight and gain (g/100 g).
Abw, Ap = mature (asymptotic) body weight (kg) and mature protein mass (g).
t = time (day); t_{28} = 28 days of age.
t' = time at point of inflection of the Gompertz growth function (day).
E = egg production (g/bird per day).

In all cases protein refers to N × 6.25. This convention has been followed to facilitate comparisons, although it is probably not appropriate for detailed consideration of protein growth. Håkansson, Eriksson and Svensson (1978b) have defined tissue 'protein' as (dry matter − (fat + ash)), but this also involves questionable assumptions.

Protein deposition in poultry production

The average rates of protein deposition during commercial production cycles of different types of poultry are shown in *Table 14.1*. The data used are not necessarily indicative of the maximum possible levels of animal performance, but are typical.

Among growing birds the average daily rate of protein deposition is highest in turkeys grown to t_{84}, followed by ducks (t_{49}) and broiler chickens (t_{56}). The values are probably affected both by species and the time interval considered. In the laying hen the daily rate of protein deposition is lower, but similar to that in the growing chicken.

If these results are expressed on a metabolic body weight basis ($\Delta P/BW^{0.75}$), then a value of about 6 g/kg per day is found for all the growing birds. This result is to be expected in view of the relatively small differences in $P\%$ and the fact that roughly comparable stages of maturity are involved. It is interesting to note however that, at similar stages of growth, this generalization also applies to a much larger animal, the pig, according to the equation relating ΔP to BW in that species, as given by Whittemore (1977). In the laying hen the rate of

Table 14.1. AVERAGE RATES OF PROTEIN DEPOSITION IN POULTRY PRODUCTION

Type (period)	*Sex*	*P* (g/day)	$P/BW^{0.75}$ (kg)	*P%* at slaughter
Broiler* (t_0–t_{56})	M	8.1	6.3	17
	F	5.8	5.6	16
Turkey† (t_0–t_{84})	M	12.4	6.3	21
	F	9.2	5.8	22
Duck‡ (t_0–t_{49})	M/F	9.4	6.2	14
Layers**	—	5.1	3.0	(12)

* Based on unpublished data supplied by Dr S. Leeson. Body weights at t_{56}, 2.675 kg (males) and 1.967 kg (females).
† Based on unpublished experiment of author. Body weights at t_{84}, 4.830 kg (males) and 3.586 kg (females).
‡ Approximate data from a variety of sources. Body weight at t_{49}, 3.4 kg.
** Assumes 45 g E/day, body weight 2 kg.

protein deposition per unit metabolic body weight is much lower. This presumably reflects the stage of maturity of the animal and/or the fact that protein synthesis is restricted largely to two organs of the body – the liver and oviduct. It cannot, of course, be implied from the values in *Table 14.1* that total protein synthesis is lower in the laying hen than in growing birds; there is an almost complete absence of data on protein turnover in avian species.

Protein growth in chickens

Five experimental studies have been found in which the growth of total body protein mass in chickens is reported. These will be referred to by the names of the originators: Wilson (Dr B.J. Wilson, unpublished data); Leeson (Dr S. Leeson, unpublished data); Edwards (Edwards *et al.*, 1973, Experiment 2); Mitchell (Mitchell, Card and Hamilton, 1931); Håkansson (Håkansson, Eriksson and Svensson, 1978a, 1978b). A detailed study of growth in Fayoumi fowl by Shebaita *et al.* (1977) should also be mentioned but cannot be included in the present survey as the results are presented only in graphical form.

From a statistical point of view it was found that the data could be accurately described by the Gompertz function (described as a growth curve by Winsor, 1932). It should be noted however that A_p was not well defined in any of the experiments and there is therefore uncertainty about the use of any growth function to interpret these data. The 'closeness of fit' with the Gompertz function was, however, extremely high (R^2 ranging from 0.952 to 0.999) even when the curve was fitted by the approximate graphical methods of Ricklefs (1967).

The eight protein growth curves found in these studies are shown in *Figure 14.1*. The parameters of the curves are given in *Table 14.2*, together with some data on body composition at a single age, t_{56}. There is extremely close agreement between the studies of Wilson and Leeson in every respect except that of fat

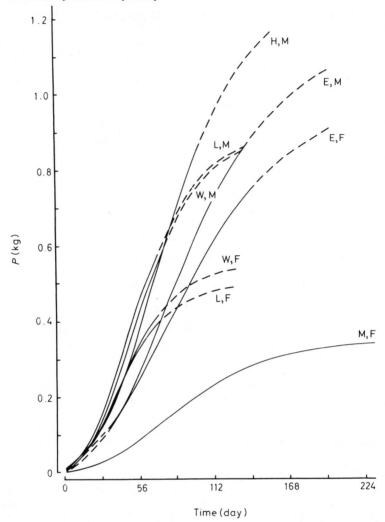

Figure 14.1   Protein growth in chickens, Gompertz curves. Data, for males (M) and females (F) from Wilson (W). Leeson (L), Edwards (E), Mitchell (M) and Håkansson (H). (Details of data sources and a statement concerning closeness of 'fit' to original data will be found in the text)

growth, which was much higher in Leeson's experiment, leading to higher BW and lower $P\%$ values at very similar levels for P. Håkansson used the same strain of commercial broiler chickens as Wilson but obtained a slightly slower rate of protein growth and a considerably higher mature protein mass, as implied by the fitted curve. In the study by Edwards, a modern broiler strain was also used, but a much lower rate of protein growth and a relatively high mature protein mass are again implied by the fitted curve. In this last case this is consistent with the higher protein and very low fat content of the birds at t_{56}, which are also suggestive of relative immaturity.

Males and females of a given strain differ, as expected, in A_p but the results for rate of maturity are inconsistent. Females either mature more quickly than

Table 14.2. CONSTANTS OF THE GOMPERTZ GROWTH CURVES FOR BROILERS SHOWN IN *FIGURE 14.1* AND BODY COMPOSITION AT t_{56}

| Data source* | Sex | Constants | | | Body composition, t_{56} | |
		A (g)	k	t' (day)	protein (g/kg)	fat (g/kg)
Wilson	M	900	0.032 59	47.3	187.4	136.5
Wilson	F	550	0.039 14	37.8	178.8	183.9
Leeson	M	900	0.033 36	45.5	169.0	211.0
Leeson	F	500	0.040 16	35.3	164.0	233.0
Edwards	M	1150	0.020 24	74.6	225.3	51.4
Edwards	F	1000	0.018 27	73.9	227.6	76.5
Mitchell	F	350	0.021 74	70.9	218.8	25.2
Håkansson†	M	1300	0.023 68	64.6	ca. 210	—

* See text for origins of data.

† Body composition at t_{56} cannot be quoted as birds killed according to weight not age. Approximate protein content as shown; fat contents varied from ca. 200 g/kg to 350 g/kg according to energy content of the diet.

males (Wilson, Leeson) or at the same rate (Edwards). Thus the age at which the maximum rate of protein growth occurs, t', which with this model occurs when $P = 0.368A_p$, ranges from about 36 days in females of fast-growing broiler strains, through 45–47 days for males of the same strains, to 70+ days in slower growing birds of both broiler and light breeds.

The general interpretation of these data is confounded by the fact that each study represents one strain in one set of environmental conditions, both of which would be expected to affect protein growth. The slower growth rate obtained by Edwards is probably nutritional in origin, as it seemed unlikely that strain differences of such magnitude would exist among commercial stocks. Håkansson also used at least one diet of very low energy content. However, the higher A_p values implied by the analysis of these two sets of results appears to be real when compared with those of Wilson and Leeson. The asymptote of 900 g protein estimated for male broilers in the Wilson study is consistent with the observed asymptote in body weight for birds of the same strain grown under the same conditions (Wilson, 1977), and yet is clearly exceeded by the observations in the Edwards and Håkannson studies.

The implied difference in rate of maturity between the results of Wilson and Leeson for broiler strains and those of Mitchell for White Leghorns is also surprising, in view of the absence of any breed effect on this parameter for total body weight when similar strains were compared on a single diet (Wilson, 1977 and unpublished). Nutritional effects may again be involved.

In general, it is to be expected that rate of maturity will be relatively constant among strains of chickens grown under the same conditions and, since $P\%$ also varies little, different strains should give approximately constant values for $\Delta P/BW^{0.75}$ when compared over the same stage of growth. This is found in the data of Edwards and Denman (1975), who compared five widely differing strains from t_0 to t_{28}. On an adequate diet, ΔP ranged from 1.70 to 2.89 g/day, $P\%$ from 19.2 to 20.8 and $\Delta P/BW^{0.75}$ from only 8.13 to 9.53 g/kg, thus confirming this expectation.

The data in *Figure 14.1* do emphasize the need for further studies of protein growth in poultry and the opportunity which exists for manipulating genetic and nutritional factors to achieve differences in growth and in the chemical composition of that growth.

The essentially similar results obtained by Wilson and Leeson provide the best available estimates of protein deposition in modern broilers. In the sense that the liveweight gains observed are rarely exceeded it may be assumed that the nutritional limitations to growth were unimportant in these experiments. Calculation of the rate of protein gain, ΔP, for these two studies, using the curves in *Figure 14.1*, gives the results shown in *Figure 14.2*. The maximum rates are about 8 g/day in females and 11 g/day in males, occurring as already noted (*Table 14.2*) at approximately 5 and 6/7 weeks of age, respectively. For the purpose

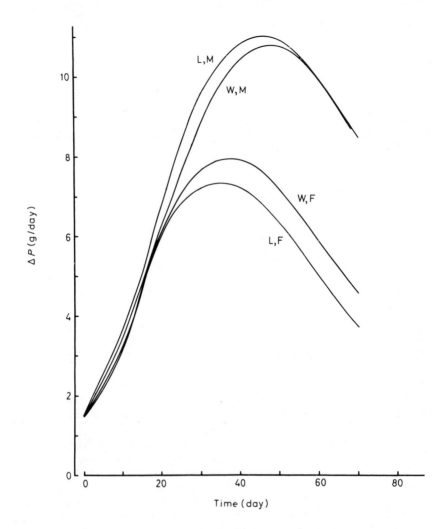

Figure 14.2   Rates of protein growth (ΔP,g/day) in broiler chickens. (Derived from curves in *Figure 14.1* for Wilson and Leeson data only)

of model construction, Whittemore and Fawcett (1976) have assumed that *potential* ΔP is constant over the approximate range $P = 0.06$ to $0.36A_p$. From the results presented here, such an assumption is unlikely to be appropriate for broiler growth.

When ΔP is expressed as a function of BW, $BW^{0.75}$ or P there is a gradual decline throughout growth (*Table 14.3*). Without greater knowledge of protein

Table 14.3. PROTEIN DEPOSITION AS A FUNCTION OF BODY WEIGHT, METABOLIC BODY WEIGHT AND PROTEIN MASS IN BROILER CHICKENS AT DIFFERENT AGES*

Age (day)	(kg BW)	*Protein deposition* (g) (kg $BW^{0.75}$)	(kg P)
0	24.8	12.1	152.1
7	18.8	11.8	120.7
14	14.7	11.2	95.3
21	11.6	10.3	75.6
28	9.3	9.2	59.9
35	7.5	8.1	47.4
42	6.1	7.1	37.5
49	4.9	6.0	29.7
56	4.0	5.1	23.5
63	3.2	4.2	18.6
70	2.6	3.5	14.7

* Calculated from Gompertz growth functions fitted to protein mass and body weight of broiler males (Leeson data).

turnover it is impossible to state the implications of these figures. Whittemore and Fawcett (1976) have proposed that ΔP and total protein synthesis (ΔP_{total}) are related in a manner that is proportional to maturity. Using the data of Millward, Nnanyelugo and Garlick (1974) for rats they suggest the following relationship:

$$\Delta P/\Delta P_{total} = 0.23 \, (A_p-P)/A_p$$

When applied to the Leeson data for males at t_{28} (when $BW = $ ca. 900 g) the ratio obtained is about 0.2. This is much lower than the value of 0.57 suggested by Buttery, Boorman and Barratt (1973) for muscle protein alone in chickens of similar liveweight. Clearly, more data on total body protein turnover in poultry are required to resolve this issue.

Consideration of the composition of growth at different stages shows that $\Delta P/\Delta BW$ increases slightly during the early stages of growth and then declines as maturity is approached. The relationship could not be derived satisfactorily from the Gompertz growth curves, but for the Wilson and Leeson studies the data could be described by a quadratic equation which was significant although not a close fit. For the pooled data for both sexes the equation was

$$\Delta P/BW = 0.1480 + 0.002\,567t - 0.000\,044\,t^2 \; (r = 0.527, P<0.01)$$

This describes an increase in $\Delta P/BW$ from about 0.15 during the first week of life to a maximum of 0.185 at 4–5 weeks of age (about 7 days prior to maximum ΔP), followed by a decline of 0.135 during the 9–10 week period of growth.

Finally we may consider the distribution of protein in the chicken body. Here the recently reported experiment by Håkansson, Eriksson and Svensson (1978a, 1978b) provides the most adequate description.

In this experiment broilers were fed three diets differing in energy content and were killed for carcass evaluation at approximately 0.5, 1.0, 1.5, 2.0, 3.0

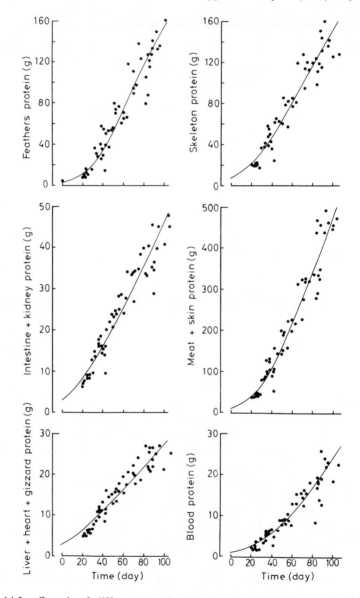

Figure 14.3   Growth of different protein components in broiler chickens, Gompertz curves. The data points (●) refer to individual birds, but the dietary treatments applied are not shown. From (Håkansson, Eriksson and Svensson, 1978b)

and 4.0 kg body weight ($t_{21} - t_{105}$). Although the different diets affected fat deposition they had only a minor influence on protein growth and the data have been combined. Values for 54 individual birds are given and have been interpreted by fitting the Gompertz function.

Body protein was split into the following six components: (a) blood; (b) feathers; (c) meat + skin; (d) skeleton; (e) liver, heart and gizzard; (f) other 'offal' including neck, lungs, kidneys, gastro-intestinal tract, skin from legs, head, etc., internal fat and other minor organs — this group will be referred to as 'intestines plus kidneys'. The data for each of these and the fitted Gompertz curves are shown in *Figure 14.3* and the parameters in *Table 14.4*.

Table 14.4. CONSTANTS OF THE GOMPERTZ GROWTH CURVES FOR DIFFERENT COMPONENTS OF PROTEIN GROWTH IN BROILERS*

Component †	A (g)	Constant k	t (day)
Meat and skin	960	0.018 69	81.9
Skeleton	300	0.016 67	76.7
Feathers	260	0.022 57	70.0
Intestines + kidneys	80	0.017 15	67.6
Liver + heart + gizzard	45	0.017 01	60.3
Blood	55	0.016 42	91.4
Total	1300	0.023 68	64.6

* From the data of Håkansson, Eriksson and Svensson (1978b), as illustrated in *Figure 14.3*.
† The exact definition of the components, especially intestines + kidneys, is given in the text.

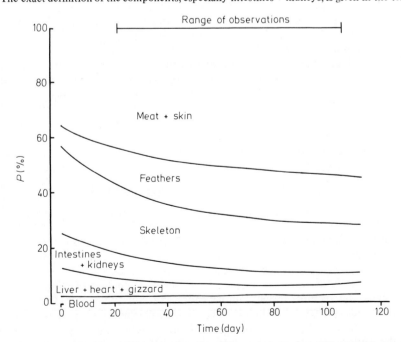

Figure 14.4   Percentage distribution of total body protein in broiler chickens. (Constructed from the Gompertz curves in *Figure 14.3*)

The fitted curves again provide an excellent description of the data but the upper asymptotic values are not entirely satisfactory, since the sum of the values for individual components exceeds the total. However, the ratios between the estimated asymptotes are reasonable and this fact, together with the high correlations, justifies the use of the curves to summarize the data, but not for extrapolation. The small differences in rate of maturity of the different components implied by the values in *Table 14.4* are probably not established with sufficient accuracy to merit further comment.

In *Figure 14.4* the data are expressed as percentage of total body protein in the various components in birds at different ages. As expected the proportion of protein in meat and skin and in feathers tends to increase during growth at the expense of skeletal and 'organic' protein. The main edible components, meat and skin, increase from about 35% to 55% of the total protein, while a further 20% of the total is in feathers after about 40 days of age. Edwards *et al.* (1973) found 23.6% of the protein of 8-week-old broiler males in the feathers, which is in reasonable agreement.

Protein deposition in laying hens

A typical egg weighs about 58 g and contains just under 7 g protein. Of this about 42% is yolk protein, synthesized in the liver, about 54% egg white protein, synthesized mainly in the magnum region of the oviduct, and the remaining 3—4% in the shell and its associated membranes which originate from the isthmus and uterus regions of the oviduct (figures from Svensson, 1964). One component of egg white protein, ovotransferrin, which accounts for about 14% of the total, may be formed from serum transferrin (Gilbert, 1971) and not in the oviduct. The values given will vary slightly according to breed, age, diet and the presence of disease.

Egg production, and therefore protein deposition, in laying birds is a reproductive activity and represents an unusually high level of material turnover (Gilbert, 1971). Detailed reviews of the initiation and control of egg production and associated ovarian and oviductal activity are available (Gilbert, 1971; McIndoe, 1971; Aitken, 1971; Gilbert and Wood-Gush, 1971) and in this volume Gilbert has considered the transfer of yolk material to the ovarian follicle. For the discussion in this paper it is only necessary to consider one aspect of this topic, whether egg protein deposition is a continuous or discontinuous process.

In the modern hen, through genetic selection and environmental manipulation, a continuous reproductive state is maintained for at least one year. Eggs are typically laid in clutches, one egg per day, each clutch being separated by an interval of one day. Such a sequence is described by the term 'closed cycles'. It is obvious that the minimum rate of egg production (rate = number of eggs/ number of days) that can be sustained in closed cycles is 0.50 or 50%, i.e. continuous one-egg clutches and one-day intervals. Lower rates of lay must involve inter-clutch intervals of greater length. Examination of clutch patterns in populations of birds reveals, among those with low (below 50%) egg production, a wide variety of combinations of normal laying and pausing, but one general feature is that the proportion of such birds shows a gradual increase as the laying year progresses (Overfield, 1969). Strains with low average egg production potential, such as breeders for broiler production and turkeys, may also be expected to show a higher incidence of such birds. When a protein-deficient diet is fed, birds tend to lay in short closed cycles for a time followed by a relatively long (4–10 days) period of non-laying. Complete cessation of lay is not generally found in short-term experiments, even with severe deficiencies (Fisher, 1970).

The pattern of protein deposition underlying this phasic production of eggs cannot be described in detail but an attempt has been made in *Figure 14.5* to describe the major features which seem most probable. It must be emphasized, however, that knowledge of this topic is very tentative.

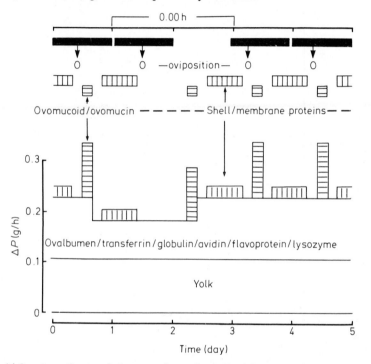

Figure 14.5   An estimate of the rate of protein synthesis/deposition in laying hens. The upper part shows the approximate timing of oviposition, egg white and egg shell secretion on successive days, assuming a sequence: 4-egg clutch, 1-day interval, 3-egg clutch. The lower part shows possible average rates of protein deposition, assuming that ovomucoid/ovomucin and egg shell/membrane proteins are synthesized and deposited simultaneously. The basis of this assumption and the values used to construct the figure are from Smith (1978a, 1978b)

There is good reason to argue, for birds laying in closed cycles, that yolk protein deposition, and presumably synthesis, is a continuous process. The ovary will typically contain between 4 and 6 follicles in the 'rapid' stage of development (ca. 8 days prior to ovulation), and during this period the mass of individual follicles increases almost linearly with time (Warren and Conrad, 1939). It seems to be a reasonable tentative conclusion that yolk protein deposition also occurs at a roughly constant rate, although minor diurnal rhythms may exist.

The evidence for periodicity in the synthesis of egg white proteins has been reviewed by Smith (1978a, 1978b) and *Figure 14.5* is based on the most extreme case that he considers likely. This is that the ovomucoids and ovomucin (about 12% of egg white protein) and shell and shell membrane proteins (about 4.5% of the total) are synthesized and deposited simultaneously as the egg passes through the different regions of the oviduct. In all cases, rates have been averaged over the relevant time periods. The remaining egg white protein components are assumed to be synthesized and stored prior to deposition on the egg at a constant rate over each interval between the passage of successive eggs. From a

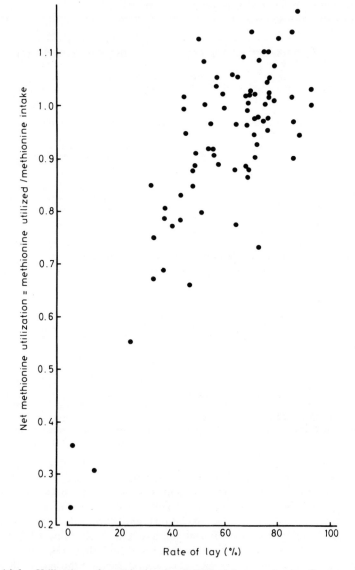

Figure 14.6   Utilization of methionine by individual hens and rate of egg production. All birds received a deficient diet containing 0.156% methionine, which was shown to be the first limiting nutrient. Net methionine utilization = methionine utilized/methionine intake, where methionine utilized = $4E + 25BW + 3\Delta W$. (From Fisher, 1970)

detailed consideration of the available evidence Gilbert (1971) concludes that 'the bulk of the (egg white) protein required for one egg is produced on that day', which is in agreement with *Figure 14.5*.

If *Figure 14.5* is reasonably close to the truth it suggests that egg protein synthesis and deposition in regularly laying hens may vary from just under 0.2 to just over 0.3 g/h. Such variations seem rather small when compared with the mechanisms available to buffer the supply of amino acids to the synthetic sites.

There is indirect evidence which suggests that the efficiency with which dietary protein is utilized declines when the rate of egg production falls below 50% and closed cycles cannot be maintained. This is taken to imply that variations in protein deposition greater than those shown in *Figure 14.5* occur, either because the deposition of egg white is more discontinuous or because yolk protein synthesis itself becomes phasic. It is implicit in the observed reduction in efficiency that dietary supply is not adjusted in phase with utilization. *Figure 14.6* (from Fisher, 1970) shows some evidence for this effect. The same argument can be derived from studies of protein utilization under different circumstances (Fisher, 1976), when it is found that average efficiency only declines in those conditions (e.g. aging, adverse lighting patterns) which increase the proportions of pausing birds in the flock. When a majority of birds are laying in closed cycles, as for example they are in flocks of young pullets, the efficiency of utilization of amino acids appears to be relatively constant in birds fed *ad libitum* (Fisher, 1976).

Dietary protein supply and protein deposition in poultry

The supply of dietary protein, or of essential amino acids, is obviously a major determinant of protein deposition, and is the factor most amenable to short-term manipulation. In practical terms the nutritional objective is normally to supply sufficient protein to meet the needs for maximum productivity. In more general terms protein supply can be considered as a determinant of production and a level of supply chosen so as to optimize some function (ideally an economic function) of productivity. In either case the essential information required is knowledge of the relationship between the dietary supply of protein (amino acids) and productivity.

This relationship can be investigated and described in two ways which will be called the *empirical* and *causal* methods. The empirical method consists of fitting a response function to experimental data, typically using 'goodness of fit' as the criterion of success. The elements of the function do not usually have any biological meaning. The causal method involves consideration, at some level of analysis, of the biological components of the required relationship and is typically used to construct predictive or simulation models.

The distinction between these two approaches can be illustrated by considering the data from strain A, the more productive strain, in *Figure 14.7(a)*. The curve shown in the diagram is derived from causal considerations (see below) and is based on the following relationship between methionine intake (MET, mg/day), egg production (E, g/day) and body weight (BW, kg) of an individual hen:

$$MET = 3.89E + 35.04BW$$

The residual mean square in E, after fitting the model, is 4.7 g^2.

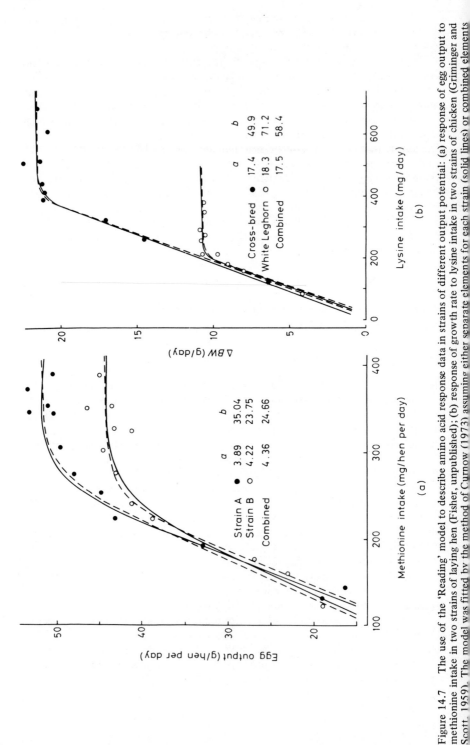

Figure 14.7   The use of the 'Reading' model to describe amino acid response data in strains of different output potential: (a) response of egg output to methionine intake in two strains of laying hen (Fisher, unpublished); (b) response of growth rate to lysine intake in two strains of chicken (Griminger and Scott, 1959). The model was fitted by the method of Curnow (1973) assuming either separate elements for each strain (solid lines) or combined elements

These same data can be equally well described by a quadratic curve which gives a residual mean square of 4.1 g². The equation is

$$E = 40.21 + 52.54 \text{ MET} - 7.47 \text{ MET}^2$$

Although both are equally effective in describing these data the causal method has properties which give it considerable advantages, both as a method of describing experimental data and in practical nutrition. These properties have been discussed in detail by Fisher, Morris and Jennings (1973) in relation to this particular model, but may be summarized as follows:

(1) The elements in the model can be assumed to apply to other populations of birds with different levels of E and BW.
(2) Since the elements have a definite biological meaning the magnitude of the estimates obtained can be compared with other independent evidence.
(3) Following from (2), it is possible to reinforce the general value of the model by direct investigation of the separate components.
(4) New elements can be introduced into the model as required to increase its descriptive and/or predictive power.

To generalize from this example, it is proposed that only by the description of the relationship between protein supply and deposition in causal terms and the development of appropriate models will our understanding of quantitative protein nutrition be advanced from either a scientific or a technological viewpoint.

There have been some attempts to choose response curves which have empirical properties but to base the choice on *a priori* biological considerations. As two of these have been widely used in poultry nutrition studies they should receive a brief mention here.

Almquist (1953) argued that the relationship between the logarithm of the amino acid supply (x) and the response (y) is linear up to a maximum asymptotic value (A) which can be estimated by examination of the data. In summary, Almquist's argument is as follows: for protein or amino acid intake, x, efficiency of utilization, c, and output, y, we have

$$x \xrightarrow{\quad c \quad} y$$

so that at the maximum $A = y = xc$. From the law of diminishing (logarithmic) returns, the rate of utilization of x when $y < A$ is $- \, dx/dt = x.c/y$, whence $(-1/x).dx/dt = c/y$ and $d\log x/dt = y/c$. Integration for $t = 1$ yields $\log x = (1/c).y + \text{constant}$.

This method has been widely used in the interpretation of amino acid requirement studies in chicks. It has the limitation that the slope $(1/c)$ cannot be independently verified and will vary according to age and rate of growth.

An alternative approach has been widely used to interpret amino acid requirement studies carried out at the University of Maryland. Combs (1960) first summarized the methionine requirements of laying hens by means of a factorial equation

$$\text{MET} = 5E + 50BW + 6.2\Delta BW$$

Similar equations for other amino acids in both laying hens and growing chicks were subsequently derived (Combs, 1968; Hunchar and Thomas, 1976; Kessler and Thomas, 1976; Thomas, Twining and Bossard, 1977; and elsewhere).

The coefficients of relationship were established by multiple regression techniques applied to performance data of either individual birds or groups for which it could be assumed that the supply of amino acid in question was the factor limiting output. Although theoretically satisfactory, this method tends to give very unstable solutions owing to the close correlations between the various forms of output when deficient diets are fed, and the results also depend on the particular part of the response curve which is considered (Fisher, 1970).

An alternative, but related, approach to the determination of coefficients in factorial models to predict amino acid requirements of laying hens, is to be found in the work of Hurwitz and Bornstein (1973) and Smith (1978a, 1978b). These authors used experimentally determined estimates of maintenance requirements (Leveille and Fisher, 1960), thus exploiting one of the advantages of the causal method as described above. Requirements for egg production were estimated by multiplying the amino acid content of egg by a utilization coefficient. This in turn was calculated by assuming that some components of egg protein were derived from body tissue and others from the food. A variety of assumptions were made which have been discussed in detail by Smith (1978a).

The final alternative to be considered, again based on the same concepts, has been developed in recent years at the University of Reading. This method was developed to describe the amino acid requirements of the laying hen (Fisher, Morris and Jennings, 1973) and has been applied to methionine (Fisher, 1970), lysine (Pilbrow and Morris, 1974) and tryptophan (Morris and Wethli, 1978). Here the general properties of the method will be briefly described; in principle they should be applicable to protein deposition in any animal.

The Reading model is based on three very simple propositions about the level of egg production from an individual bird receiving a given level of amino acid intake. These are:

for $AAI > bBW$, $E = 1/a \, (AAI - bBW)$

for $AAI < bBW$, $E = 0$

for $E > E_{max}$, $E = E_{max}$

where AAI is the amino acid intake; E, BW are as described above; and a, b are constants representing, respectively, the amino acid requirement per g E (mg/g) and for maintenance per kg BW.

E_{max} is the potential upper limit to E for the individual bird and as a simple concept represents the genetic potential.

This basic causal model refers only to individual animals. The response model for a flock is simply the average of individual values and is derived by defining the variation in W and E_{max} and the co-variation between them; the coefficients a and b are assumed to be invariate. If the distributions of W and E_{max} are normal, a fully specified curve with seven parameters is obtained. The parameters are E_{max}, σE, BW, σ_{BW}, ρ_{EW}, a and b, and the statistical properties of the curve have been defined (Curnow, 1973). A full description of this model can be found in Fisher, Morris and Jennings (1973), with further discussion in Fisher (1976).

There is no theoretical reason why the same model should not be used to describe growth responses, either of the total body or of protein. In *Figure 14.7*, two sets of data are shown, both of which compare the responses to amino acid intake in breeds with different potential output. *Figure 14.7(a)* shows the response of egg production to methionine intake (Fisher, 1970) and *Figure 14.7(b)* the response of growth rate to lysine intake (Griminger and Scott, 1959). In both, the model clearly provides a very satisfactory description of the data, and the responses of widely different breeds can be described by the same model elements. For the laying hen, Fisher (1976) has argued that the general body of published information on methionine, lysine and tryptophan can be accommodated by the model, although with the data available the case cannot be closely argued. As used in *Figure 14.7(b)* the model for growing birds would not be expected to be generally applicable because of differences in body composition; a minor extension to the model could accommodate this however. Finally it should be noted that there is no conceptual difficulty about adding further output parameters to the model (e.g. growth in laying hens) or in considering non-normal distributions, although the facility with which the model could be used would be reduced.

In its present form the model discussed here is based on extremely simple concepts, although for practical nutrition it may be adequate. From such a technological point of view the addition of further complexity would be justified on the pragmatic grounds of increased utility. For scientific purposes the model concepts can be extended as required to illuminate the particular aspect of protein nutrition which is under investigation. One of the values of this approach to protein nutrition is that it provides a means of relating detailed studies of parts of metabolism to whole animal requirement experiments of a more practical type.

Efficiency of protein utilization in poultry

In the method of calculating amino acid requirements that has been outlined here, the model elements a and b are coefficients of utilization and it is important to consider whether they are constant under all conditions.

The coefficient b, the maintenance requirement, is poorly defined, both conceptually and quantitatively. Apart from pointing out that the values obtained by fitting the Reading model are reasonably consistent with other direct evidence (Fisher, 1976) there is little basis for further discussion.

The efficiency with which amino acids are utilized for egg production or growth, however, can be readily determined by comparing the estimates of the coefficient a with the amount of amino acid deposited. For example, in *Figure 14.7(a)* it is suggested that 4.36 mg 'available' methionine are required per g E. If it is assumed that eggs contain 11.25% protein and the protein 3.26% 'available' methionine (Carpenter and Anantharaman, 1968) then the efficiency of utilization is 3.67/4.36 = 0.84. Values between 0.80 and 0.85 are also obtained by similar calculations for lysine and tryptophan (Fisher, 1976).

Efficiency in this sense appears to be relatively constant when different breeds of laying fowl are compared at the same age. This is illustrated for methionine in *Figure 14.7*, and has also been shown for lysine (Pilbrow and Morris,

1974) and tryptophan (Morris and Wethli, 1978). (There is some uncertainty about this conclusion in the latter two papers which should be referred to in order to assess the degree of approximation inherent in this general statement.) While it would be expected that the magnitude of a would vary with egg composition, this has not been demonstrated experimentally.

As pointed out previously (Fisher, 1976) the only situations in which large differences in protein utilization can be demonstrated are when birds are compared at different stages of lay (see Jennings, Fisher and Morris, 1972; Bray, 1968, 1971; Wethli and Morris, 1978) and when adverse lighting patterns are used during lay (Bray, 1968, 1971). In both of these circumstances the birds under comparison differ widely in rate of lay and in the incidence of pausing. Wethli and Morris (1978) have shown that the stage of lay effect is reversed if egg production levels are restored by moulting. In general this evidence, and the converse absence of differences in utilization under other conditions, can be accommodated within the hypothesis that the utilization of amino acids for egg production by laying hens is constant except in those circumstances where egg laying patterns show a high incidence of pausing, i.e. when a significant proportion of the flock have egg production levels below 50%, presumably leading to discontinuous protein synthesis. This may also be important when extrapolation is made from laying hens to other species, e.g. turkeys, with characteristically lower rates of egg production.

No similar analysis of the data on growing birds has been carried out. The one example of a breed comparison shown in *Figure 14.7(b)* can obviously be reconciled with common values for a and b. The birds in this experiment were grown from t_{14} to t_{28} and the equation for composition of gain given above (page 257) indicates an average protein content of 18% during this period. If this protein contains 6.9% lysine (Block and Weiss, 1956), efficiency is 12.42/16.45 = 0.71. Critical comparison of this figure with the rather higher values suggested for laying hens would require more careful consideration of dietary lysine levels and their availability, and of carcass composition.

The coefficient a, when applied to protein gain, might still be expected to vary in birds of different ages because of the effects of protein turnover; the greater the amount of protein synthesis and breakdown associated with a given rate of deposition the less efficient utilization would be expected to be. This concept is incorporated in the growth model of Whittemore and Fawcett (1976) but its quantitative importance in poultry feeding remains to be established.

The high efficiencies of net amino acid utilization suggested here contrast with the rather low efficiencies of gross protein utilization found, in common with other animal systems, in poultry production. Values for gross efficiency depend heavily on the assumptions made but are of the order of 0.25–0.30 for laying hens (egg protein/feed protein), 0.30–0.40 for total protein growth (body protein/feed protein) and 0.25–0.35 for carcass (saleable) protein growth (carcass protein/feed protein).

The most important single factor accounting for the differences between gross protein and net amino acid utilization is the imperfect balance of amino acids in dietary proteins. For example, the gross utilization of the first-limiting amino acid in laying hens (egg amino acid/feed amino acid) is approximately 0.45–0.50 (Fisher, 1976), and there is evidence that perfectly balanced proteins (e.g. egg protein) are utilized with these efficiencies (Shapiro and Fisher, 1965). The simple model proposed here to describe responses to amino acids attributes the remaining differences to maintenance costs and to variability between individuals. Improved efficiency through manipulation of either of these appears to be a very slight prospect.

The final possible way of achieving overall improvements in efficiency is to increase the net utilization of amino acids. Suffice it to say that evidence about the potential in poultry for such improvements is almost entirely lacking, although there is little reason to hope that major advances can be used.

Acknowledgements

I am grateful to Dr B.J. Wilson and Dr S. Leeson for making unpublished data available to me.

References

AITKEN, R.N.C. (1971). In *Physiology and Biochemistry of the Domestic Fowl* (Bell, D.J. and Freeman, B.M., Eds), Academic Press, London and New York

ALMQUIST, H.J. (1953). *Archs Biochem. Biophys.,* **44**, 245–247

BLOCK, R.J. and WEISS, K.W. (1956). *Amino Acid Handbook,* Charles C. Thomas, Springfield, Illinois

BRAY, D.J. (1968). *Poult. Sci.,* **47**, 1005–1013

BRAY, D.J. (1971). *Proceedings 14th World's Poultry Congress,* Vol. 2, pp. 629–667

BUTTERY, P.J., BOORMAN, K.N. and BARRATT, E. (1973). *Proc. Nutr. Soc.,* **32**, 80A

CARPENTER, K.J. and ANANTHARAMAN, K. (1968). *Br. J. Nutr.,* **22**, 183–197

COMBS, G.F. (1960). *Proceedings of the Maryland Nutrition Conference, 1960,* pp. 28–45

COMBS, G.F. (1968). *Proceedings of the Maryland Nutrition Conference, 1968,* pp. 86–96

CURNOW, R.N. (1973). *Biometrics,* **29**, 1–10

EDWARDS, H.M. and DENMAN, F. (1975). *Poult. Sci.,* **54**, 1230–1238

EDWARDS, H.M., DENMAN, F., ABOU-ASHOUR, A. and NAGURA, D. (1973). *Poult. Sci.,* **52**, 934–948

FISHER, C. (1970). *Ph.D. thesis,* University of Reading

FISHER, C. (1976). In *Protein Metabolism and Nutrition* (Cole, D.J.A., Boorman, K.N., Buttery, P.J., Lewis, D., Neale, R.J. and Swan, H., Eds), Butterworths, London

FISHER, C., MORRIS, T.R. and JENNINGS, R.C. (1973). *Br. Poult. Sci.,* **14**, 469–484

GILBERT, A.B. (1971). In *Physiology and Biochemistry of the Domestic Fowl* (Bell, D.J. and Freeman, B.M., Eds), Academic Press, London and New York

GILBERT, A.B. and WOOD-GUSH, D.G.M. (1971). In *Physiology and Biochemistry of the Domestic Fowl* (Bell, D.J. and Freeman, B.M., Eds), Academic Press, London and New York

GRIMINGER, P. and SCOTT, M.M. (1959). *J. Nutr.,* **68**, 429–442

HÅKANSSON, J., ERIKSSON, S. and SVENSSON, S.A. (1978a). Report No. 57, Swedish University of Agricultural Sciences, Department of Animal Husbandry

HÅKANSSON, J., ERIKSSON, S. and SVENSSON, S.A. (1978b). Report No. 59, Swedish University of Agricultural Sciences, Department of Animal Husbandry

HUNCHAR, J.G. and THOMAS, O.P. (1976). *Poult. Sci.,* **55**, 379–383

HURWITZ, S. and BORNSTEIN, S. (1973). *Poult. Sci.,* **52**, 1124–1134

JENNINGS, R.C., FISHER, C. and MORRIS, T.R. (1972). *Br. Poult. Sci.,* **13**, 279–281

KESSLER, J.W. and THOMAS, O.P. (1976). *Poult. Sci.,* **55**, 2379–2382

LEVEILLE, G.A. and FISHER, H. (1960). *J. Nutr.,* **72**, 8–15

McINDOE, W.M. (1971). In *Physiology and Biochemistry of the Domestic Fowl* (Bell, D.J. and Freeman, B.M., Eds), Academic Press, London and New York

MILLWARD, D.J., NNANYELUGO, D.O. and GARLICK, P.J. (1974). *Proc. Nutr. Soc.,* **33**, 55A

MITCHELL, H.H., CARD, L.E. and HAMILTON, T.S. (1931). *Agricultural Experiment Station Bulletin 367*, University of Illinois

MORRIS, T.R. and WETHLI, E. (1978). *Br. Poult. Sci.,* **19**, 455–466

OVERFIELD, N.D. (1969). *NAAS Qly Rev.,* **86**, 84–91

PILBROW, P.J. and MORRIS, T.R. (1974). *Br. Poult. Sci.,* **15**, 51–73

RICKLEFS, R.E. (1967). *Ecology,* **48**, 978–983

SHAPIRO, R. and FISHER, H. (1965). *Poult. Sci.,* **44**, 198–205

SHEBAITA, M.K., SALEM, M.A.I., BADRELDIN, A.L. and ARRAM, G.A. (1977). *Arch. Geflügelk.,* **41**, 49–55

SMITH, W.K. (1978a). *Wld's Poult. Sci. J.,* **34**, 81–96

SMITH, W.K. (1978b). *Wld's Poult. Sci. J.,* **34**, 129–136

SVENSSON, S.A. (1964). *Lantbr-Hogsk. Annlr,* **30**, 405

THOMAS, O.P., TWINING, P.V. and BOSSARD, E.H. (1977). *Poult. Sci.,* **56**, 57–60

WARREN, D.C. and CONRAD, R.M. (1939). *J. Agric. Res.,* **58**, 875–893

WETHLI, E. and MORRIS, T.R. (1978). *Br. Poult. Sci.,* **19**, 559–565

WHITTEMORE, C.T. (1977). In *Recent Advances in Animal Nutrition – 1977* (Haresign, W. and Lewis, D., Eds), Butterworths, London

WHITTEMORE, C.T. and FAWCETT, R.H. (1976). *Anim. Prod.,* **22**, 87–96

WILSON, B.J. (1977). In *Growth and Poultry Meat Production* (Boorman, K.N. and Wilson, B.J., Eds), British Poultry Science Ltd, Edinburgh

WINSOR, C.P. (1932). *Proc. Natn Acad. Sci., U.S.A.,* **18**, 1–8

15

PROTEIN METABOLISM IN FISH

C.B. COWEY
Institute of Marine Biochemistry, Aberdeen

Summary

Energy expenditure by fish is probably lower than that of conventional farm animals both because of their poikilothermy and of certain aspects of their aquatic habit. Food energy conversion may therefore be more efficient in fish than in terrestrial homeotherms. Many fish are, however, carnivorous and this militates against efficient conversion of food protein to tissue protein. Available data on protein metabolism in fish are examined against this general background.

Minimal dietary protein levels for optimal growth of most fish are much higher than those of omnivorous birds and mammals. This is also true of essential amino acid requirements of salmon, eel and common carp, although channel catfish seem to require levels more similar to those of terrestrial omnivores.

The main nitrogenous end product of protein metabolism is ammonia, and the excretion pattern in young Pacific salmon following a single daily meal indicates a rapid rate of amino acid assimilation and catabolism. Some control of amino acid catabolism, via changes in the tissue levels of amino acid degrading enzymes in response to changes in dietary protein intake, has been shown to occur in carp. Enzymes that deaminate non-essential amino acids in plaice did not respond to changes in protein intake.

There was no change in the relative rate of oxidation of amino acids (both essential and non-essential) as assessed by the single-pulse tracer method in marine flatfish in response to gross changes in dietary protein intake. In carp, the rate of leucine oxidation decreased as dietary protein level was reduced. These differences may reflect that omnivores adapt more readily to dietary protein restriction than do carnivores.

The overall efficiency with which food protein is deposited as tissue protein has been examined in several fish species at a gross level. The values obtained are generally lower than those pertaining in omnivorous homeotherms. However, the quantity of protein deposited in fish tissues per unit food energy consumed exceeds that of omnivorous birds and mammals. The recent application of direct calorimetry to fish has indicated that this efficient utilization of energy during growth arises not only from factors such as poikilothermy, locomotion in water requiring less energy than in air, etc., but also because specific dynamic action in fish is much less than in birds and mammals. This low specific dynamic action results from ammonotelism and means that fish derive much more energy from dietary protein than do uricotelic and ureotelic animals.

Introduction

The energy demands of fish differ from those of terrestrial homeotherms, and these differences may affect the efficiency of energy and protein transformation in the two types of organism. For instance, lack of an expenditure of energy in maintaining the body at a different temperature from that of the environment represents a potential advantage for fish over homeotherms. This advantage may, however, be offset by the likelihood that metabolic reactions function more efficiently, certainly more rapidly, at higher temperatures.

By comparison with terrestrial birds and mammals several other features of piscean life will affect, one way or the other, the energetics of growth; locomotion and maintenance of position in water require less energy than they do in air. On the other hand, respiration from water necessitates a continuous passage of water over very permeable surfaces, entailing the risk to the fish of either hydration or dehydration depending on the salinity of the water, so that some energy must be expended in maintaining osmotic equilibrium. Again, since diffusible metabolic end products may readily be disposed of across permeable surfaces into the surrounding water, energy does not have to be expended in converting potentially toxic substances such as (in the context of protein metabolism) ammonia to harmless compounds such as urea.

The relative efficiency of energy conversion to growth in fish and in warm-blooded vertebrates will be determined by the totality of these adaptations and differences. Blaxter (1975) quoted data by Nijkamp, van Es and Huisman (1974) to show that a comparison of food utilization by carp and poultry did not suggest any great advantage for the poikilotherm. It is intended in the present chapter to examine various facets of protein metabolism in fish and to assess the present status of fish as converters of protein and energy.

Finally it is apposite to point out here that most of the studies of protein metabolism in warm-blooded animals have been made on species that are omnivorous or herbivorous. The preponderance of fish in the natural environment are carnivorous, as are several of the species under cultivation. Thus, it must be borne in mind that many of the data referred to in succeeding pages concern carnivores and in these instances differences between fish and homeotherms may be as much a reflection of the carnivore/omnivore contrast as of the cold-blooded/warm-blooded states.

Dietary protein requirement

Dose—response (growth) curves have been used to determine the minimal dietary protein level giving optimal weight gain in a number of fish. Some of the results obtained are shown in *Table 15.1*. In making the measurements most authors claim to have used isoenergetic diets; generally, the different protein levels were obtained by substituting carbohydrate for protein, it being assumed that the two materials supply equal amounts of metabolizable energy on a weight basis. Recent studies have shown that this assumption may not be valid (Smith, Rumsey and Scott, 1978b) and some adjustment of some of the values may eventually be shown to be necessary.

With the possible exception of channel catfish the values shown in *Table 15.1* are uniformly high. This is to be expected of carnivores but the values obtained for omnivorous (*Cyprinus carpio*) and herbivorous (grass carp) fish are surprisingly high compared with the requirements of omnivorous birds and mammals. In view of the doubts raised about the metabolizable energy content of some of the experimental diets used, further evidence seems desirable to establish that a real difference in protein requirement exists between warm-blooded and cold-blooded herbivores.

Several environmental factors may apparently affect the protein requirement

Table 15.1. ESTIMATED DIETARY PROTEIN REQUIREMENT OF CERTAIN FISH

Species	Crude protein level in diet for optimal growth (g/kg)	Reference
Rainbow trout (*Salmo gairdneri*)	400–460	Satia (1974)
		Zeitoun *et al.* (1976)
		Tiews, Gropp and Koops (1976)
Carp (*Cyprinus carpio*)	380	Ogino and Saito (1970)
Chinook salmon (*Oncorhynchus tschawytscha*)	400	De Long, Halver and Mertz (1958)
Eel (*Anguilla japonica*)	445	Nose and Arai (1972)
Plaice (*Pleuronectes platessa*)	500	Cowey *et al.* (1972)
Gilthead bream (*Chrysophrys aurata*)	400	Sabaut and Luquet (1973)
Grass carp (*Ctenopharyngodon idella*)	410–430	Dabrowski (1977)
Channel catfish (*Ictalurus punctatus*)	220–320	Garling and Wilson (1976)
Tilapia aurea	360	Davis and Stickney (1978)

of fish; among those that have been shown to do so in certain species are temperature and salinity. De Long, Halver and Mertz (1958) showed that the optimum dietary protein level for chinook salmon was 400 g/kg at 8.3 °C and 550 g/kg at 14.4 °C. Zeitoun *et al.* (1973) found that when rainbow trout were grown at a salinity of 10 p.p.t. the minimum dietary protein level for optimal growth was 400 g/kg; the corresponding value for trout grown in a salinity of 20 p.p.t. was 450 g/kg. The minimum protein requirement of coho salmon (*Oncorhynchus kisutch*), on the other hand, was the same at both these salinities (Zeitoun *et al.*, 1974).

Amino acid requirements

Quantitative requirements of three species of fish for essential amino acids are shown in *Table 15.2*, where they are compared with those of the rat. While the

Table 15.2. ESSENTIAL AMINO ACID REQUIREMENTS OF CERTAIN SPECIES OF FISH AND OF RAT (g/kg dry diet)

Amino acid	Chinook salmon[1]	Japanese eel[2]	Carp[2]	Rat[1]
Arginine	24	14	16	2
Histidine	7	8	8	4
Isoleucine	9	15	9	5
Leucine	16	20	13	9
Lysine	20	20	22	10
Methionine		12(Cys= 0)	12(Cys= 0)	6(Cys= 0)
	6(Cys=10)	9(Cys=10)	8(Cys=20)	
Phenylalanine		22(tyr= 0)	25(tyr= 0)	9(tyr= 0)
	17(tyr= 4)	12(tyr=20)	13(tyr=10)	
Threonine	9	15	15	5
Tryptophan	2	4	3	2
Valine	13	15	14	4

[1] Mertz (1972).

[2] Nose (1979)

expression of requirement (g/kg dry matter) is not ideal for a strict comparison, nevertheless the requirement for most essential amino acids by all three species (including the omnivorous carp) is markedly greater than that of the rat.

Recent measurements on channel catfish, however, have given values that are of a similar order to those of homeotherms. Harding, Allen and Wilson (1977) showed that, in the absence of cysteine, the methionine requirement of this species is 5.6 g/kg diet; the requirement for lysine amounts to 12.3 g/kg (Wilson, Harding and Garling, 1977), while those for threonine and tryptophan are 5.3 and 1.2 g/kg, respectively (Wilson *et al.*, 1978). These values indicate that homotherms and poikilotherms of similar dietary habit are not greatly different in their requirements.

Nitrogen excretion

The main nitrogenous end product of protein metabolism in fish is ammonia, virtually all of which is excreted extrarenally through the gills. Thus in marine teleosts 80% of total waste nitrogen is excreted as ammonia (Wood, 1958). In rainbow trout, Fromm (1963) showed that urinary nitrogen made up only 3% of excreted nitrogen, while 60% of the total waste nitrogen was excreted as ammonia. Several groups (Goldstein, Forster and Fanelli, 1964; McBean, Neppel and Goldstein, 1966; Pequin and Serfaty 1966; Kenyon 1967) using hepatectomy or perfusion techniques demonstrated that the liver is the main site of ammonia formation.

The pattern of nitrogen excretion has been examined by Brett and Zala (1975) using young sockeye salmon (*Oncorhynchus nerka*). Fish were given a single meal daily, approximating the maintenance requirements of the animal, and the excretory products were monitored at frequent intervals. Before the single daily meal, ammonia excretion was relatively constant at 8.2 mg N/(kg h) (15 °C); following the feed, excretion rose sharply reaching a peak, 4½ h after the meal, of 35 mg N/(kg h); it then fell in exponential fashion to the base level. Urea excretion meanwhile was constant at 2.16 mg N/(kg h).

From earlier studies (Brett and Higgs, 1970) it was possible to superimpose a plot of gastric evacuation at 15 °C over that for ammonia excretion; a general parallelism existed between the two curves, from which it could be inferred that, at the time of peak ammonia excretion, the stomach would be approximately half empty. Brett and Zala (1975) conclude that the large pulse in ammonia excretion is entirely a product of food intake.

The surprising feature of these results is the speed with which ingested protein is assimilated and catabolized, the overall rate being of the same order as that of warm-blooded animals but at a much lower temperature.

Amino acid catabolism

The rapidity with which dietary protein nitrogen is excreted is indicative both of high rates of proteolysis in the gut and high catabolic (or at least deamination) rates. The rapid assimilation of amino acids means that they are transported to the tissues at a rate greater than that at which they are utilized for protein synthesis. Consequently tissue concentrations of amino acids increase leading directly to enhanced rates of deamination and of lipogenesis, gluconeogenesis or oxidation of the carbon skeletons.

In omnivorous mammals two main factors are seen as controlling the rate of degradation of amino acids (Krebs, 1972). These are (1) the tissue concentrations (or activities) of amino acid degrading enzymes, and (2) the K_m values of these enzymes relative to tissue concentrations of the amino acid substrates.

The first factor is seen as a coarse control (Krebs, 1972). In omnivorous mammals grown on a high protein diet, the tissue levels of many amino acid degrading enzymes are several-fold higher than in animals given a low protein diet. This is especially true of enzymes that degrade essential amino acids; very low concentrations of these enzymes are found under conditions of dietary protein restriction when conservation of essential amino acids is necessary. In addition, the total liver content of all urea cycle enzymes is directly proportional to daily protein consumption in rats (Schimke, 1962). Alanine aminotransferase showed a similar response to the urea cycle enzymes but, interestingly, glutamate dehydrogenase which is considered necessary to supply one of the two nitrogen atoms present in the urea molecule was not affected by a change in protein intake.

Few comparable data are available for fish; in fact, for several essential amino acids the catabolic pathways involved have not yet been established. However, responses similar to those of mammals have been described for enzymes catabolizing two essential amino acids in carp. Sakaguchi and Kawai (1970) found that the activities of histidine deaminase and urocanase in carp given a diet containing 800 g casein/kg were, respectively, 19.2 and 21.8 μmol/(h/g) tissue, but the activities in carp given diets containing 50 g casein/kg were reduced to 2.0 and 4.8 μmol/(h/g), respectively. Addition of histidine (2.7%) to the low casein diet resulted in intermediate enzyme activities of 4.6 and 12.9 μmol/(h/g). The latter observation is a little surprising because enzymes of protein metabolism are not normally induced in animals given a restricted protein intake, and a diet containing 50 g crude protein/kg would be a very restricted protein intake for a fish. In a later paper (Sakaguchi and Kawai, 1974) it was shown that liver histidine deaminase and urocanase could be induced by intraperitoneal injection (three doses at 12 h intervals) of histidine; similar injections of urocanate were without effect on urocanase activity. It appears probable that the induction of both histidine deaminase and urocanase in carp liver are dependent mainly on histidine intake. This contrasts with the induction of threonine dehydrase in rats which requires the presence in the diet of four essential amino acids, threonine, valine, methionine and tryptophan (Mauron, Mottu and Spohr, 1973).

The enzymic pathway of leucine breakdown in carp has also been shown (Zébian and Créach, 1978) to respond to variations in protein intake. This pathway (or the first two enzymes in it, leucine aminotransferase and α-ketoisocaproic dehydrogenase) has been the subject of several studies in rats (Sketcher, Fern and James, 1974; Sketcher and James, 1974). Control of leucine catabolism does not seem to reside in the initial (transaminase) reaction; under conditions of low protein intake, muscle largely regulates leucine oxidation in rats limiting it by a decrease in the α-ketoisocaproic dehydrogenase activity at a time when no change occurred in the activity of either of the liver enzymes.

The levels of these enzymes in carp tissues are shown in *Table 15.3* (Zébian and Créach, 1978). Although red muscle contains highest activities of both enzymes the amounts present in liver are relatively much greater than in the rat. As with the rat, no change in activity of either of the liver enzymes occurred in response to alterations in dietary protein level but, by contrast with the rat, both the aminotransferase and the dehydrogenase activity fell in the red muscle in response to this dietary change, while in the white muscle only the amino-transferase activity was affected. The significance of these changes is not easy to interpret at the present time mainly because it is not clear which of the tissues examined by Zébian and Créach is quantitatively the most important in leucine oxidation. The animal contains vastly more white muscle than it does either of the other tissues but the vascularization of this tissue is very poor compared with that of liver or red muscle.

Table 15.3. LEUCINE-α-OXOGLUTARATE AMINOTRANSFERASE AND α-KETOISOCAPROIC ACID DEHYDROGENASE ACTIVITIES IN TISSUES OF CARP GIVEN HIGH OR LOW PROTEIN DIETS FOR 7 WEEKS (from Zébian and Créach, 1979)

Tissue	Diet regime	Activity of enzymes (nmol leucine or keto acid oxidized/100 mg tissue per h)	
		leucine-α-oxoglutarate aminotransferase	α-ketoisocaproic acid dehydrogenase
Liver	HP (320 g protein/kg)	331	35
	LP (200 g protein/kg)	296	38
Red muscle	HP	1109	176
	LP	671	88
White muscle	HP	245	37
	LP	168	33

Enzymes deaminating non-essential amino acids showed little change in activity in response to changes in dietary protein intake. Nagai and Ikeda (1973) were unable to show any effect on aspartic aminotransferase or alanine amino-transferase in carp liver. Cowey *et al.* (1974) did not find any significant changes in total hepatic glutamate dehydrogenase or either of these aminotransferase activities in plaice fed high or low protein diets for several weeks. These amino-transferases have, of course, important functions (such as the transport of four carbon units between mitochondria and cytosol) other than amino acid degrada-tion.

The lack of effect of gross differences in dietary protein intake on glutamate dehydrogenase activity is surprising because ammonia formation is considered to be mainly by transdeamination (Forster and Goldstein, 1969; Watts and Watts, 1974). Glutamate dehydrogenase is an allosteric enzyme and it may be that enhanced activity in both mammals and fish given high protein diets is due to allosteric effects (high ADP concentration or altered pyridine nucleotide concen-trations related to changes in mitochondrial activity). One other possibility is that glutamate dehydrogenase is not primarily involved in ammonia production — McGivan and Chappell (1975) suggested that the main function of the enzyme concerns nitrogen storage. The latter possibility implies that most of the ammonia excreted by fish is formed by some other pathway such as the purine nucleotide cycle (Lowenstein, 1972). There is little evidence that this pathway is active in fish liver, the major site of amino acid degradation; the tissue distribution of ammonia-forming reactions in rainbow trout is shown in *Table 15.4*. The inference from these data is that amino acid deamination in the liver occurs through the mediation of glutamate dehydrogenase; this conclusion has been verified by studies on trout liver mitochondria (Walton and Cowey, 1977) which showed that glutamate is oxidized mainly by the glutamate dehydrogenase pathway, comparatively little aspartate being formed.

Table 15.4. ACTIVITY OF AMMONIA-FORMING ENZYMES IN TISSUES OF RAINBOW TROUT (μmol NH_3 formed/g wet weight per min at 15 °C ± S.E.)

Enzyme	Liver	Kidney	Gill	Muscle
Glutamate dehydrogenase*	0.95 ± 0.17	0.78 ± 0.13	0.31 ± 0.06	not detected
Glutaminase	3.37 ± 0.99	1.93 ± 0.15	2.07 ± 0.14	not detected
AMP deaminase	not detected	not detected	9 ± 4	226 ± 45

* Glutamate deamination in the absence of adenosine diphosphate.

Fine control of amino acid catabolism (control by K_m) has been illustrated in omnivorous mammals (Krebs, 1972). The tissue concentrations of most amino acids are normally less than 1 mM and K_m values of enzymes that degrade amino acids are millimolar or more. These relative levels militate against amino acid breakdown but any increase in amino acid concentrations (e.g. post-absorptive or by increasing protein intake) will lead to a rapid increase in amino acid degradation. Amino acid concentrations in systemic venous blood plasma of rainbow trout given diets containing either 600 g or 100 g crude protein/kg (Cowey *et al.*, 1977) are shown in *Table 15.5*. The plasma samples are taken 18 h after the trout had last been given food. For most amino acids, concentrations, even on the high protein diet, are less than 1 mM, the exceptions among the

Table 15.5. CONCENTRATIONS (mmol/l) OF THE AMINO ACIDS IN BLOOD PLASMA OF RAINBOW TROUT GIVEN DIETS OF DIFFERENT PROTEIN CONTENT (mean values with standard errors for four fish/group)

Amino acid	High protein diet (600 g crude protein/kg)	Low protein diet (10 g crude protein/kg)
Lysine	0.69 ± 0.06	0.23 ± 0.03
Histidine	0.17 ± 0.04	0.09 ± 0.01
Arginine	0.26 ± 0.03	0.09 ± 0.01
Aspartic acid	0.04 ± 0.007	<0.01
Threonine	0.76 ± 0.05	0.12 ± 0.02
Serine	0.47 ± 0.14	0.27 ± 0.04
Glutamic acid	0.25 ± 0.02	0.05 ± 0.008
Proline	0.36 ± 0.04	<0.01
Glycine	1.40 ± 0.22	0.74 ± 0.18
Alanine	0.68 ± 0.13	0.38 ± 0.04
Cysteine	0.05 ± 0.01	0.04 ± 0.003
Valine	1.61 ± 0.33	0.24 ± 0.01
Methionine	0.31 ± 0.07	0.12 ± 0.01
Isoleucine	1.07 ± 0.14	0.14 ± 0.01
Leucine	1.41 ± 0.29	0.20 ± 0.01
Tyrosine	0.13 ± 0.04	0.09 ± 0.009
Phenylalanine	0.15 ± 0.04	0.12 ± 0.001
Tryptophan	<0.01	<0.01

essential amino acids being the branched chain amino acids. Much lower concentrations were present in plasma of trout given the low protein diet, the total quantity of essential amino acids being reduced by 80% from 6.43 mmol/l to 1.35 mmol/l and those of non-essential amino acids by about 55% from 3.77 mmol/l to 1.35 mmol/l. Few data are available on K_m values of enzymes that deaminate amino acids in fish; those for which values are available are in excess of the plasma concentrations shown in *Table 15.5*; alanine aminotransferase (trout liver cytosol) 2.2 mM, alanine aminotransferase (trout liver mitochondria) 2.5 mM, serine pyruvate aminotransferase (trout liver mitochondria) 5–10 mM.

These K_m values were all measured at 15°C; temperature may affect the value of this parameter markedly. For many enzymes from fish, the commonest effect of a decrease in environmental temperature is a reduction in apparent K_m (Hochachka and Somero, 1973; *Figure 15.1*) although, to date, most of the results concern glycolytic enzymes. The resultant increase in enzyme—substrate affinity tends to compensate at non-saturating substrate concentrations for the fall in reaction rate with temperature. The net effect of a change in environmental temperature on the metabolic fate of amino acids depends on whether

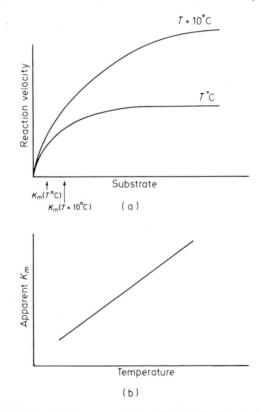

Figure 15.1   (a) Relationship between reaction velocity and substrate concentration for an enzyme at two assay temperatures. The affinity for substrate approximately doubles as the temperature is decreased by 10 °C. Rate-compensating effects of positive thermal modulation occur only at non-saturating substrate concentrations. (b) Relationship between apparent K_m and assay temperature, illustrating positive thermal modulation. (Courtesy Drs P.W. Hochachka and G.N. Somero and W.B. Saunders Company, Philadelphia)

changes in K_m of amino acid deaminating and amino acid activating enzymes are in a similar direction and of a similar magnitude. The optimal temperature for protein deposition in a given species will clearly be influenced by these effects.

The ability of the rat to adapt metabolically to a restriction in dietary protein intake has been demonstrated by experiments in which the oxidation of isotopically labelled essential and non-essential amino acids has been followed. McFarlane and van Holt (1969) thus showed that glutamate and alanine are oxidized rapidly irrespective of dietary protein intake, while leucine and phenylalanine oxidation was markedly reduced in animals subject to dietary protein restriction.

Since then many similar experiments have been performed to examine the manner in which the animal adapts to dietary stress. The validity of this type of experiment, in which a single pulse is injected, has been questioned because of both the heterogeneity of the free amino and pool and of short-term fluctuations in rates of protein synthesis (Neale and Waterlow, 1974). Even when

Table 15.6. INCORPORATION OF RADIOACTIVITY FROM L-[1-^{14}C]-LEUCINE, L-[1-^{14}C]-PHENYLALANINE, L-[1-^{14}C]-ALANINE AND L-[1-^{14}C]-GLUTAMIC ACID INTO LIVER PROTEIN, CARCASS PROTEIN AND CARBON DIOXIDE OF TURBOT GIVEN EITHER A HIGH OR LOW PROTEIN DIET

Amino acid	Protein level in diet (g/kg)	Oxidation as $^{14}CO_2$ (% dose given)	Incorporation into liver protein (% dose given)	Incorporation into carcass protein (% dose given)
Leucine	60	23.5	1.1	13.6
	500	28.9	0.9	13.7
Phenylalanine	60	24.1	0.9	11.4
	500	19.9	1.1	13.7
Glutamic acid	60	56.3	0.08	1.1
	500	59.1	0.08	1.0
Alanine	60	56.6	0.11	1.2
	500	48.5	0.11	1.8

constant infusion over a period of 4 h was used in rats, isotopic equilibrium within the precursor pool was not achieved (Neale and Waterlow, 1974).

Despite the criticisms made of the single-pulse method, it has been used subsequently. Sketcher and James (1974) claim that while the method does not give an absolute measure of amino acid oxidation nevertheless it does, if 1-^{14}C-amino acids are used, reflect the rate of catabolism of the carbon skeleton because recycling of the label is minimal once it has been released as $^{14}CO_2$.

When the method was applied to marine flatfish, Cowey (1975) was unable to detect any relative alterations in rates of oxidation of essential and non-essential amino acids in plaice given diets of grossly different protein content. Similarly results obtained with turbot (*Table 15.6*) did not reveal any ability of turbot to regulate amino acid oxidation in response to dietary protein restriction.

Since then Zébian and Créach (1979) have measured $^{14}CO_2$ production from carp given diets of different protein content for 7 weeks and then injected intra-peritoneally with $[1\text{-}^{14}C]$-leucine. The results (*Table 15.7*) obtained were some-what similar to those of Sketcher and James (1974) in that a reduction of protein in the diet led to a fall in $^{14}CO_2$ production from the substrate.

Table 15.7. *IN VIVO* OXIDATION OF L-$[1\text{-}^{14}C]$-LEUCINE IN CARP GIVEN HIGH OR LOW PROTEIN DIETS — mean values with standard errors for 6 groups of fish (from Zébian and Créach, 1979)

Dietary regime	Radioactivity in expired CO_2 (% of dose injected)
H.P. (320 g protein/kg)	13.3 ± 3.8
L.P. (200 g protein/kg)	7.2 ± 2.8

It is hazardous at this stage to make any conclusions about regulation of protein metabolism in fish. Little adaptation to a decrease in protein intake seems to occur in carnivorous species, although the general lowering of plasma and tissue amino acid concentrations resulting from such a regime will lead to some reduction in amino acid catabolism. Omnivorous species on the other hand, may yet be shown to possess some metabolic regulatory ability to enable them to adapt to low protein intakes.

Protein synthesis

A number of tracer incorporation studies have been made and have given some information but this has necessarily been restricted because of the inherent limitations of the method. In goldfish given $[U\text{-}^{14}C]$-leucine intraperitoneally (Das and Prosser, 1967) the ratio, incorporation of radioactivity into protein : radioactivity in the free amino acid pool, 1 h after the injection was similar in liver and gill but the level in muscle was very much lower than in these tissues. Narayansingh and Eales (1975) found that incorporation of a mixture of $[U\text{-}^{14}C]$-amino acids was highest in liver followed by gill and small intestine; incorpora-tion into muscle was barely measurable. Incorporation of leucine, phenylalanine, glutamic acid and alanine into liver and carcass protein of plaice and turbot was not significantly affected by gross differences in protein level of diets given to the fish (Cowey, 1975).

Jackim and La Roche (1973) examined incorporation of intraperitoneally injected $[1\text{-}^{14}C]$-leucine into the muscle proteins of *Fundulus heteroclitus*. Fish were acclimatized to different temperatures 17—21 days before being studied. Amino acid incorporation into proteins increased with temperature up to a critical point at 26—29 °C, beyond which it decreased sharply. Amino acid incorporation was decreased by fasting, by reducing dissolved oxygen concentra-tion in the water below 2.5 mg/l, and by stress (confinement in small volumes of water). Insulin treatment increased amino acid incorporation but exercise, darkness, fish size and sex had no significant effect. The radioactive leucine used in this work was injected, together with large amounts of carrier leucine, to off-set possible changes in amino acid pool size during various treatments. Pool sizes, however, were not measured directly and alterations in the pool sizes of essential amino acids other than leucine could alter the observed rate of incorporation of leucine, especially if such amino acids were rate limiting for protein biosynthesis.

Haschemeyer and co-workers have carried out a number of studies on protein synthesis in the liver of toadfish, *Opsanus tau*. The object was primarily to determine whether 'a control over protein synthesis might provide a common basis for the increased levels of enzymes of respiratory metabolism and other pathways responsible for physiological adaptation to low temperatures' (Haschemeyer, 1969b). During the course of these studies, methods permitting simultaneous measurement of amino acid uptake, activation and incorporation into growing polypeptide chains by liver *in vivo* were elaborated. As a result, Haschemeyer and Persell (1973) were able to show that the overall rate of protein synthesis in toadfish liver measured at 24 °C after fish had been acclimatized at sensibly the same temperature (22°C) was 1.6 mg/(h/g) liver, about one-fifth of the mammalian rate; the authors infer that 'when temperature and ribosome concentration differences are taken into account, it would appear that the protein synthetic system of toadfish liver is comparable to the mammalian liver with respect to levels of substances involved in control of elongation rates'.

Concerning temperature compensation, it is evident that when measurements of protein synthesis were made at a given temperature (either 24 °C or 11 °C) fish acclimatized at 11 °C had approximately 50% greater rates of protein synthesis in their livers than fish acclimatized at 22 °C irrespective of the temperature selected for measurement (Haschemeyer and Persell, 1973). Earlier Haschemeyer (1969a) had demonstrated that the rate of protein biosynthesis is controlled at the stages of polypeptide chain elongation and release; and that the rate increase in cold-acclimatized fish was paralleled by elevated levels of elongation factor I, the enzyme that promotes binding of amino acyl tRNA units at codon recognition sites (Haschemeyer, 1969b). Finally it has been shown that neither the rate of accumulation of amino acids into the liver nor the rate of formation of amino acyl tRNA units limited the rate of protein biosynthesis (Haschemeyer and Persell, 1973).

It is known that the levels of plasma proteins in fish do not change seasonally with temperature changes. In line with this, Haschemeyer (1973) has shown that the levels of circulating plasma proteins in cold- and warm-acclimatized fish do not differ. Likewise the rate of turnover of plasma proteins is not significantly altered by cold-acclimation ($t_{1/2}$=9.2±2.1 days at acclimation temperature 20 °C; $t_{1/2}$=8.0±1.0 days at acclimation temperature 10 °C). Haschemeyer (1973) concluded that, under conditions of an increased rate of protein biosynthesis in liver, control over the levels of circulating plasma proteins is exercised at a stage between completion of biosynthesis and secretion of the protein.

Retention of dietary energy and protein

Much of the data on energy metabolism in fish has been obtained by indirect methods. Nijkamp, van Es and Huisman (1974) compared energy retention in growing chicks and carp. Groups of carp were given daily rations (amounting to between 0% and 9% of their body weight per day) for four weeks of a diet containing 530 g crude protein/kg. Collection of faecal material was not possible so that utilization of nitrogen and gross energy was estimated from analysis of the food and of the carcass. Maintenance energy was obtained from a plot of energy gain against gross energy intake. The results obtained indicated that the daily maintenance requirement of the chick was about six times that of carp at 23 °C; above maintenance, however, conversion of food energy by the chick was several-fold more efficient than in carp, so that on balance gross efficiency of energy conversion was of a similar order in the two species. Energy retention as a proportion of energy intake was maximally 26.6% for carp and averaged 30% for chicks. These values led Blaxter (1975) to conclude that poikilothermy conferred no great advantage in terms of energy conversion by fish.

A fuller account of this data was later given by Huisman (1976) and the data was extended to rainbow trout. The value obtained for maintenance ration of carp as 23 °C was 720 $kJ/kg^{0.8}$ per week. From measurements of oxygen consumption, heat production at the maintenance ration was estimated as 462 $kJ/kg^{0.8}$ per week. Huisman then claims that as 'the total amount of available energy in the maintenance food ration is released as heat' the percentage available energy at maintenance is $462/720 \times 100 = 64\%$. Similar calculations indicated that available energy in trout rations given at maintenance level amounted to 67%. With increasing ration size there was apparently a relative decrease in available energy (which appeared to vary between 30% and 60% of the diet energy).

These values for available energy in the food are much lower than those reported by other workers who estimated available energy mainly by subtracting energy in faeces from that in food. Menzel (1960) obtained values of 91—97% for a Bermuda reef fish; Warren and Davis (1967) values of between 82% and 89% for a variety of fishes; Kelso (1972) 82—98% for the walleye. Huisman (1976) claims that these data are in error because they ignore 'soluble excrements', but until such soluble excrements have been shown to exist, identified and quantified then the weight of the evidence that well over 80% of dietary energy is normally assimilated and utilized cannot be ignored. The conclusion that an appreciable proportion of dietary energy is unavailable to trout and carp raises doubts about the experimental approach. The assumption mentioned above, for example, relating heat production and energy intake at a maintenance ration, seems to ignore energy requirements for locomotion, osmotic balance and so on. Consequently these values may not provide an ideal basis for comparison of the relative efficiency of food energy conversion between fish and birds and mammals.

Studies of energy transformation in fish have suffered from two main drawbacks. First, indirect measurements of heat production may have been inaccurate because the oxycalorific equivalents used were inexact and, secondly, the weight exponent in the equation relating metabolism to body weight may also have been erroneous.

Recently Smith, Rumsey and Scott (1978a) applied direct calorimetry to several species of salmonid. Over the temperature range tested, heat production increased linearly. The Q_{10} for heat production was less than 2, from which the authors deduced that fish have a mechanism which alters enzyme functioning and reduces fluctuations in metabolic rate to less than that expected in a biochemical system.

With rainbow trout of less than 4 g in weight, heat production was directly proportional to body weight; with trout of body weight 4—57 g, heat production was proportional to body weight to the exponent 0.63. These values are quite different from those obtained by indirect methods.

Smith, Rumsey and Scott (1978b) went on to study the increase in heat production (specific dynamic action) in trout, associated with the ingestion of protein, lipid and carbohydrate. Heat increment with all three components was less than 5% of the metabolizable energy fed. Values obtained were 1.36 soybean oil, 3.5 dextrin and 3.26 protein. The value for protein is not significantly different from that for dextrin. This contrasts markedly with mammals, where ingestion of dietary protein leads to an increased heat production of between 25% and 30%. This also contrasts with previous values for the specific dynamic action of protein for fish obtained by indirect methods.

Smith, Rumsey and Scott (1978b) ascribe these differences to the ammonotelism of fish, on the one hand, and uricotelism of birds and ureotelism of mammals, on the other. Specific dynamic action of proteins is seen as deriving from three main causes: (1) metabolism of the carbon skeleton of the amino acid molecule – this would be of similar magnitude to heat production from carbohydrate and fat metabolism and is perhaps partly due to the occurrence of futile or substrate cycles; (2) synthesis of the waste product that is excreted; (3) the energy cost of concentrating and excreting via the kidney (in mammals and birds) the waste product.

The energetics of ammonia, urea and uric acid excretion, as evaluated by Smith, Rumsey and Scott (1978b), is reproduced as *Table 15.8*. They assume a

Table 15.8. CALCULATED ENERGY DISTRIBUTION OF PROTEIN CATABOLIZED BY AMMONOTELIC, UREOTELIC AND URICOTELIC ANIMALS (from Smith, Rumsey and Scott, 1978b, courtesy American Institute of Nutrition and of Dr R.R. Smith)

Fraction	*Excretory product*		
	ammonia (kJ/g)	urea (kJ/g)	uric acid (kJ/g)
Gross energy	23.8	23.8	23.8
Digestion loss (8%)	1.9	1.9	1.9
Digestible energy	21.9	21.9	21.9
Metabolic loss	3.0	3.6	5.5
Metabolizable energy	18.9	18.3	16.4
Heat increment			
Waste product synthesis	0.0	2.1	1.8
Waste product concentration and excretion	0.0	0.9	1.2
Metabolism of non-nitrogen	1.2	1.2	1.2
Total	1.2	4.2	4.2
Net energy	17.7	14.1	12.2

digestion loss of 8%; the metabolic loss reflects the heats of combustion of ammonia, urea and uric acid; the energy costs of urea and uric acid synthesis/g protein are theoretical costs and the portion of heat output arising from concentration and excretion of these compounds has been obtained by subtraction of these values from the measured heat production. *Table 15.8* demonstrates clearly the contribution of ammonotelism to the economy of the fish; it means that net energy derived from dietary protein by the fish is greater than that derived by omnivorous birds and mammals.

In line with this are values obtained for energy retention as a percentage of energy intake in rainbow trout (Lee and Putnam, 1973) and channel catfish (Garling and Wilson, 1976). In both sets of experiments a large number of diets

Table 15.9. EFFECT OF VARIATION IN DIETARY PROTEIN ENERGY/TOTAL ENERGY RATIO ON PROTEIN UTILIZATION BY RAINBOW TROUT (from Lee and Putnam, 1973)

Total energy in diet (MJ/kg)	Protein level in diet (g/kg)	Protein energy/ total energy ratio	Protein retention* (%)
13.1	350	0.46	37
13.2	440	0.57	37
13.4	530	0.67	35
16.1	350	0.38	40
16.2	440	0.46	40
16.4	530	0.55	37
19.1	350	0.32	46
19.2	440	0.39	42
19.4	530	0.46	37

* Protein retained/protein intake.

that varied widely in protein and energy content were used. In their best diets, Lee and Putnam (1973) obtained values between 45% and 48% for energy retention in rainbow trout. Several of the diets of Garling and Wilson (1976) gave values for energy retention by channel catfish in excess of 60%, the maximum value being in excess of 68%.

Retention of dietary protein by carp was examined by analyses of food and of growth increments (Huisman *et al.*, 1979). Their data for carp of different sizes and at different temperatures given diets containing over 50% protein are shown in *Figure 15.2*. In fingerling carp, retention of protein is enhanced by a rise in temperature from 23 °C to 27 °C at high intake levels. At lower intake levels, retention at 27 °C is less than at 23 °C due to the increased metabolic rate. In two-year-old carp, protein retention at 27 °C is lower than at 23 °C at both high and low intake levels. In one-year-old carp, this temperature change was without effect on protein retention. Optimal values for efficiency of protein retention were generally in the range 20–25%.

This relatively low value for protein retention by carp is in line both with expectation and with earlier results, given the high protein content of the ration (cf. *Table 15.1*). Protein retention (net protein utilization, NPU) was examined in mammals by Miller and Payne (1961); highest values for NPU were obtained at the maintenance level of intake and decreased as the proportion of protein in the diet was increased. Cowey *et al.* (1972) showed that a similar relationship between NPU and dietary protein level held good for plaice. As the diets of carnivorous fish contain much higher levels of protein than do those of omnivorous mammals, then the ratio protein retained/protein intake will generally be superior in the latter animals to that in carnivorous fish. By the same token, protein conversion efficiencies of 40% have been routinely obtained with the grass carp (*Ctenopharyngodon idella*) – Domaniewski (1977).

The effect of variation in the ratio total energy/protein energy in the diet on protein retention and other characteristics of growth has been examined in several species of fish (Ringrose, 1971; Lee and Putnam, 1973; Page and Andrews, 1973; Cowey *et al.*, 1975; Takeda *et al.*, 1975). Certain trends or findings were common to most of these studies, thus an increase in the ratio digestible energy/protein energy in the diet led to an increased deposition of lipid in the fish; an increase in dietary energy level at constant dietary protein level resulted in improved feed efficiency; protein efficiency ratio was positively correlated with the ratio digestible energy/protein energy in the diet.

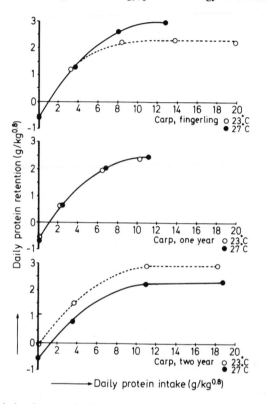

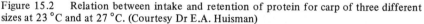

Figure 15.2   Relation between intake and retention of protein for carp of three different sizes at 23 °C and at 27 °C. (Courtesy Dr E.A. Huisman)

The object of studies on protein energy/total energy ratio in the diet has been to attain a ratio that, while permitting rapid growth and economic feed conversion, would not result in a gross alteration in carcass composition. Few investigators have been able to include sufficient different total energy and protein energy levels in their experiments to permit a useful prediction of the consequences of any particular ratio for protein retention. Perhaps the results of Lee and Putnam (1973) are most complete (*Table 15.9*). The energy contents of their diets have been recalculated using a calorific value of 18.8 kJ/g protein but retaining the factors applied by Lee and Putnam (1973) to the carbohydrate content of starch and the protein content of casein—gelatin. Values for protein retention shown in *Table 15.9* were not given in Lee and Putnam's paper and have been calculated from other data they supply.

Within the confines of their experiments (350—530 g protein/kg; 12.5—18.8 MJ/kg), Lee and Putnam (1973) found no significant differences in weight gain unless both protein and energy in the diet were low. The relationship between protein retention and protein energy/total energy ratio is given by: $N=51.76-0.27P$, where N is protein retention and P is dietary protein energy as a percentage of total dietary energy.

Garling and Wilson (1976), using the warm-water channel catfish, showed that a diet containing 11.5 MJ/kg produced optimal growth with 241 g crude protein/kg; a diet containing 14.3 MJ/kg produced optimal growth at a dietary protein level of 280–320 g/kg. From their data the relationships between protein utilization and protein energy/total energy ratio are given by:

$$N = 76.94 - 0.82P \text{ (diets containing 11.5 MJ/kg)}$$
$$N = 63.11 - 0.65P \text{ (diets containing 14.3 MJ/kg)}$$

For these species these equations permit the prediction of protein utilization within reasonably wide limits of protein concentration and energy density; they provide an acceptable approach to reaching a desirable protein energy:total energy ratio in both experimental and practical diets.

References

BLAXTER, K.L. (1975). *Proc. Nutr. Soc.*, **34**, 51–56

BRETT, J.R. and HIGGS, D.A. (1970). *J. Fish. Res. Bd Can.*, **27**, 1767–1779

BRETT, J.R. and ZALA, C.A. (1975). *J. Fish. Res. Bd Can.*, **32**, 2479–2486

COWEY, C.B. (1975). *Proc. Nutr. Soc.*, **34**, 57–63

COWEY, C.B., ADRON, J.W., BROWN, D.A. and SHANKS, A.M. (1975). *Br. J. Nutr.*, **33**, 219–231

COWEY, C.B., BROWN, D.A., ADRON, J.W. and SHANKS, A.M. (1974). *Mar. Biol.*, **28**, 207–213

COWEY, C.B., KNOX, D., WALTON, M.J. and ADRON, J.W. (1977). *Br. J. Nutr.*, **38**, 463–470

COWEY, C.B., POPE, J.A., ADRON, J.W. and BLAIR, A. (1972). *Br. J. Nutr.*, **28**, 447–456

DABROWSKI, K. (1977). *Aquaculture*, **12**, 63–73

DAS, A.B. and PROSSER, C.L. (1967). *Comp. Biochem. Physiol.*, **21**, 449–467

DAVIS, A.T. and STICKNEY, R.R. (1978). *Trans. Am. Fish. Soc.*, **107**, 479–483

DE LONG, D.C., HALVER, J.E. and MERTZ, E.T. (1958). *J. Nutr.*, **65**, 589–599

DOMANIEWSKI, J.C.J. (1977). *Nature, Lond.*, **267**, 102

FORSTER, R.P. and GOLDSTEIN, L. (1969). In *Fish Physiology*, Vol. 1, pp. 313–350 (Hoar, W.S. and Randall, D.J., Eds), Academic Press, New York

FROMM, P.O. (1963). *Comp. Biochem. Physiol.*, **10**, 121–128

GARLING, D.C. and WILSON, R.P. (1976). *J. Nutr.*, **106**, 1368–1375

GOLDSTEIN, L., FORSTER, R.P. and FANELLI, G.M. (1964). *Comp. Biochem. Physiol.*, **12**, 489–499

HARDING, D.E., ALLEN, O.W. and WILSON, R.P. (1977). *J. Nutr.*, **107**, 2031–2035

HASCHEMEYER, A.E.V. (1969a). *Proc. Natn Acad. Sci. U.S.A.*, **62**, 128–135

HASCHEMEYER, A.E.V. (1969b). *Comp. Biochem. Physiol.*, **28**, 535–552

HASCHEMEYER, A.E.V. (1973). *J. Biol. Chem.*, **248**, 1643–1649

HASCHEMEYER, A.E.V. and PERSELL, R. (1973). *Biol. Bull. Mar. Biol. Lab. Woods Hole*, **145**, 472–481

HOCHACHKA, P.W. and SOMERO, G.N. (1973). *Strategies of Biochemical Adaptation,* Saunders, Philadelphia

HUISMAN, E.A. (1976). *Aquaculture,* 9, 259–273

HUISMAN, E.A., KLEIN BRETELER, J.G.P., VISMAN, M.M. and KANIS, E. (1979). In *Finfish Nutrition and Fishfeed Technology*, Vol. I, pp. 175–188 (Halver, J.E. and Tiews, K., Eds), H. Heenemann GmbH & Co., Berlin

JACKIM, E. and LA ROCHE, G. (1973). *Comp. Biochem. Physiol.,* 44A, 851–856

KELSO, J.R. (1972). *J. Fish. Res. Bd, Can.,* 29, 1181–1192

KENYON, A.J. (1967). *Comp. Biochem. Physiol.,* 22, 169–175

KREBS, H.A. (1972). *Adv. Enzyme Reg.,* 10, 397–420

LEE, D.J. and PUTNAM, G.B. (1973). *J. Nutr.,* 103, 916–922

LOWENSTEIN, J.M. (1972). *Physiol. Rev.,* 52, 382–413

MAURON, J., MOTTU, F. and SPOHR, G. (1973). *Eur. J. Biochem.,* 32, 331–342

McBEAN, R.L., NEPPEL, M.J. and GOLDSTEIN, L. (1966). *Comp. Biochem. Physiol.,* 18, 909–920

McFARLANE, J.G. and VON HOLT, C. (1969). *Biochem. J.,* 111, 557–563

McGIVAN, J.D. and CHAPPELL, J.B. (1975). *FEBS Letts,* 52, 1–5

MENZEL, D.W. (1960). *J. Cons. Perm. Explor. Mer,* 25, 216–222

MERTZ, E.T. (1972). In *Fish Nutrition*, pp. 106–143 (Halver, J.E., Ed.), Academic Press, New York

MILLER, D.S. and PAYNE, P.R. (1961). *Br. J. Nutr.,* 15, 11–19

NAGAI, M. and IKEDA, S. (1973). *Bull. Jap. Soc. Scient. Fish.,* 39, 633–643

NARAYANSINGH, T. and EALES, J.G. (1975). *Comp. Physiol. Biochem.,* 52B, 399–405

NEALE, R.J. and WATERLOW, J.C. (1974). *Br. J. Nutr.,* 32, 11–25

NIJKAMP, H.J., VAN ES, A.J.H. and HUISMAN, A.E. (1974). *Eur. Ass. Anim. Prod. Publ.,* 14, 277–280

NOSE, T. (1979). In *Finfish Nutrition and Fishfeed Technology,* Vol. I, pp. 145–156 (Halver, J.E. and Tiews, K., Eds), H. Heenemann GmbH & Co., Berlin

NOSE, T. and ARAI, S. (1972). *Bull. Freshw. Fish. Res. Lab. Tokyo,* 22, 145–155

OGINO, C. and SAITO, K. (1970). *Bull. Jap. Soc. Scient. Fish.,* 36, 250–254

PAGE, J.W. and ANDREWS, J.W. (1973). *J. Nutr.,* 103, 1339–1346

PEQUIN, L. and SERFATY, A. (1966). *Comp. Biochem. Physiol.,* 18, 141–149

RINGROSE, R.C. (1971). *J. Fish. Res. Bd, Can.,* 28, 1113–1117

SABAUT, J.J. and LUQUET, P. (1973). *Mar. Biol.,* 18, 50–54

SAKAGUCHI, M. and KAWAI, A. (1970). *Bull. Jap. Soc. Scient. Fish.,* 36, 783–787

SAKAGUCHI, M. and KAWAI, A. (1974). *Daigaku Shokuryo Kagaku Kenkyusho Hokoku,* 37, 28–31

SATIA, B.P. (1974). *Progve Fish Cult.,* 36, 80–85

SCHIMKE (1962). *J. Biol. Chem.,* 237, 459–468

SKETCHER, R.D. and JAMES, W.P.T. (1974). *Br. J. Nutr.,* 32, 615–623

SKETCHER, R.D., FERN, E.B. and JAMES, W.P.T. (1974). *Br. J. Nutr.,* 31, 333–342

SMITH, R.R., RUMSEY, G.L. and SCOTT, M.L. (1978a). *J. Nutr.,* 108, 1017–1024

SMITH, R.R., RUMSEY, G.L. and SCOTT, M.L. (1978b). *J. Nutr.,* 108, 1025–1032

TAKEDA, M., SHIMENO, S., HOSOKAWA, H., KAJIYAMA, H. and KAISYO, T. (1975). *Bull. Jap. Soc. Scient. Fish.,* 41, 443–447

TIEWS, K., GROPP, J. and KOOPS, H. (1976). *Arch. Fisch Wiss.,* 27, 1–29

WALTON, M.J. and COWEY, C.B. (1977). *Comp. Biochem. Physiol.,* 57B, 143–149

WARREN, C.E. and DAVIS, G.E. (1967). In *The Biological Basis of Freshwater Fish Production*, pp. 175–214 (Gerking, S.D., Ed.), Blackwell, Oxford

WATTS, R.L. and WATTS, D.C. (1974). In *Chemical Zoology*, Vol. 8, pp. 369–446 (Florkin, M. and Scheer, B.T., Eds), Academic Press, New York

WILSON, R.P., ALLEN, O.W., ROBINSON, E.H. and POE, W.E. (1978). *J. Nutr.,* **108**, 1595–1599

WILSON, R.P., HARDING, D.E. and GARLING, D.L. (1977). *J. Nutr.,* **107**, 166–170

WOOD, J.D. (1958). *Can. J. Biochem. Physiol.,* **36**, 1237–1242

ZÉBIAN, M.F. and CRÉACH, Y. (1979). In *Finfish Nutrition and Fishfeed Technology*, Vol II, pp. 531–544 (Halver, J.E. and Tiews, K., Eds), H. Heenemann GmbH & Co., Berlin

ZEITOUN, I.H., HALVER, J.E., ULLREY, D.E. and TACK, P.I. (1973). *J. Fish. Res. Bd Can.,* **30**, 1867–1873

ZEITOUN, I.H., ULLREY, D.B., HALVER, J.E., TACK, P.I. and MAGEE, W.T. (1974). *J. Fish. Res. Bd, Can.,* **31**, 1145–1148

ZEITOUN, I.H., ULLREY, D.E., MAGEE, W.T., GILL, J.L. and BERGEN, W.G. (1976). *J. Fish. Res. Bd, Can.,* **33**, 167–172

LIST OF PARTICIPANTS

Armstrong, Prof. D.G.	Dept of Agricultural Biochemistry, University of Newcastle-upon-Tyne, Newcastle-upon-Tyne
Arnal, Dr M	Laboratoire du Métabolisme Azote, CRZV, Theix, 63110, Beaumont, France
Ash, Dr R.	School of Applied Biology, University of Bradford, Bradford
Baird, T.	Institute of Physiology, University of Glasgow, Glasgow
Baker, K.R.	Tate and Lyle Feeds, 215 Tunnel Avenue, Greenwich SE10
Bamgbose, M.	Olagun Ltd, PO Box 3304, Ibadan, Nigeria
Bardsley, Dr R.G.	Dept of Applied Biochemistry and Nutrition, University of Nottingham School of Agriculture
Bassett, Dr J.M.	Nuffield Institute for Medical Research, University of Oxford, Headley Way, Headington, Oxford
Beckerton, Dr A.	Colborn Group Ltd, Heanor Gate Industrial Estate, Heanor, Derbyshire DE7 7SG
Beever, Dr D.E.	Grassland Research Institute, Hurley, Maidenhead, Berkshire
Bergman, Prof. E.N.	Dept of Physiology, Biochemistry and Pharmacology, New York State College of Veterinary Medicine, Cornell University, Ithaca, New York 14850, U.S.A.

Blair, Dr T.

Imperial Chemical Industries Ltd, Jealott's Hill Research Station, Bracknell, Berkshire RG12 6EY

Boekholt, Dr Ir H.A.

Dept of Animal Physiology, Agricultural University, Haarweg 10 – 6709 PJ Wageningen, The Netherlands

Boorman, Dr K.N.

Dept of Applied Biochemistry and Nutrition, University of Nottingham, School of Agriculture

Box, P.G.

Glaxo-Allenburys Research, Breakspeare Road South, Harefield, Uxbridge, Middlesex

Breeuwsma, Dr A.J.

Intervet International B.V., PO Box 31, 5830 AA, Boxmeer, The Netherlands

Broom, Dr J.

Dept of Surgery, University Medical Buildings, Foresterhill, Aberdeen AB9 2ZD

Buraczewski, Prof. S.

IFZZ PAN, 05–110 Jablonna, k/Warszaw, Poland

Burgess, R.J.

Dept. of Biochemistry, Queen's Medical Centre, University of Nottingham

Burleigh, Dr I.G.

ARC Meat Research Institute, Langford, Bristol

Burns, Dr R.A.

Dept of Applied Biochemistry and Nutrition, University of Nottingham School of Agriculture

Buttery, Dr P.J.

Dept of Applied Biochemistry and Nutrition, University of Nottingham, School of Agriculture

Capper, B.S.

Tropical Products Institute, 56/62 Gray's Inn Road, London WC1X 8LU

Carter, T.J.

Frank Wright International Ltd, Misbourne House, Chiltern Hill, Chalfont St. Peter, Buckinghamshire

Chalmers, Dr M.I.

Rowett Research Institute, Aberdeen

Champredon, C.

Laboratoire du Metabolisme Azote, CRZV, Theix, 63110, Beaumont, France

Cheeseman, Mrs C.

Animal Nutrition Research, Pfizer Ltd,
Sandwich, Kent CO13 9NJ

Christopherson, Dr R.J.

Dept of Animal Science, University of
Alberta, Edmonton, Canada

Clague, M.B.

Dept of Surgery, University of Newcastle-
upon-Tyne, Newcastle-upon-Tyne

Close, Dr W.H.

ARC Institute of Animal Physiology,
Babraham, Cambridge CB2 4AT

Cole, Dr D.J.A.

Dept of Agriculture and Horticulture,
University of Nottingham School of
Agriculture

Cowdy, P.E.M.

BOCM Silcock Ltd, Basing View,
Basingstoke RG21 2EQ

Cowey, Dr C.B.

Institute of Marine Biochemistry,
St. Fittick's Road, Aberdeen AB1 3RA

Cuthbertson, Dr W.F.J.

Glaxo-Allenbury's Research, Sefton Park,
Stoke Poges SL2 4DZ

Davis, C.

Elanco Products Ltd, Kingsclere Road,
Basingstoke, Hampshire

Demeyer, Dr D.I.

Lab. Voeding E.W. Hygiene, University of
Ghent, Proeofhoevestraat, 10,
9230 Melle, Belgium

Donaldson, Dr I.A.

Biological Laboratory, University of Kent,
Canterbury, Kent CT2 7NJ

Duckworth, Dr J.E.

Meat and Livestock Commission,
Queensway House, Bletchley, Milton Keynes

Edmunds, B.K.

Pauls and Whites Food Ltd, New Cut West,
Ipswich 1PZ 8HP

Edwards, Dr R.A.

Edinburgh School of Agriculture, King's
Buildings, West Mains Road, Edinburgh
EH9 3JS

Eeckhout, Ir W.

Ryksstation voor Feervoeding,
Melle Gomtrode 9231, Belgium

Eenaeme, Dr C. Van

Faculte de Medecine Veterinaire,
Universite De Liege, 45, Rue Des Veterinaires,
1070, Brussels, Belgium

Es, Dr A.J.H. Van   Institut Voor Veevoedingsonderzoek,
                                    'Hoorn' Runderweg 2, Lelystad,
                                    The Netherlands

Fau, Dr D.   Centre de Recherches sur la Nutrition,
                                    CNRS, Rue Hetzel, 92190 Meudon, France

Fauconneau, B.   Laboratoire d'Etude du Métabolisme Azote,
                                    CRZU, Theix, 63110, Beaumont, France

Filmer, D.G.   BOCM Silcock Ltd, Basing View,
                                    Basingstoke RG21 2EQ

Finot, Dr P.A.   Nestle Products Technical Assistance Ltd,
                                    CH1814, La Tour De Peilz, Switzerland

Fisher, Dr C.   ARC Poultry Research Centre, King's
                                    Buildings, West Mains Road, Edinburgh
                                    EH9 3JS

Fuller, Dr M.F.   Rowett Research Institute, Greenburn Road,
                                    Bucksburn, Aberdeen

Gaetani, Dr S.   Istituto Nazionale Nutrizione,
                                    Via Lancisi 27 — Rome, Italy

Garlick, Dr P.J.   Clinical Nutrition and Metabolism Unit,
                                    Dept of Human Nutrition, London School
                                    of Hygiene and Tropical Medicine,
                                    London WC1

Gilbert, Dr A.B.   ARC Poultry Research Centre, King's
                                    Buildings, West Mains Road, Edinburgh
                                    EH9 3JS

Grimble, Dr R.F.   Nutrition Dept, School of Biochemical
                                    and Physiological Sciences, University of
                                    Southampton, Bassett Crescent East,
                                    Southampton

Halliday, D.A.   Ulster Polytechnic, Jordanstown,
                                    Co. Antrim, N. Ireland

Hardy, Dr B.   Dalgety Crosfields Ltd, Dalgety House,
                                    The Promenade, Clifton, Bristol

Harris, Dr C.I.   Rowett Research Institute, Greenburn Road,
                                    Bucksburn, Aberdeen

Hart, Miss D.   Biochemistry Dept, May and Baker Ltd,
                                    Dagenham, Essex RM10 7XS

Haselbach, C.

Inst. f. Tiereproduktion, Gr. Ernahrung, ETH, Zentrum, Zurich, Switzerland

Heitzman, Dr R.J.

ARC Institute for Research on Animal Diseases, Compton, Nr. Newbury, Berkshire RG16 0NN

Henderickz, Prof. H.K.

Dept of Nutrition, Faculty of Agricultural Science, University of Ghent, Proefhoevestr. 10, 9230 Melle, Belgium

Hodgson, Dr J.C.

Animal Diseases Research Association, 408 Gilmerton Road, Edinburgh EH19 7JH

Hoffman, Prof. B.

Institut für Veterinärmedizin des Bundesgesundheitsamtes, Bundesgesundheitsant, Postfach D.1000, Berlin 33, W. Germany

Holm, Dr H.

Institute for Nutrition Research, School of Medicine, University of Oslo, PO Box 1046, Blindern, Oslo 3, Norway

Holme, Dr D.W.

Pedigree Petfoods, Animal Studies Centre, Melton Mowbray, Leicestershire

Holsheimer, Ing. J.P.

Spelderholt Institute for Poultry Research, Beekbergen, The Netherlands

Homan, Ir G.W.

Hendrix' Voeders B.V., Veerstraat 38, 5831 JN Boxmeer, The Netherlands

Hudson, K.A.

Beecham Animal Health, Broadmead Lane, Keynsham, Avon

Jackson, Dr A.

The Tropical Metabolism Research Unit, University of the West Indies, Mona, Kingston 7, Jamaica

Jeacock, Dr M.K.

Dept of Physiology and Biochemistry, University of Reading, Whiteknights, Reading RG6 2AJ

Jones, Dr C.T.

Nuffield Institute for Medical Research, University of Oxford, Oxford OX3 9DS

Just, Dr A.

Dept Research in Pigs and Horses, National Institute of Animal Science, Rolighedsvej 25, DK 1958, Copenhagen, Denmark

Keith, Dr M.C.

Unilever Research Laboratory, Greyhope Road, Aberdeen

Kendall, Dr P.T.

Animal Studies Centre, Freeby Lane, Waltham-on-the-Wolds, Leicestershire

Knudsen, K.E.B.

National Institute of Animal Science, Animal Physiology and Chemistry, Rolighedsvej 25, 1958 Copenhagen, Denmark

Krieg, Dr R.

Institut für Kleintierzucht, Bundesforschungsanstalt für Landwirtschaft, Braunschweig − Völlerade, Dörnbergstrabe, 25/27, 3100 Celle

Lamming, Prof. G.

Dept of Physiology and Environmental Studies, University of Nottingham School of Agriculture

Lardeux, B.

Centre de Recherches sur la Nutrition, CNRS 9, rue Jules Hetzel, 92 190 Meuden Bellevue, France

Laurent, Dr G.

c/o Professor Turner Warwick, Cardiothoracic Dept, Brompton Hospital, London

Lawrie, Prof. R.A.

Dept of Applied Biochemistry and Nutrition, University of Nottingham School of Agriculture

Lewis, Prof. D.

Dept of Applied Biochemistry and Nutrition, University of Nottingham School of Agriculture

Lindsay, Prof. D.B.

Dept. of Biochemistry, ARC Institute of Animal Physiology, Babraham, Cambridge

Lobley, Dr G.E.

Rowett Research Institute, Greenburn Road, Bucksburn, Aberdeen AB2 9SB

Lodge, Prof. G.A.

School of Agriculture, 581 Kings Street, Aberdeen AB9 1UD

Lomax, Dr M.A.

Dept of Animal Physiology and Nutrition, University of Leeds

MacRae, Dr J.C.

Rowett Research Institute, Greenburn Road, Bucksburn, Aberdeen AB2 9SB

Mathers, Dr J.C.

Dept of Applied Biology, University of Cambridge, Pembroke St, Cambridge

Mayer, Dr R.J.	Dept of Biochemistry, Queen's Medical School, University of Nottingham
Mayes, Dr R.W.	Hill Farming Research Organization, Bush Estate, Penicuik, Midlothian EH26 0PY
Mepham, Dr T.B.	Dept of Physiology and Environmental Studies, University of Nottingham School of Agriculture
Motagally, Dr Z.Z.	National Research Centre, Dokko, Cairo
Mul, Ir A.J.	Intervet International BV, PO Box 31, 5830 AA, Boxmeer, The Netherlands
Musharaf, N.A.	Gezira University, PO Box 365, Wad Medani, Sudan
Neale, Dr R.J.	Dept of Applied Biochemistry and Nutrition, University of Nottingham School of Agriculture
Niess, Dr E.	Institut für Tierernährung Endenicher, Allee 15, D-5300, Bonn, W. Germany
Noda, Dr K.	Clinical Nutrition and Metabolism Unit, Hospital for Tropical Diseases, 4 St. Pancras Way, London NW1
Obled, Dr C.	Laboratoire du Métabolisme Azote, CRZV, INRA Theix, 63110 Beaumont, France
Oldham, Dr J.D.	NIRD, Shinfield, Reading RG6 2QZ
Orr, Dr R.M.	Seale Hayne College, Newton Abbot, Devon
Pain, Dr V.M.	Dept of Human Nutrition, London School of Hygiene and Tropical Medicine, London W1
Palmer, F.G.	MAFF, ADAS, Block 2, Government Buildings, Lawnswood, Leeds LS16 5PY
Perry, F.G.	BP Nutrition (UK) Ltd, Stepfield, Witham, Essex CM8 3AB
Petersen, Dr V.	Institut für Tierernährung, Bundesforschungsanstalt für Landwirtschaft, Bundesallee 50, D-3300, Braunschweig

Pickford, J.R.   BP Nutrition (UK) Ltd, Stepfield, Witham,
                                         Essex CM8 3AB

Pion, Dr R.   INRA, Theix, 63110, Beaumont, France

Pisulewski, Dr P.   Institute of Animal Production,
                                         32—D83 Balice, Poland

Porter, P.D.   The Lord Rank Research Centre, Lincoln
                                         Road, High Wycombe, Buckinghamshire

Potthast, Dr V.   Institut für Tierernährung, Endenicher Allee
                                         15, 5300 Bonn, W. Germany

Reeds, Dr P.J.   Rowett Research Institute, Greenburn Road,
                                         Bucksburn, Aberdeen AB2 9SB

Roberts, Dr C.J.   ARC, Institute of Animal Diseases,
                                         Compton, Newbury, Berkshire

Robins, Dr S.   Rowett Research Institute, Greenburn Road,
                                         Bucksburn, Aberdeen AB2 9SB

Rosochacki, Dr S.   CNM Unit, Hospital for Tropical Diseases,
                                         4 St. Pancras Way, London NW1

Ross, M.T.   278 Centennial Street, Winnipeg, Manitoba,
                                         Canada R3N 1P3

Russell, Dr S.M.   Dept of Biochemistry, University of
                                         Nottingham, Queens Medical Centre

Sanderson, Dr B.   Biochemistry Dept, May and Baker Ltd,
                                         Dagenham, Essex RM10 7XS

Schreurs, Dr V.V.A.M.   Dept of Animal Physiology, Agricultural
                                         University, Haarweg 10 — 6709 PJ,
                                         Wageningen, The Netherlands

Schryver, R. De   Laboratory of Animal Nutrition, University
                                         of Ghent, Heidestraat 19, 9220 Merelbeke,
                                         Belgium

Shepherd, Dr D.A.L.   Dept of Physiology and Biochemistry,
                                         The University, Whitenights, Reading

Slater, Dr J.S.   Animal Disease Research Association,
                                         408 Gilmerton Road, Edinburgh EH17 7JH

Smith, Dr R.W.

Biochemistry Dept, National Institute for Research in Dairying, Shinfield, Reading RG2 9AT

Steel, Dr G.T.

Central Toxicology Laboratory, Alderley Park, Cheshire SK10 4TJ

Strachan, Mrs P.J.

Unilever Research Laboratory, Colworth House, Sharnbrook, Bedford

Swan, Dr H.

Dept of Agriculture and Horticulture, University of Nottingham School of Agriculture

Taylor, Dr A.J.

Unilever Research Laboratory, Colworth House, Sharnbrook, Bedford

Taylor, Dr S.S.

Animal Breeding Research Organisation, West Mains Rd, Edinburgh EH9

Thomas, Dr P.C.

The Hannah Research Institute, Ayr, Scotland KA6 5HL

Thompson, A.P.

Hill Farming Research Organisation, Bush Estate, Penicuik, Midlothian, EH26 0PY

Turner, Miss R.

Pauls and Whites (Food) Ltd, Mill Road, Radstock, Bath, Avon

Tyler, R.W.

Nutrition Chemistry Dept, MAFF, ADAS, Olantigh Road, Wye, Ashford, Kent

Ulyatt, Dr M.J.

Applied Biochemistry Division, DSIR, Palmerston North, New Zealand

Unsworth, E.F.

Dept of Agriculture, Queen's University of Belfast, Newforge Lane, Belfast BT9 5PX

Varley, Dr M.A.

Seale Hayne College, Newton Abbot, Devon

Vernon, Dr B.G.

Dept of Applied Biochemistry and Nutrition, University of Nottingham School of Agriculture

Waterworth, D.G.

Imperial Chemical Industries Ltd, Jealotts Hill Research Station, Bracknell, Berkshire RG12 6EY

Wilkinson, Dr J.I.

Lilly Research Centre Ltd, Erle Wood Manor, Windlesham, Surrey

Wilson, Dr B.J.

Cherry Valley Farms Ltd, Rothwell,
Lincoln LN7 6BR

Wilson, C.J.

Imperial Chemical Industries Ltd,
Agricultural Division, Billingham,
Cleveland

Wilson, Prof. P.N.

BOCM Silcock Ltd, Basing View,
Basingstoke, Hampshire

Wise, Dr D.R.

School of Veterinary Medicine, Madingley
Road, Cambridge

Wiseman, Dr J.

Dept of Agriculture and Horticulture,
University of Nottingham

Young, Dr P.W.

Beecham Pharmaceuticals, Walton Oaks,
Dorking Road, Tadworth, Surrey

Young, Prof. V.R.

Dept of Nutrition and Food Science, and
Clinical Research Center, Massachusetts
Institute of Technology, Cambridge,
Mass. 02139, USA

INDEX